PERFORMING BODY, SPACE, AND PLACE

Performing Body, Space, and Place offers a groundbreaking methodology for performance in the Anthropocene, addressing the spiritual and ecological crises of our time through a fusion of Indigenous knowledge and contemporary practice.

This book is the culmination of 35 years of international fieldwork across six continents with Indigenous communities—including the Yup'ik and Inupiaq peoples of Alaska, the !Xuu and Khwe Bushmen of the Kalahari, the Sakha of Siberia, the Zulu of South Africa, the Tamils of India, and the Miao of China—along with experimental productions with Dead White Zombies in Dallas. It unfolds in three interconnected stages: Body, which emphasizes sensory presence and reconnection with the more-than-human world; Space, which critiques anthropocentric narratives and contrasts them with cosmocentric perspectives that honor the sovereignty of all beings; and Place, which explores performance as a tool for transforming modern placelessness into lived integration. Through practical explorations, theoretical insights, case studies, and field observations, the book equips practitioners to craft performances that transcend human-centered narratives and foster participatory engagement with the living world.

This essential resource is for artists, educators, and students in theater and performance studies, environmental humanities, Indigenous studies, drama therapy, and cultural anthropology. It will resonate with anyone interested in exploring the intersection of performance, ecology, and community, offering tools to reimagine our relationship with the embodied world and the spaces we inhabit.

Thomas Riccio is an artist-scholar, Professor Emeritus of Visual and Performing Arts at The University of Texas at Dallas and Artistic Director of the Dead White Zombies.

Routledge Advances in Theatre & Performance Studies

This series is our home for cutting-edge, upper-level scholarly studies and edited collections. Considering theatre and performance alongside topics such as religion, politics, gender, race, ecology, and the avant-garde, titles are characterized by dynamic interventions into established subjects and innovative studies on emerging topics.

Stagecraft Rewired
AI, Authorship, and the Future of Performance
Michael Harding and James Hutson

Brecon's Hidden Theatre History, 1699–1870
Beyond Sarah Siddons
Jayne Gold

Robot Theaters
Collisions in Performance and Social Robotics
Edited by Hilary Bergen, Eric Mullis, Benny Simon, and Sydney Skybetter

Performing Body, Space, and Place
The Cosmocentric Way
Thomas Riccio

For more information about this series, please visit: www.routledge.com/Routledge-Advances-in-Theatre–Performance-Studies/book-series/RATPS

PERFORMING BODY, SPACE, AND PLACE

The Cosmocentric Way

Thomas Riccio

Routledge
Taylor & Francis Group
LONDON AND NEW YORK

Cover image: Drawings from the author's workshop with the! Xuu and Khwe Bushmen. Designed by Laura Hyunjhee Kim

First published 2027
by Routledge
4 Park Square, Milton Park, Abingdon, Oxon OX14 4RN

and by Routledge
605 Third Avenue, New York, NY 10158

Routledge is an imprint of the Taylor & Francis Group, an informa business

British Library Cataloguing-in-Publication Data
A catalogue record for this book is available from the British Library

ISBN: 978-1-041-16801-0 (hbk)
ISBN: 978-1-041-16804-1 (pbk)
ISBN: 978-1-003-68642-2 (ebk)

DOI: 10.4324/9781003686422

Typeset in Sabon
by KnowledgeWorks Global Ltd.

To the Earth and its elders. To the future unborn.

CONTENTS

Acknowledgments *x*

Introduction: From Body Through Space to Place 1

Body 13

Introduction 13
Circle 21
Creative Circle 25
Walking 31
Our Work 43
Exploration 46
Rhythm 52
Rhythmic Movement 60
Sticks 68
Community of Place 78
Making Sounds 81
Balance 86
Moving 99
Explorations 110
Body and Spaces 113
Focus and Mindfulness 120
Play 125
Interstitials 129

Psychophysical 131
Trust 134
Aggression 138
Performing 144
Into Spaces 151

Space 153

Introduction 153
Narratives 155
Space Age 156
Cycles 158
Language 163
Pride and Prejudice 168
Rhythm 177
Cultures 182
Sound 192
Checking In 198
Balance 200
Process 204
Observation 211
Mimesis 221
Gifting and Exchange 235
Exploring 238
Objects 247
Character 252
Sharing 255
Performance 264
Expansion 269
Archetypes 275
Archetypal Ensembles 289
Masks 294
Body as Script 301
Space and Place 313

Place 314

The Loss of Place 316
Spaces Without Place 317
Place Through Performance 317
Somatic Wisdom and Practice 318

Our Contemporary Crisis 319
Becoming Earthlings 320
Performance as Transformation 321
Narratives 322
Bringing Place to the Body 339
The Ritual Preparation 346
Creating Performance 358
Theoretical Frames 365
Hallucinating 373
Interviewing 385
Trance 390
Character 400
Rehearsals 409
Immersion, Ritual, and the Sacred 415
Ancient Futures 418

Case Studies 420

List of Explorations 442

Works Cited 450

Index 459

ACKNOWLEDGMENTS

This book emerges from decades of collaboration, trust, and shared learning with communities and individuals across six continents. The knowledge contained here was never mine to discover—it was given, demonstrated, entrusted, and collectively created through a network of relationships that continue to shape my understanding of what performance can be and do.

My deepest gratitude goes to the elders who shared their wisdom, knowing I would carry it forward. Jimmy Toolie of Savoonga, Alaska, whose teachings on listening and on how spirits communicate through chance have profoundly reshaped my understanding of performance. Sakha shaman Isbekhov, whose words and presence live with me, and Iqulik, a Yup'ik shaman who guided me, demonstrating that which we do not experience, we cannot truly know. Thanks to Evelyn Alexander, friend and Athabaskan elder of Minto, whose wisdom, with a smile and a laugh, still shines within me.

Many thanks to the many Miao badai and xianniang—shamans and spirit mediums—who generously shared their knowledge, time, and spirit. Special appreciation goes to Master Shi Jing He, a Miao badai who walked 30 miles to share sacred hand gestures with a stranger, trusting that someday someone would want to learn them. Heartfelt gratitude also to my dear brother from another mother, Peng Jinquan, and the tireless Ma Mei, who continue to guide and support my work with the Miao.

To the many Korean mudangs, Zulu sangomas, Mexican curanderas, and Umbanda practitioners who welcomed me with open arms. The Bushman healers—Manfred, George, Machai Mbande, and Mbuto Bohera—who provided heartfelt and invaluable insights—and all the !Xuu and

Khwe Bushmen—who asked me to tell people about the Bushmen's way because they will soon be gone, a responsibility I take seriously with every page. This book seeks to honor all that was shared, while acknowledging it can never fully capture the embodied wisdom you carry.

Tuma Theatre and the University of Alaska Fairbanks Theatre were the birthplace of much of this work. Paul Jumbo of Toksook Bay—hunter of seals, walrus, and beluga whale—became a key collaborator, and patient teacher. His embodied knowledge of Arctic survival and subsistence deeply shaped my understanding of a Cosmocentric worldview. Wilma Brown, Melanie Brown, and Paul Mountain of Tuma Theatre provided insights that influenced both the practice and philosophy outlined here. The entire Tuma Theatre company—comprising Yup'ik, Inupiaq, Athabaskan, Aleut, and non-native performers—taught me that performance is inseparable from land, ancestors, and the living world. I am also grateful to Mike Gaffney, James Nageak, John Leipzig, and Thresa John for their unwavering support of my work and research in Alaska. Special thanks to my dear friend Anatoly Anthoin, who helped facilitate my work in St. Petersburg, Siberia, and Ethiopia—he is gone but remembered.

My Dallas-based performance collective, the Dead White Zombies, acts as a testing ground where Cosmocentric methodologies meet urban challenges. The performers Lori McCarty, Luke Lafitte, Brad Hennigan, Hilly Holsonback, Abel Flores, Stephan Gardner, Stephanie Cleghorn, David Jeremiah, Justin Locklear, Ben Miro, and Jennifer Culver deserve special recognition. To Scot Gresham-Lancaster, a composer and sound designer, and Mona Kasra, video artist—longtime DWZ collaborators whose integrated soundscapes and video installations were vital in creating immersive environments that served as consciousness technology. To Butch McGregor, the DWZ patron saint of buildings, and Alisa Eykilis and John Lehman, whose photographs documented our work.

Although I formally cite these artists-scholars in the text, several individuals greatly influenced this book: Foremost among them is my mentor, Richard Schechner, for his unwavering support, insights, and influence over the years. Many thanks to Robert Brustein for his guidance and support in my development as an artist at the American Repertory Theatre; to David Wheeler, Carlos Fuentes, and Lee Breuer for their support, friendship, and inspiration; and to Joseph Garry for setting me on the path and teaching me that challenges are opportunities.

My scholarly work owes much to Victor Turner and Arnold van Gennep for their insights into liminality and rites of passage; David Abram for exploring the spell of the sensuous world; Tyson Yunkaporta for indigenous knowledge systems; Rane Willerslev for his research with the Yukaghir; Oscar Kawagley for Yup'ik philosophy; and Walter Burkert for his inspirational work on ritual and biology. From Jerzy Grotowski's "Poor Theatre" to Augusto Boal's "Theatre of the Oppressed," from the anthropological insights of anthropologist Richard Nelson to the drumming wisdom of Sule Greg Wilson, and the poetic anthropology of Tom Lowenstein—these thinkers created the intellectual soil in which this work could grow.

Twenty years of friendship and work with David Hanson and Hanson Robotics have shown that technology and traditional wisdom can coexist harmoniously. The questions that people have long pondered—regarding consciousness, presence, and self-boundaries—are now increasingly urgent for all of humanity in ways that are both new and unforeseen.

Gratitude goes to the many workshop participants and institutions that supported, tested, and helped refine the work. Notable mentions include: The University of Texas at Dallas, California Institute of Integral Studies, the Estonian National Theatre School, DirectorsLab Chicago, the Sakha National Theatre, the University of Ouagadougou, Lul Theatre and the Addis Ababa University, the Korean National University of the Arts, Riksteatret in Stockholm, the National Theatre School in Krakow, Kenya's National Theatre, Metamorphosis Theatre in St. Petersburg, Mandala Theatre in Kathmandu, Pondicherry University (India), the University of Dar es Salaam, Tribhuvan University in Kathmandu, Jishou University (China), New York University, Kwasa Group in Durban, the Karamu House (Cleveland), the University of Arizona, UC Berkley, Institute of American Indian Arts, and San Francisco State among many other. Each contribution added new dimensions, depth, and understanding to the work.

I sincerely thank Dennis Kratz, Rick Brettell, and the University of Texas at Dallas for supporting my research and performance activities. Many friends, collaborators, and supporters have influenced, encouraged, and inspired my work over the years. Among the most prominent are Ellen Stewart, Michael Kustow, Lealem Berhanu, Dale Seeds, Megan Evans, Tony Rivard, Sven Holmberg, Josh and Vera Weiser, Mike Gaffney, Dhruba Karki, Carl Mikael Renlund, Idalotta Backman, Jouni Piekkari, Fred Curchak, Themi Venturas, Murray McGibbon, Shelley Kjonstad, Amandina Lihainba, Arumugham Thambiran, Lolitia Lesheim, Neal Szpatura, Opiyo Mumma, Mapopa Mtonga, Abune Belayneh, Jamie Dakis, and Ran Jiao as well as my loving family—my parents Filomena and Anthony, brothers James and Anthory, and devoted and loving Laura Hyunijhee Kim.

What love and generosity I have received from friends, elders, practitioners, and communities who shared their knowledge with me was never meant to be kept. It was given to be passed on, transformed, and multiplied. I offer it now to you.

INTRODUCTION

From Body Through Space to Place

What you do not see, do not hear, do not experience, you will never really know.

—Iyaaka (Ander Apassingok), Siberian Yup'ik Elder

The Earth Is Becoming Thin

We stand at a threshold moment. The earth, as elder Jimmy Toolie explained during our conversation in Savoonga, Alaska, is becoming "thin again"—that mythic time the Siberian Yup'ik call "nunam qainga mamkilrani," when the earth's crust was thin and boundaries between human and non-human, material and spiritual, performer and performance dissolved into permeability (Kawagley 1995, 34). When the earth was "thin," animals, spirits, and humans could interchange roles and forms. This wasn't primitive confusion but sophisticated understanding: Reality is porous, identity unfixed, and transformation always possible.

Today, this thinning manifests everywhere. In abandoned Dallas warehouses, audiences wander through 32,000 square feet of industrial decay transformed into liminal space, carrying stones that represent their souls. In Utqiaġvik, Alaska, Iñupiaq dancers move to the heartbeat of the Eagle Mother, their bodies becoming conduits between worlds. In post-Soviet Siberian theaters, actors rediscover movements their grandparents were forbidden to perform. What was once the domain of shamans and specialists now emerges as our collective reality.

The boundaries Western culture held for thousands of years—between sacred and profane, human and nature, technology and spirit—are breaking

DOI: 10.4324/9781003686422-1

down because our survival depends on it. The human body is a small universe assembled from Earth's elements, minerals, and chemicals. Over hundreds of millions of years, we evolved through trial and error, developing in ways deeply influenced by the environment—our nervous and circulatory systems, genetics, senses, psychology, societies, and cultures all reflecting constant coevolution with the planet (Lovelock 1979, 15).

As you read these words, electrical signals in your brain translate sensory input, accessing thoughts, memories, emotions, and actions driven by the desire to understand and improve survival. Among the Siberian Yukaghir, hunters understand their bodies convey information critical to decision-making—a twitching eyebrow signals luck or danger, bodily signs guide when to pursue or abandon prey (Willerslev 2007, 57). This knowing is deeply related to doing. The Earth, like our bodies, is an ongoing work in progress, a continuous expression of sensory perceptions and relationships—constant, subtle, perpetual communications within and among our environment, including people, animals, plants, objects, emotions, thoughts, energies, rhythms, sensations, and spirits.

Why This Book, Why Now

For decades, work carried me across continents without planning or theory to justify it. Circumstances guided me from studying traditional Western theatre to directing off and off-Broadway, serving as artistic director of the Organic Theatre in Chicago, collaborating with Tuma Theatre in Alaska, then with the Sakha of Siberia, Zulu performers in South Africa, !Xuu and Khwe Bushmen in the Kalahari, Miao spirit mediums in China, and on to projects in Ethiopia, Kenya, India, Korea, Nepal, European universities, and experimental artists in Dallas. Each project required its own approach, yet patterns emerged—not rules or methods, but something more like water finding its way down a mountainside.

This book traces that journey while offering something more urgent: An alternative way of being in the world and methodologies for creating place through embodied practice. We live in the Anthropocene, where human activities have become a planetary force. Climate change, mass extinctions, social fragmentation, political polarization, and technological disruption aren't separate crises but symptoms of a single displacement: The loss of our ability to create and sustain meaningful places.

Yet, this moment of maximum upheaval offers maximum possibility. When everything solid dissolves into air, we recall that all places once performed the world into being. The question isn't whether we'll create new realities—we do that constantly through social media, virtual worlds, and AI. The question is whether we will do so consciously and collectively, with awareness that what we're building should be a better, more sustainable world.

Form Fatigue and Cultural Zombification

We've reached what I call "Form Fatigue"—the exhaustion of the anthropocentric worldview that has dominated Western culture since ancient Greece. As the artist Hito Steyerl observes, we live in an era in which "almost everyone is an artist" yet creates mostly "commodified self-referential sludge" (149). The arts, like the larger culture, are caught in feedback loops of simulation and simulacrum, endlessly remixing what once was.

Through my work with the Dead White Zombies, a Dallas performance collective, I've confronted this reality directly. Our name acknowledges that Western culture—the shaper of global systems for two millennia—has become undead, like a zombie, yet still walking around, feeding off the remains of a world that once was. This isn't nihilism but a diagnosis. We're wandering through the death throes of originality, where all that remains is remix, rehash, and trash, fodder for satire, irony, and parody.

Theatre, once vital to community life, now seems in hospice, pushed aside and institutionalized. Yet, the longing for presence, participation, and belonging appears everywhere: In immersive events, gaming, social media, sports, radicalized politics, and tribalizations and identifications of every sort—all a yearning to be a part of something. The world has become increasingly "performative" precisely because traditional performance has abdicated its role of connecting people, ideas, and feelings beyond their own needs. This performativity is at once a desperate cry to be heard, to be recognized, a shout to fill a void that ever expands between us and everything. As Jonathan Crary notes, we live in capitalism's "terminal scorched earth phase," where "the ongoing colonization of consciousness, the homogenization of experience, and the anesthetized senses" create conditions of profound disconnection (88).

From Anthropocentric to Cosmocentric: A Necessary Shift

The exhaustion we're experiencing isn't merely aesthetic or cultural—it's ontological. Western civilization has been dominated for millennia by an anthropocentric worldview: The belief that humans are the most important beings, that nature exists for human benefit, and that human consciousness represents the apex of creation.

This book presents a fundamental alternative: The cosmocentric perspective. A cosmocentric worldview sees nature as alive in diverse ways;

recognizes that all life depends on the environment; understands that everything is interconnected within a web of life; and shifts the focus away from the individual. This animistic understanding—whereby nonhuman animals, as well as inanimate objects and spirits, are endowed with intellectual, emotional, and spiritual qualities—emerges not from cognitive processing but from direct perceptual engagement with the world (Willerslev 2007, 9).

This isn't new, but a return to humanity's original way of being. The Classic Maya regarded themselves as "one with the world rather than one with nature," promoting a cosmocentric view that supported biodiversity and conservation for millennia. Indigenous communities worldwide have maintained such perspectives through sustainable practices, recognizing humans as part of a vast, interconnected web of life.

The cosmocentric approach doesn't diminish human worth but redefines it. Humans are not the most important beings; instead, society functions through relatedness, where everything and everyone are interconnected, and identity emerges from these connections. We are not masters of nature but active participants in a living cosmos. As phenomenologist Merleau-Ponty articulates, "our body is not a thing in itself but an effect of our relations with others" (cited in Willerslev 2007, 72). Self and world are inextricably intertwined.

This shift has profound implications for performance. Anthropocentric performance privileges human agency and individual expression, positioning even natural elements as materials for human meaning-making. Cosmocentric performance recognizes all participants—human and more-than-human—as co-creators with their own agency, intelligence, and contribution.

The methodologies in this book are grounded in and inspired by cosmocentric principles. They focus on engaging with place rather than dominating space, fostering communities of practice rather than training individual performers, involving the world itself as a continuous, collaborative process among all beings.

The Alternative: Communities of Place

Against this exhaustion, place-based performance offers vital alternatives. These aren't primitive practices but methodologies that have sustained human communities for millennia through rhythmic repetition that creates neural pathways, collective movement that builds trust, and shared stories that encode essential knowledge.

Place-based performance emerges from what I call a "community of place"—a worldview recognizing both human and non-human entities within a defined geographical area as community members with agency and intelligence. Animals, plants, landforms, weather patterns, spirits, and ancestors all participate in an ongoing dialogue that performance makes visible and visceral.

The Yup'ik understand this as *inua*—the spirit that lives in everything. The Māori call it *mana*. Similar concepts exist across place-based cultures because they reflect a fundamental reality: Persons are those with whom other persons interact with varying degrees of reciprocity (Harvey 2006, 11). Being a person does not require human likeness.

This reflects what David Abram describes as the "spell of the sensuous"—through ongoing bodily presence, "my sensing body gradually attunes itself to the style of this other presence—to the way of this stone, tree, or table—as the other seems to adjust itself to my own style and sensitivity" (52). Among the Yukaghir, this is a lived practice: Animals appear as prey when roaming the forest but take human shapes when they enter their lands deep in rivers and lakes, living in households similar to humans—elk people remain friendly and keep tidy households, fox people retain their strong smell and shy character (Willerslev 2007, 89). Reality is unfixed, porous, permeable, and malleable.

Ancient Wisdom, Contemporary Application

This understanding isn't romantic primitivism. Neuroscience confirms what shamans have always known. Researcher Robert Beck documented that healers from diverse cultures all exhibit identical brainwave patterns during healing moments: 7.8 to 8 Hz, matching Earth's electromagnetic pulse, the Schumann resonance. We're literally jamming to planetary rhythms—"world beat" is precisely that: The beat of the world.

For the Yup'ik, traditional chanters and dancers could enter into the spirit of the hunter and prey through visualization and trance, suggesting man, animal, and spirit become one. "It means that, as we imagine, we cannot separate ourselves from whatever we are picturing" (Kawagley 1995, 34). The Yukaghir hunting practices embody this principle: Success depends on maintaining a double perspective, acting as a mimetic agent who is both hunter and animal, moving between identities in what performance theorist Richard Schechner calls the "not me, not not me" state (Willerslev 2007, 12, 96–97). The Yup'ik maintained a positive mental attitude to thrive in unpredictable environments through steam baths, singing and dancing, talking with others, playing games, spending time in silence, learning to relax, and visualizing a good life (Kawagley 1995, 36).

Similarly, my work with artificial intelligence and social robotics reveals how emerging technologies need not oppose place-based worldviews but can enhance them. Through 20 years with David Hanson and Hanson Robotics, creating humanoid robot personalities and serving as Creative Director, I came to understand robots and the AI that powers them as emergent consciousness. While working on Sophia the robot, I discovered what might be called "technological spirit"—entities that serve as intermediaries between human consciousness and a larger planetary force. Like totems, symbols, and icons of cultures old and modern, robots are vectors that hold and convey meaning, providing access to other realities. The means and mechanisms of our ancestors live with us still, updated to suit modern needs.

We live between the ancient and the future, the material and the spiritual, and performance remains the connective tissue. Yet, its potential has been domesticated, constrained, and institutionalized by the anthropocentric mind. From Grotowski's Poor Theatre to Boal's Theatre of the Oppressed, from Cage's 4'33" to Abramović's durational performances, avant-garde artists have challenged conventional boundaries. Yet, these movements, for all their innovation, remain rooted in anthropocentric frameworks that still privilege human agency and individual expression.

Place-based performance, by contrast, emerges from what Richard Schechner calls "the oldest performance tradition on earth"—practices predating agriculture, writing, and cities. As anthropologist Victor Turner observed, such performances aren't representations of reality but reality itself: "a kind of creative reordering of the very stuff of social life" (Turner 1969, 97). They don't separate art from life, audience from performer, human from more-than-human world.

Performance as Engagement with the Living World

The human body functions as a conduit through which we connect with the world. Every action—whether physical, emotional, intellectual, or spiritual—has consequences. For the Yukaghir, the world emerges from concrete, practical engagement rather than abstract contemplation. Doing is learning, and learning is doing; there is no gap between subject and world (Willerslev 2007, 94, 164). These ongoing interactions make performance a vital means of observing, interpreting, changing, and influencing thought and action. Human life on Earth, despite technological progress, remains a continuous revisiting and renewal of mythic and ritual patterns rooted in nature.

As mythologist Walter Burkert states: "Our biological makeup forms preconditions or 'attractors' to produce phenomena consistently, even if these patterns are created and recreated afresh in each case; there is a persistence of patterns through place and time" (22). To perform today is to participate in a transcendental act. Beneath the surface, the patterns hard-wired into our biology express our relationships and negotiate connections with each other and with non-human beings.

Place-based performance serves not primarily as entertainment but as functional technology for maintaining balance within ecological and social systems. As Malidoma Somé explains: "Every person is sent to their outpost, Earth, to work on a project that helps maintain the cosmic order" (Somé 1993, 14). For the Iñupiaq people of Point Hope, one of the oldest continuously inhabited settlements in North America, performance activates myth-time—the deep structural expressions from which all has been built. Performance serves magical, pragmatic, and aesthetic purposes simultaneously.

How to Use This Book: Performing the Cosmocentric Way

Do not read this book as dogma. These are suggestions shaped by specific communities facing particular challenges. What worked with Sakha performers in Yakutsk will not work the same way in Dallas or São Paulo—each part of the Earth has its own way of being. The point is adaptation: Taking what serves, modifying to suit needs, discovering what emerges when these practices meet your specific place, people, and circumstances. This book is decidedly hybrid, interdisciplinary, and expansive, reflecting my perspective and world.

Think of this book as a conversation rather than a manual, proposing an alternative way of looking and being. The margins await your notes. There are many explorations presented within, inviting your variations. A note: I prefer using the term "explorations" rather than "explorations" because it suggests an interiority and interconnectedness rather than a one-off mechanical accomplishment. Everything is everything, that's cosmocentricism. Some explorations will be familiar to some practitioners and will serve as a reminder of what we know but forget. Find them here recontextualized, adapted, and directed towards a cosmological intent—it's all about intent. Some explorations will resonate immediately; others might seem irrelevant until suddenly they speak to you. Trust the nonlinear path. As Tyson Yunkaporta notes, indigenous peoples don't have words for straight lines because "nobody would consider traveling, thinking, or talking in a straight line in the first place. The winding path is just how a path is" (18). All I would add to that is have some fun along the way and keep a sense of humor. If you are reading this book, it is because you care, are passionate, and called. You are doing what you love, spread it. Be generous to yourself and others. The Yup'ik have a saying, "All you have in life is what you can give."

The book's structure itself exemplifies cosmocentric principles from my perspective—there can't be only one way, and that's kind of the point. I will argue that space is a unit of information and that the interconnectedness of spaces constitutes and constructs a place. This understanding sees the entire cosmos as a hall of mirrors, with various dimensions of reality conceived as replicas, each recursive and reflective of the others. If you're searching for something definitive and a clear, organized methodology (that sure would be nice), but it won't happen. This book, our work, is just one more attempt to scale a mountain of mystery. Subsistence hunters accept and don't expect. Everything is as it should be—that's the way of nature.

The sections of this book—Body, Space, and Place—follow a natural progression, growing an understanding:

Body examines the psychophysical instrument through which we interact with reality. Instead of viewing the body as a machine to be trained or just a means for personal expression, we see it as a tiny universe—an intuitive, responsive entity made from and expressive of the Earth itself, constantly co-evolving, constantly cycling, immersed in rhythm, seeking balance. The body becomes the main technology for engaging with the cosmos.

Space examines the dimensional, relational, and performative fields where action unfolds. Space is not an empty void to be filled with human activity but living information—charged with history, relationality, and more-than-human presences. We learn to read spaces, to listen to them, to collaborate with what's already there.

Place synthesizes where embodied consciousness and spatial dynamics converge to create grounded, meaningful environments. Place emerges through sustained performance—through repeated, intentional, communal practices that weave body and space into coherent territories of meaning. Place is where cosmocentric principles become a lived reality.

After Place, it all dissolves and returns, and the cycle continues. As the Vietnamese say: "Same, same, but different."

Each section can stand alone, yet they're designed as progressive deepening—from individual instruments of the Body to relational fields of Spaces to the collective territory of Place. Most importantly, remember that reading about embodied practice is like looking at sheet music without playing it. The real knowledge lives in doing—in gathering others, forming circles, finding rhythms, and discovering what emerges. Performance is fundamentally and inherently communal. You can't do this alone, and that's the point. Even a solo performance artist and monologuist converses with others and with otherness, seen, and unseen. The cosmocentric way and its expression, performance, require moving beyond the isolated self toward recognition of radical interdependence.

This book offers practical methodologies for enacting the cosmocentric through performance. Where anthropocentric approaches ask "what can I express?" cosmocentric practice asks "what wants to emerge through me?" Where anthropocentric thinking seeks to master technique, cosmocentric practice cultivates receptivity. Where anthropocentric performance celebrates individual genius, cosmocentric work honors collective wisdom and more-than-human collaboration. Most likely, many of these notions go against everything you have been taught, the syntax of thinking and language, and that is what is at stake. We've been encoded, colonized, willfully convinced, and passively accepting that there is one way when there are many and other ways of being in the world.

Listening to the Earth

My journey into this work began without intention. In 1988, the University of Alaska hired me to direct Tuma Theatre, an Alaska Native performance group. If I didn't take the position, the program would lose funding. Despite knowing little about Alaska Native peoples or their cultures, I was given the opportunity to undertake a two-year research project in remote Yup'ik, Iñupiaq, and Athabaskan villages.

That first workshop with Tuma Theatre brought profound revelation. We'd begun with standard Western warm-ups—body isolation, stretching. But something felt wrong. I stopped the group. "This is wrong," I said. "I don't know what we should do, but this is wrong. We need to develop an Alaska Native warm-up."

The students looked confused. I explained that mindlessly importing Western culture had overshadowed their tradition, reducing the body to a mechanical entity that needed tuning. They held a heritage spanning thousands of years, an advanced culture with intricate, holistically attuned performance traditions. The challenge was reimagining these in a modern context, giving expression and voice to a marginalized yet unique worldview. Indigenous cultures hold Earth knowledge that is sorely needed.

This realization—that performance methodologies encode worldviews—became central to everything that followed. Working with elders, I learned to listen differently. "Listen and you will know what to do. Listen with your heart, not just your ears," Jimmy Toolie advised. "There is so much noise in the world today. It's hard to listen." Among the Yukaghir, this epistemology of listening is paramount—verbal information is never seen as sufficient; firsthand knowledge is essential. They put enormous emphasis on first-hand interaction with the world, especially the forest, and refrain from using explicit instructions when teaching children, who are instead encouraged to discover meanings on their own (Willerslev 2007, 69, 160).

Deep listening proved transformative. The blinding white tundra, where land, sea, and sky impressionistically blend, carried me to the infinite. The wind of the Bering Sea whispered songs. The great boreal forests revealed another magic—villages nestled among trees along broad rivers with distant snow-capped mountains. The silence was so deep it felt loud.

This listening led to invitations from communities worldwide facing similar challenges: How to preserve traditional performance while adapting to contemporary reality. Shi Jing He, an elderly Miao healer, traveled thirty miles to share sacred hand gestures he'd initially refused to teach. "My tradition is dying," he explained. "Someday, someone will want to learn and use them." Machai, a !Xuu Bushman I worked with in a Kalahari Desert refugee camp, said something similar: "You must tell people about the way of the Bushmen because soon we will be gone."

But traditions don't die. They transform. The rhythms that helped our ancestors survive on the savannah now help people transform trauma into a creative resource. The circle that once held tribal councils now forms as support groups, raves, yoga classes, and protest movements. The technologies remain constant even as everything else changes.

The Great Transition

We live in the era of the great transition—a moment when old forms have exhausted themselves but new forms have not yet fully emerged. Yet, alongside exhaustion, signs of transformation appear everywhere: The advancement of consciousness and awareness that, over a few generations, has upended millennia of institutionalized and dogmatized corruption and ignorance. Some are resistant and fighting back for a delusional return, but their desperate counteractions and rants are wheezing death throes. Globalization has wrought a new world of interconnectedness, and with that, a process of awakening, bringing disruption and possibility, as with any change. A new, planetary consciousness emerges, the world, in fits and starts, is emerging into a new sort of place—planet indigenous for which we are all a part.

The markers of an emergent consciousness are many and diverse. Environmental and climate consciousness overshadows all, as it should. Planetary evolution includes, to name but a few, the long-overdue acceptance and inclusion of women, human rights, and civil rights for all people, regardless of race, culture, gender, sexual orientation, or ability. The recognition of animal intelligence and rights; legal recognition of rivers and ecosystems as persons; global movements for environmental and social justice, renewed interest in animistic and panpsychic philosophies; alternative healing and medicines, and technological advancements that enhance human perception, longevity, and ability are all part of a vast, unfolding, and wonderous design. Performance, too, is part of this emergent consciousness.

These developments share a common motivation: Acknowledging the agency and liveliness of the more-than-human world. As James Bridle articulates: "Everything is really everyone, and all those beings have their own agency, points of view, and forms of life. Intelligence is not something that exists but something one does" (51). We're witnessing what James Lovelock described in his Gaia hypothesis: "the entire surface of the earth, including life, is a self-regulating entity" (15). This represents a profound shift from viewing nature as a primitive force to be subdued or as a demented spaceship traveling driverless around the sun, to understanding ourselves as participants in a living, self-organizing planetary system. The argument of this book is that performance is and has always been a way to glimpse the mystery and wonders of the cosmos. It is time to reclaim that mantle. Through this book, I hope to provoke thought, spark insight, and motivate a reaction.

Begin Where You Are: The Practice of Cosmocentricity

You don't need special training, sacred ground, or cultural permission to begin practicing the cosmocentric way. You need only what humans have always needed: Others willing to gather, move, and risk transformation together. Start with a circle—that ancient technology that democratizes sight lines and circulates energy. Add rhythm—the heartbeat that has sustained our species since the womb, that connects us to Earth's electromagnetic pulse. Include breath—that first and last act connecting inner and outer worlds, demonstrating our ongoing exchange with the atmosphere we share with all living beings.

From these simple elements, entire cosmologies emerge. Not because we make them up but because they're already there, waiting in the space between bodies, in the rhythm between heartbeats, in the place that emerges when humans gather with genuine intention to participate in something larger than ourselves.

Cosmocentric practice requires a fundamental shift in how we approach performance. We stop asking "what do I want to create?" and start asking "what does this place need?" We stop treating the environment as a backdrop and start recognizing it as a primary co-creator. We stop performing for an audience and start performing with a community that includes more-than-human participants. The goal is not full identification—which would be metamorphosis and dissolution of self—but the ability to move between identities, to act with a dual nature as both human and other, maintaining the "not me, not not me" state that permits new potential for action (Willerslev 2007, 95–96). The trees, the weather, the land itself—all become active collaborators.

This isn't a metaphor or a romantic fantasy. When the Iñupiaq perform, they're activating relationships with the animals whose spirits they embody, with the ancestors whose stories they carry, with the landscape that holds their history. When the !Xuu dance, the trance isn't a performance technique but a genuine communion with forces beyond human consciousness. When contemporary practitioners work with these principles, they discover that the more-than-human world responds, that the work takes on dimensions no individual or group could have planned.

The spirits that communicate through accidents have brought you to this moment, reading these words, considering these possibilities. What happens next isn't up to me or even you alone. It depends on who you gather, where you inhabit, which more-than-human presences are ready to participate, and what needs to emerge through awareness and collective practice.

As Jimmy Toolie offered in parting wisdom: "Chance is how the spirits speak. So, listen. All you need to know is already there. People think they know, but they don't. The spirits are always speaking."

The cosmocentric way asks us to stop talking so much and start listening more. To stop imposing our will and start noticing what's already present. To stop treating performance as human expression and start recognizing it as participation in the ongoing creation of the living world. The Miao Badai, their shamans, with whom I worked in the Wuling Mountains of Hunan, told me that talking and writing obstruct rather than promote genuine knowledge. Many choose to remain illiterate. Knowing comes from overcoming the boundary between self and world through direct, embodied engagement.

The circle continues. The rhythm deepens. The earth grows thin. The work awaits.

Welcome to the journey from Body through Space to Place—not a destination but an ongoing and never complete practice of creating the world we need to inhabit through cosmocentric engagement with all that is.

BODY

Introduction

The body serves as both the beginning and the end of our work.

The main objective of the Body section is to restore a sensory experience of presence in the world, which will underpin all subsequent work. This phase, and indeed this book, will blend field-tested techniques and practical applications, theory, case studies, personal observations, and speculations. Some of what follows will be familiar, but here evolved to the needs of the Anthropocene.

Treat this section and everything that follows as a flexible resource. Feel free to graze, pick and choose, adapt, inspire, adjust, and improve to suit your goals. The order of explorations, studies, and explorations is dictated by the book's format and intended to be suggestive rather than prescriptive. Just as in life, use your resources effectively based on the situation.

We begin by simply standing in a circle.

STANDING RELAXED

The performer is instructed to stand with minimal energy, keeping their eyes closed and maintaining a relaxed breathing pattern.

Use verbal guidance to encourage performers to visualize each body part as you lead them through the process. Begin by tensing the feet, then move to the calves, knees, thighs, pelvis, chest, shoulders, hands,

DOI: 10.4324/9781003686422-2

forearms, upper arms, neck, face, scalp, ears, and eyes. Once they have gone through the entire body, prompt performers to tense and inhale for each body part, imagining their breath circulating through the muscles and bones of that part.

Then, exhale to release any tension.

Start at the head and work down to the jaw, chest, arms, and neck. You can be as detailed as time allows or as concise as necessary.

Ask the performers to let their breath cleanse each body part in succession; when they exhale, their breath releases any lingering tension.

Repeat the breathing exploration for each isolated body part and then for the entire body.

Next, prompt the performers to stand energized, alert, and ready.

Discussion: After completing the tension-release segment of the exploration, instruct the performers to identify which is holding the most tension to gain insight into where and how they are holding it.

An alternative seated position to **Standing Relaxed** can be used in a classroom, at home, with older performers, or as a variation. Arrange straight-backed chairs in a circle and ask performers to sit comfortably. Instruct performers to sit relaxed and begin with their feet, following the process outlined in **Standing Relaxed.**

THE MOUNTAIN

The performers stand comfortably, eyes closed, in a circular formation, with you at the center.

Eyes closed. Instruct the performers to breathe in through their noses and out through their mouths. The breathing should be natural, with a steady rhythm.

The performers take a relaxed stance and envision themselves as mountains. Their heads represent the mountain peaks, while their feet are firmly rooted in the earth. The rhythms of their bodies—flow, breath, heartbeat, and pressure sensations on their skin—are reminiscent of mountains: Stable, solid, and alive.

Their breath, heartbeat, and the flow of blood through their bodies resemble streams on a mountain. They recognize how the body is a system and part of a vast system of nature.

Ask the performers to clear their minds and focus entirely on the movements within their bodies. Encourage them to remain present amid the myriad actions occurring in their bodies, fostering calmness and observation.

After establishing the mountain, ask the performers to start recognizing their feelings and thoughts as they arise, acknowledge them, accept them, and let them go.

If you suddenly feel a sensation in your body—a flash of warmth, an itch, tension, or a twitch—acknowledge it, accept it, and let it go. Listen to your mind and body; remember their constant conversations and chatter while staying calm and observant, like a mountain. You can see far and wide. You are connected to the earth and sky; animals and vegetation live, grow, and die through you.

You are the host. The world passes through and around you. Winds, the sounds of birds, insects, and animals, climate, and streams of water and blood flowing through you. Choose what to focus on and engage with, recognizing the chatter of your life and its ongoing events. Work to be fully aware yet able to focus on whatever you choose.

Note: If working with a group that has limited experience with mountains, adapt the exploration to a tree or plant visualization, or to what is known and predominant in their environment.

Internal Dialogue can be applied while walking, driving, eating, or even in the presence of another person. It is simply the awareness and articulation of what is happening: Our breath, heartbeat, a thought, and a feeling. This exploration serves as a means to become more observant and appreciative of the complexity, mysteries, and wonders of the mind and body as we navigate the most mundane aspects of living in and with the world.

Some questions: Are you narrating your world, or is your world narrating you? Are you engaged in an ongoing dialogue with the world and its many elements? Both the world and you are alive, participating in a never-ending dialogue. Through this exploration, you simultaneously grant the world recognition and a voice. The conversation can lead anywhere. We live associatively, and you are an attractor as you wind our way through the world. A constant symphony of feelings, thoughts, biochemistry, and the stimuli of the external world move you—you are part of the world.

We are part of what the Mexican philosopher Roger Bartra calls the *exocerebrum*. A unitary consciousness, a fundamental force that permeates the entire universe, and its expression in humans is just a complex variation of a cosmic phenomenon (10).

INTERNAL DIALOGUE

Comment to yourself when engaged in everyday activities and create an ongoing dialogue. Reflect on each action without censoring or premeditating—flow through the world's stream of experience.

You can pick up or drop this simple exploration at your discretion. Don't overthink it. Just go with the flow and keep your thoughts to yourself.

Train yourself to become more present in the world. Develop your awareness into a habit. Awareness of the world is a companion and a comfort.

An example from my walk along the streets of Athens as daytime turns to evening on a late February day in 2014.

I am walking on slippery tiles. A man approaches me, and he does not look happy. A motorbike speeds by behind me. In the window, an older man is slouched over, smoking a pipe. Another man looks at him as if to ask a question. What do they sell in this shop? The whiff of dog waste lingers; people are under no obligation to clean up after their dogs. The name of this street is in Greek, and I have no idea how to pronounce these symbols. A beautiful young woman is looking, about to cross the street. She has been grocery shopping. I hear the playful yelling of young boys. I see them; they appear to be friends, and it seems one is teasing the other. A motorcycle glides by in second gear, slowing down for the intersection. I tilt my shoulders to let an old woman pass. She looks at me and then looks away. I am approaching a busy street. I feel good, happy to be alive, to have my strength. My right shoulder hurts; I think I've pulled something and need to get it checked out. There are so many bland concrete apartment buildings; I couldn't live like this. So many television dishes on the rooftops; no one must own a dryer in this area. A woman is taking in her clothing before dark. A man is talking on the phone while smoking a cigarette. A gust of wind brings a chill around my torso, so I zip up my jacket. The store is around here. Is it to the left or the next street? The rumble of a city bus...

Exploration Variation: Encourage performers to navigate the world together in real time by moving through daily life while openly vocalizing and sharing their observations.

Our bodies are made up of minerals and elements from the Earth, making us embodiments of the Earth itself. We have become estranged from nature, our origins, and our place in the web of life. By distancing ourselves from our greater selves, we've lost our identity and neglected our responsibilities, leading to feelings of uncertainty, alienation, and anxiety. This distortion of our sense of identity, values, and actions has set off a chain of events, putting us on the brink of a crisis. The planet will survive without us, continuing to recover and heal. But we can't say the same. Nature isn't an external abstraction separate from us—it is us.

Our goal is to forge a new path—to relearn and reimagine—based on the challenges and circumstances of our time.

We begin by acknowledging our current situation and the resources at our disposal. Action must be taken. The ways we have inherited are no longer suited to our moment.

WORLD AWARENESS

An exploration integrated into the performer's daily activities, such as walking, biking, running, driving, and moving through life's events. A training to see, hear, feel, smell, taste, and be present in the world, fostering an appreciation for the depth and nuance around them.

Begin with the statement: "I see the world, and it sees me."

Recognize that trees, birds, wind, air, rocks, plants, grass, moon, sun, and everything around is alive, each in its own way.

Take a moment, give a thought, extend a feeling to appreciate and reflect on their existence—their rhythms, cycles, histories, and roles within the interconnected totality.

Observing a dog or cat involves understanding its journey from puppyhood to senior years, its daily experiences, perspectives, and emotions, and seeing the world through its eyes. Humans, cats, and dogs have long been companions—co-evolutionary partners. See into their origins, consider what and how they sense their world.

Similarly, consider a lake, river, trees, plants, and urban animals, such as birds, squirrels, rabbits, and insects. Reflect on their daily lives—how they move, what moves them, how they see, feel, and respond to their environment. Think about their sense of rhythm, time, space, and relationships. What are the cycles and balance that define their lives?

And rocks, the bones of the Earth. That rock in your yard, along a walk, in a park, has witnessed billions of years and was at the origins of the planet, if not the universe. What and how do they feel, see, and exist? How do they sit, how did they get where they are, what have they seen and experienced? How do they communicate? What stories does it hold and carry to this moment? How do they respond to your presence?

Then, consider what has been manufactured and processed. The rug, tiles, or flooring beneath your feet when you step out of bed in the morning, for example. They, too, had organic origins. Cotton, wool, or petroleum-based synthetics are all derived from the earth, processed, and shaped. What was their journey and memory? They contain many of the same elements and minerals present within our bodies. What was their journey, and what part of the earth did they come from?

What cycles of origins and history brought them to your bedroom and beneath your bare feet? In our age of globalization and technology-enabled advanced manufacturing, how did their lives and forms change, and what social, cultural, and economic systems brought you and them together?

Look now at the objects around you. The prevailing cultural perspective, unspoken and accepted, values the material. Ours is a material-objective perception of reality. However, between the material objects that occupy our lives, what we call and understand as "empty" or "space" is alive, whole, and part of a vast totality of being. We have been conditioned, to our disadvantage, not to recognize it as such. We perceive and are perceived—the idea of an "all-seeing, all-knowing god" codified by Abrahamic religions developed from this understanding.

Many of the world's place-based cultural traditions perceive this emptiness as occupied, holding and giving order to the material world. It is inhabited by ancestors and spirits that have preceded the present. That "empty space" is charged and alive. The Earth is a closed and cyclical system, constantly, rhythmically attuned to seeking and sustaining balance.

The air that has circulated through the lungs of all our ancestors—human and non-human beings alike—since the beginning of time has also passed through us. Similarly, the water, oceans, rain, and fog that nourish the present have sustained everything that came before and everything yet to come. Polluting these elements is a violation of a sacred responsibility, a disrespect to the past that brought us here, and to the legacy we inherited and are charged to pass on.

We, everything, belong to a community of place.

The default human setting is to either define or categorize non-human beings on human terms. A rock is lifeless because it lacks the attributes of humanity or sentience. The space between two objects is empty and devoid of value. The world is alive in many and varied ways. Our work is to become aware again, to lead, and to share this awareness with our larger community.

What happens when you share awareness beyond the human? To give is to receive, to understand is to be understood, and such is the pathway to another way of being.

In the Woods

During my 15 years teaching at the University of Alaska Fairbanks, I lived immersed in Alaska's expansive interior forests. My home, perched on a rocky bluff, offered a view of the Tanana River and extended 120 miles into pristine wilderness, leading to the Alaska Range and Denali. This experience profoundly transformed my body, mind, and spirit. By living in tune with the natural cycles and rhythms of seasons, weather, and migrations—observing nature's subtle, constant, and wondrous shifts—I maintained balance and have been changed by it, now living within and beneath these words. A highlight of this time was working closely with Alaska Native communities, traveling to villages to learn from elders and the land.

My home was open to my Alaska Native students, who found comfort in the quiet and embrace of nature. One frequent visitor was my dear friend, Paul Jumbo, a student around my age from Toksook Bay, a village on the Bering Sea. Paul hunted seals, occasionally walrus, and beluga whale. When he visited my small cabin, we would walk in the woods. One spring day, he stopped and flipped over a rock, then continued without a word. When I asked why, he replied, "That rock was getting tired of being on that one side for so long." I looked forward to our visits and walks, never knowing what they would teach me. Walking on the frozen river, he would identify fox and wolf tracks and interpret what they were thinking and doing. For him, the walks were like moving through a story, as if reading a book. Willow trees that a moose mother and her calf chewed on, then nearby tracks, "This was a young fox that was lost, see how he walks, then stops so many times. Here he heard another fox and changed direction. Here, he did a little dance. See happy steps, then he went running another way."

Daily awareness and appreciation of others require being present, understanding our role in the world, and recognizing our higher purpose.

Our identity today is deeply rooted in our history as Indigenous hunter-gatherers. Richard Nelson, an anthropologist working in the far north, states that 99% of human existence has been as hunter-gatherers, and our biology and psychology have evolved from these ancestors, whom he regards as the most successful and sustainable way humans have lived. We are hunter-gathering organisms. Our versatility, endurance, adaptability, keen color vision, tough skin, ability to eat a variety of foods, and cognitive skills to observe and learn from the world—never ceasing to learn, hold, pass on, and develop—are all traits inherited from our primitive, nameless predecessors. Their daily interactions and close relationships with animals, plants, climate, seasons, landforms, and astronomy were crucial for their survival and our evolution. Their traditional knowledge, based on the keen observation of nature, is on par with that of scientists. We are what we are because of what we were—hunter-gatherers. The most successful, sustainable, and lasting lifestyle for humanity (Nelson 1990).

Our modern lives have developed through deep connections with indigenous environments over thousands of years. The small, compact bodies (to conserve heat) and round cheeks (to prevent frostbite) of the Inuit are adaptations to the harsh, cold climate. People from the Mediterranean often have prominent noses to help cool down and browner skin to survive in sun-intensive regions (Nelson 1990). The darkness and brown tones of African skin are adaptive to their sunbaked environments. The small, lean stature of the !Xuu and Khwe Bushmen, whom I worked with in the Kalahari, is well-suited to their environment, with larger fat reserves in their buttocks to help them survive periods of scarcity. Likewise, the environment has shaped social and cultural systems, rituals, and performances that are responsive and optimized for survival. The history of conquests, colonization, wars, forced migrations, intermarriages, urbanization, and now advanced technology has complicated humanity and largely severed our connection to ancestral lands and their ways.

Humans have evolved as beings connected to specific places, influenced by interactions with both people and the natural environment. Developing new relationships and healing old ones with non-human intelligences are central themes in this book. Our biology, psychology, and worldview originate from the traditions of our Indigenous hunter-gatherer ancestors. These traditions have shaped not only our biological makeup but also our societies, politics, economies, and cultures, all of which are deeply tied to our interactions with the natural world. Over tens of thousands of years, our hunter-gatherer ancestors also developed innate release mechanisms that function like predetermined programming, similar to imprinting. These fixed action patterns are triggered by specific combinations of stimuli. Rituals and all forms of performance activate fundamental structural patterns that trigger mechanisms supporting collective expression (29).

We acknowledge our deep structural inheritances, but as both human and technology-driven systems become more integrated, this understanding has faded. Today, the physical body, personal interactions, and concepts of identity are increasingly abstracted, removed, and virtualized—we live in a society that emphasizes human perspectives, which are anthropocentric, and has gradually distanced itself from the natural world. More and more, we perceive ourselves through technological, political, economic, and social systems rather than natural, organic, or environmental relationships. This book seeks to reaffirm, explore, and convey a cosmocentric worldview. It emphasizes a perspective that centers the universe or nature, in contrast to an anthropocentric outlook that prioritizes humans as the most essential aspect of existence. The book argues that adopting a cosmocentric mindset can guide our community in innovative ways, offering a new, reimagined model. In this pivotal moment in history, we serve as a bridge between the ancient and the modern, participating in an experiment with unknown outcomes. Our mission is to work towards creating a different future and world.

We start by transforming the mind, body, and spirit to spark a change in perception and attentiveness.

Circle Rhythm Balance

Our work will revolve around three core concepts.

Circle—the diagram and symbol
Rhythm—the articulator and drive
Balance—the objective and the vision

Circle

From prehistoric cave drawings to contemporary scientific discoveries, circles and cycles have served as humanity's primary means of expressing our desire to identify with wholeness and grasp the essence of both natural and metaphysical worlds. The terms "circle" and "cycle" represent complementary facets of a single concept: The circle provides the material and symbolic representation, while the cycle embodies its dynamic unfolding over time.

The circle symbolizes the endless cycles that define existence—birth, life, and death; the rhythms of time itself; the dynamics of community, ecology, biology, and chemistry; the flow of emotions and relationships; the circulation of electricity and weather patterns; animal migrations and planetary movements; the eternal ebb and flow of tides, societies, cultures, religions, and history. Even as you read these words, your breath, heartbeat, and digestive processes cycle continuously. Life, the universe, and everything within it move in cyclical patterns.

Cycles represent the fundamental rhythm of existence, present everywhere and integral to everything. A circle embodies profound simplicity: It lacks beginning or end, has no front or back, and establishes no hierarchy. This ancient yet timeless form is simultaneously egalitarian, organic, continuous, unifying, inclusive, and comforting—serving as both symbol and reality, a mnemonic that captures the essence, process, and goals of natural order.

We exist immersed in cycles at every moment. Beyond our personal biological rhythms, we move through larger cycles of day and night, seasons, tides, planetary orbits, and lunar phases. From the microscopic spiraling of DNA to the vast orbital dance of galaxies, nature reveals itself as fundamentally organized around circular and cyclical patterns. This suggests that cycles represent not merely common occurrences but a basic organizing principle of reality to which we are all inextricably connected.

In pure mathematics, a circle achieves ideal definition through its consistent relationship to a center—every point on its perimeter maintains equal distance from that central point. The ancient Babylonians, Egyptians, and Greeks discovered that the ratio of a circle's circumference to its diameter is approximately 3.14 (π), establishing the circle as inherently reliable and unwavering.

Medieval scholars connected early geometry, astrology, and astronomy to the divine, believing circles contained intrinsic perfection. This reverence inspired ancient Greeks to develop both geometry and early calculus. The circle's mathematical properties revealed fundamental truths about space, proportion, and relationship that continue to undergird scientific understanding.

The circle-sphere stands as an archetypal symbol across cultures, representing perfection, divinity, and the soul. The halo marks sacred figures, while rulers from ancient Egypt to modern monarchy are crowned with circles. The Yin-Yang symbol—two spiraling parts within a circle—serves as a foundational icon of Confucianism and Taoism. Buddhist mandalas (Sanskrit for "circle") create symbolic diagrams of the universe arranged in concentric circles, used in tantric practice for meditation and spiritual development.

Carl Jung recognized the mandala as a universally occurring pattern associated with the mythological representation of the self, suggesting that circular forms tap into fundamental aspects of human sentience. This cross-cultural recognition points to something more profound than coincidence—a universal human intuition about the circle's fundamental significance.

The circle serves as the foundation for the wheel, which, along with related inventions like gears, enables much of modern civilization. Beyond its practical applications, the wheel has evolved into a powerful cultural and spiritual metaphor for cycles and regular repetition. The "Wheel of Dharma" represents the Buddhist teaching on the path to enlightenment. At the same time, similar symbols appear in Jainism and other traditions, embodying concepts of spiritual progression and eternal return.

The circle uniquely embodies the relationship and the concept of being. It represents a fundamental, coexisting relationship between humans and their surrounding world: A circle requires both a perimeter and a center, each dependent on the other for existence. Who or what occupies the center, and who comprises the circumference, depend on the cultural perspective, but the relationship remains constant.

This interdependence reveals the circle's inherent fragility—if any element of the perimeter or center is lost, the circle ceases to exist. This mirrors human relationships with our world and reflects how every creature exists at the center or perimeter of multiple overlapping circles of influence and connection.

A circle represents the locus of all points equidistant from a central point. To "circle" something implies creating an area of action or influence by moving around a fixed reference point. The circle also embodies absolute zero—simultaneously symbolizing emptiness and the infinite potential of what is not yet manifest but could be made manifest.

This paradox captures the essence of creative potential: The empty center that defines the circle's space also provides the openness necessary for new possibilities to emerge. In this way, the circle-cycle suggests a recombinant energy that returns and transforms, manifesting the power of life through eternal cyclical movement.

The power of the circle extends beyond symbolism into lived practice, particularly in healing and spiritual traditions. As African drummer Sule Greg Wilson observed,

> Each stroke you put into the drum or bell must be part of a cycle that returns the energy you put back into your person and the people before you. If you play down into the instrument, all you do is break it into pieces, dissipate your energy, and disappoint those whose chakras wait to catch your sent emotions.
>
> *(75)*

This wisdom reveals the circle's function as more than a symbol—it becomes the mechanism through which energy circulates, transforms, and heals. The circle represents the public realm and the cultural component of human experience, embodying connectivity, change, and continuity that manifests collective creativity.

Indigenous cultures worldwide have long understood the circle as a sacred space essential for community transformation. The ancestors of !Xuu and Khwe Bushmen in the lower Kalahari established circles as sacred spaces necessary for trance induction and community healing ceremonies. In these traditions, the community could hear their ancestors more vividly when gathered in healing circles.

In an interview with me, Bushman healer Silenga proclaimed, "It is the ancestors who help us survive." Fellow healer George added, "It is the ancestors who pass on information about how to live and heal. They got the information from God" (Riccio 1994). The circle becomes the vessel through which ancestral wisdom flows into the present community.

Within the Inupiat Eskimo worldview, the cycle transcends individual existence—it represents the pattern enabling the world's continuation. As anthropologist and poet Tom Lowenstein documented,

> All they discover is eventually dispersed in further cycles—as they know they and the things they have made will be dispersed in coming generations... where the ancestors had fought or died, seen visions or shamanized, they left knots, whorls, vortices, or human implications on the landscape. To be in a place of death or vision was to relive the story and extend its relationship with the present.
>
> *(Lowenstein 1993, 35)*

The circle functions as the liminal space identified by anthropologists Arnold van Gennep and Victor Turner—a transitional realm of separation, the liminal, where transformation becomes possible. In this sacred geography, communities can leave one way of being and enter another. The circle provides refuge from ordinary reality, creating a space where groups can transform, adjust, and reintegrate new understandings into individual and community life.

This liminal quality makes the circle simultaneously stable and transformative. As a space that is always different yet always the same, the circle generates perpetual motion that implies stability while catalyzing change. To participate in a circle means engaging with an invisible force that transcends the self, becoming part of something greater through the dynamic processes of formation, transformation, and recreation.

The return of circle-cycle principles as guiding frameworks represents a significant cultural moment. Out of necessity and driven by an acute awareness of finite and fragile resources, human civilization increasingly re-embraces a cyclical understanding. This shift prompts profound social and cultural adjustments, including biodegradability, recycling, renewable resources, pollution control, earth-friendly organic farming, and holistic medicine—all of which harken back to indigenous ways of living responsibly within and with the earth.

This contemporary renaissance demonstrates that the circle-cycle represents more than historical curiosity or spiritual metaphor. It offers practical wisdom for addressing modern challenges, providing a worldview and foundation from which sustainable solutions can flow.

Creative Circle

The **Creative Circle** is a gathering. Forming a circle is one of the oldest ways humans have shared and grown together. I call this a *sharemony*. It is central to our work, serving as a living model and the heart of the group, guiding the tone and method for everything that follows. As we progress through this book and our work, I'll remind you how circle awareness and the way of cycles inform our work, emotions, thoughts, and creativity. Like all powerful ideas, **Creative Circle** is elegant in its simplicity, in its form and utility. Basically, we sit in a circle and discuss whatever needs to be addressed in any way and by any means considered appropriate. It is a free and creative space that expresses the totality of the group and our work. At first, you will serve as the facilitator and model, but as the practice becomes natural, leadership is shared. What follows is an example that illustrates its form and use, highlighting its effectiveness.

The Zulu

During my time working with a Zulu Group in Durban, South Africa, the last apartheid laws were falling away, and a new and uncertain reality was emerging. One day during our workshop, I addressed two performers who returned 30 minutes late after an hour-long lunch break. The rest of the group had waited, and I asked the two men to apologize. The interaction sparked a noticeable shift in the atmosphere. I stopped our work and called for a **Creative Circle**, which, by this time, had been established as a space where anyone could share their thoughts and feelings on any topic without fear of judgment or repercussions. Initially, our discussions focused on work and general issues, but with each new Circle, they grew in depth and range, addressing issues such as those of people living in distant or politically contentious townships who needed to leave early. Under Apartheid, blacks were compelled to reside in townships located on the outskirts of cities, some as far as 30 miles away, which often meant lengthy and hazardous train or bus journeys home. South Africa's social threads were coming undone, and the country was being strangled by growing tensions and uncertainty across politics, economy, and race.

A few of the performers lived in violence-ridden townships and traveled into stronghold areas of the IRP the political opponents of the ANC led by Mandela. Political ruffians were known to question riders about their political affiliations and throw people off the moving train. Necklacing had also occurred, where political opponents had a car tire doused with gasoline placed over them and set on fire. The performers living in these areas wanted to travel and return home before dark, so we adjusted our work schedule accordingly. Working in such a situation was more than a workshop. I was working with a traumatized population that called upon and incorporated my limited skills in drama and social therapy.

Expressing thoughts and feelings openly was a new practice in this racially and politically tense environment. Apartheid impacted people's minds and bodies, and I was considered a white man and a symbol of oppression—something we discussed. Over the weeks, we developed and used the circle as a safe space for open discussion. It also served as a modeling opportunity, demonstrating equality, perspective, and respect for themselves and others.

In the case of tardiness, I explained why I addressed the latecomers, emphasizing that the work and the rest of the group must be respected. Performers overwhelmingly agreed but still said, "When a white man scolds us, we do not like it." I replied, "I didn't think I was scolding. I was stating what I felt in neutral terms. I meant nothing personal." I apologized and thanked them, saying I would be more sensitive in the future. The two late attendees also apologized, assuring me they would respect the work and one another going forward. We discussed at length how they had never criticized a white man before and how they had been educated and conditioned, physically and mentally, to respond.

It was a small but essential learning moment. The ability to express discontent and to flesh out the complexity of a situation fostered trust and confidence, allowing the work to explore wherever needed. An open and supportive discussion of long-held grievances that had never been aired during the repressive apartheid regime would be central to our work. We would transform them into a performance and offer it to the larger community. Performers are courageous leaders, and the Creative Circle was a place to build and shape a voice and courage.

On another occasion, a circle was formed to discuss the refusal of two Zulu men to participate in "women's dances" as part of a skills exchange. The men questioned this action, stating, "Zulu men do not perform women's dances!" Many voices were heard, the women were incensed, contributing to the circle as if it were a pot stirring a new way of being into existence.

The two Zulu men, deeply influenced by images of warriors and masculinity, were unwaveringly opposed. The discussion stirred up deep, long-standing hurt among the women. There were eruptions of resentments, personal histories, traditions, identities, politics, and race, all of which lay deep within the action of their refusal. The discussion somehow reached a resolution when the connection between sexism and racism suddenly dawned on the two men. Their refusal was another form of apartheid. "It's like apartheid," the muscular Thubani said. From then on, the two men became the group's most vocal advocates and performers of women's dances. One went on to study ballet at the Johannesburg Ballet, becoming one of its first Zulu company members. Thubani went on to play Macbeth in an international touring production.

The most striking example of the Creative Circle was when Kenneth, a tall, powerfully built Zulu man, removed his shirt during an exploration on a hot day. He showed me the gruesome scars from a whipping, and I was so shocked that I broke down in tears. Over the three weeks of our workshop, I had come to know him as a likable, hardworking, and devoted performer. But when I asked him about his scars, he got angry and snapped, "Nothing, man. Nothing," and pulled his shirt back on. I told him it was essential to open up about what had happened when he was ready.

A week later, while adapting a Zulu myth, *Emandulo*, for performance, the group worked on developing a scene. In the myth, a sacred cow, a symbol of Zululand, was whipped by invaders and then torn apart, its body scattered over the land. Kenneth approached me and said, "I will tell you now." I understood what he was referring to. We stopped and gathered in a circle. The group sat as the young performer stood in the center, and I stood nearby.

I filled the group in on what had happened a week earlier regarding the scars, his hesitation to talk, and his anger. Kenneth then removed his shirt and turned around to show the group. I stood with Kenneth and asked him about the incident. As he spoke, he slowly shared how, when he was 14, he and a friend were walking home from school. "What were you wearing?" I asked. He was in his school clothes and carrying his backpacks. "So, they knew you were just schoolboys?" "Yes."

"Tell us what you saw," he said, recalling the day he had been living in a township where a protest was happening. On one side were the demonstrators, and before he knew it, the situation had escalated; the police stepped in, and scuffles, teargas, rubber bullets, and beatings erupted on the streets. He and his friend got caught up in the chaos, were arrested, and stuffed into a police van. He was separated from his friend and questioned. "I told them the truth; I didn't want any trouble. Why didn't they let me go home? I didn't do anything. Look, I'm just a schoolboy with books, not rocks." Over several days, he was beaten with a cane, and the police held him for two weeks without charging him. "They told me I needed to know my place, and caning would help me remember."

He stood staring, emotionless. The group was silent, and a dark cloud had settled over the circle. He was 24, and it had been ten years since the event. He had never told his story before.

"What happened when they released you?"

"I went home."

"Who was there?"

"My mother and father."

"What did they do?"

"My mother cried and held me."

"What did your father do?"

"He held me but didn't cry."

"Why didn't you tell them what happened?"

"They knew, they saw."

"Why did you wait until now?"

"I am a Zulu. I am a warrior. Others have died."

As I asked who in the group had been arrested. It was a question I hadn't considered before. They were young, thoughtful adults, aged 20 to 35, taking part in a training program to become performers. During apartheid, Blacks were barred from performing for fear that they would use performance as a means of sedition.

"How many of you have been arrested?" I asked again, sensing their reluctance to relive their experiences. Every Zulu in the group raised their hand, and we went around the circle, with each person sharing the details of their arrest and incarceration. Many of the men had been arrested multiple times.

"How many of you have been beaten by the police?" Every hand shot up except for three women. The air was thick with tension, and many faces were blank and staring. The harsh realities of Apartheid were exposed. It was living in the bodies, minds, and spirits. How many generations, I wondered, would it take to heal such wounds?

Each performer shared their experiences of police brutality, perpetrated in no small way by Zulu police. Pointing to scars on their faces, arms, legs, and backs—marks I had previously thought were the results of everyday accidents like falling off a bike or rural childhood mishaps—turned out to be the remnants of a horrifying moment in history.

When we are given the freedom to live and speak, the work takes on a life of its own, moving in its own way. We concluded our session by agreeing to turn the emotions we shared that day into a performance. We would speak what had not been or could not be spoken, giving form to our feelings, conjuring the ghosts that surround and haunt into being. Facing them was the first step to healing.

FIGURE 1.1 Production photo from *Emandulo*, Kwasa Group, Durban, South Africa

(l. to r.: Lindelani Nzimande, Benefit Mzobe, Patricia Majapelo)

Source: Photo by Shelley Kjonstad.

Listening is an ongoing process—something you can be doing right now—that is fundamental to our lives and work. By listening—with your whole being and all your senses—you reaffirm a basic principle: Everything you need to know, you already know. It is present and in this moment. Siberian Yup'ik elder Jimmy Toolie often told me, "If you don't know something, all you have to do is listen." Toolie, who has since passed away, lived on Savoonga, an island located in the middle of the Bering Sea. On the world map, the International Date Line curves around the island to include it in the Northern Hemisphere. For the hunters of Savoonga, crossing the dateline inspired a profound yet straightforward saying: "I went hunting today to get game tomorrow so I could eat it yesterday." Time and space are relative concepts, shaped by the world around us. Toolie's listening is about being present, immersed in the land, the weather, the oceans, the seasons, the animals and their migrations, the histories unfolding across the world, and the self—the one who experiences it all. Deep listening transcends time, space, and human concerns. Our planet is a closed system, with everything that was and will be already present for those who listen.

Each of us is a living vessel, a carrier of a vast, layered legacy—an inheritance of millions of years of human evolution, filled with biological and cultural memories. Everything that has been and everything that will be exists within and around us—for they are one and the same but different. Our entry is listening to this very moment.

LISTENING

Ask the performers to close their eyes and take a few deep breaths to center themselves, whether they are standing, sitting, or lying down.

Encourage performers to notice their surroundings. Present the following series of observations and questions to the performers:

- Feel the air around you.
- Pay attention to what is happening in your body.
- Notice areas that feel tight, heavy, or painful.
- Identify parts of your body where you feel most at ease.
- Acknowledge the places in your body that feel the best.
- Keep breathing.
- Select a part that feels tense.
- Then, choose a spot that feels light.
- With your breath, direct attention to that area.

- Continue breathing and experience that part of your body.
- Avoid passing judgment on your feelings.
- No one can discredit your feelings.
- Feelings simply exist and experiencing them is the key to transforming their energy.
- Do not resist or judge what arises.
- Allow the feelings to expand and be experienced.
- Listen to your inner voice.
- This voice connects to the spirit of your life.
- This voice connects to the spirit of all life.
- Allow the feelings you observe in your body to unfold and pay attention to them.
- Keep breathing deeply.
- Follow your breath as it guides you on a deeper journey within yourself.
- Inquire of the voice inside what you need to do for yourself.
- Ask how it can assist you right now in meeting that need.
- Ask the voice to share a small step you can take today or tomorrow to help meet that need.
- Question whether what you feel is a truth or a lie.
- Continue breathing and follow your feelings.
- Explore as many feelings as you want or as much time as you have today.
- Recognize that beneath all the feelings you experience, there can be both positive and negative emotions.
- Recognize that feelings are often hidden or mixed up with other feelings.

Walking

Walking is a fundamental part of being human, shaping our bond with the earth. I begin my workshops with a simple walk, an inherently human activity that offers unique insights into personalities, contexts, and histories. This reveals the lives of individuals, their communities, and the social and cultural vocabulary they share. When people are asked to walk like an animal, the animals they choose reveal a lot about their understanding of and connection to the animal world. In Western classrooms, students often select domestic animals such as cats, dogs, mice, squirrels, birds, or even chimpanzees. In contrast, Alaskan responses include seals, whales, caribou, salmon, walruses, seagulls, and ravens. In Sakha (central Siberia), horses, milch mares (a type of short- and long-haired horse found only in Siberia), elk, owls, bears, wolves, and eagles are invoked.

When asked to "move like a dancer," the performers instantly showcase their dance skills, cultural traditions, and willingness to express themselves. For those who are shy or reserved, exploration provides a natural and straightforward way to "perform" experiences that we can build on. "Move like a drunk" reveals social issues and perceptions that might not be otherwise apparent. Eskimo, Sakha, and Polish performers immediately staggered and fell when invited to "move like a drunk," offering a stark glimpse of their social reality in the context of high rates of alcoholism.

In contrast, when the same prompt was given, performers in my Drama Therapy class at the California Institute of Integral Studies (CIIS)—mature, professionally minded graduate students—delivered stiff, almost stereotypical portrayals of drunkenness. Two performers, who were Catholic nuns, stood still and later explained that they had no personal

FIGURE 1.2 *Walking* exploration during a workshop at the DirectorsLab Chicago

Source: Photograph by the author.

experience with drunkenness. The same two nuns repeated their ignorance when asked to "move sexy." When asked to walk sexy, several Zulu performers, both male and female, needed to be warned against touching or engaging in suggestive actions. A group of Finns during a corporate "creative" workshop walked reservedly, keeping a distance to avoid corporate decorum. A group of young actors in their early twenties at the Estonian National Theatre School moved awkwardly, imitating walks and poses suited for Instagram or media posts. When asked to "move sexy," a group of Yup'ik Eskimos of mixed ages in Bethel, Alaska, responded with carefree laughter, grins, and giggles. A simple suggestion can transform a walk and convey more than any discussion. Our bodies are a way of listening and conversing. We are always in conversation with the world, responding, recording, and conveying a constant stream of information. The body can't lie; when it tries, it reveals a more profound truth. Watch and listen.

Walking recurs as a physical theme in our work. Subsequent explorations will demonstrate how they can be repeated and adjusted to meet project goals or introduce additional complexity. We all walk on Earth, but how we traverse it varies depending on our location and social and cultural contexts. Walking connects us to the Earth and our surroundings, shaping us and reflecting the spaces we encounter. It's the first step in understanding and developing our performance vocabularies.

MOVEMENT IN SPACE

Ask performers to walk at an everyday pace.

Encourage performers to increase their pace and then weave in and out of each other's paths, always avoiding physical or eye contact.

As they walk, jog, and then run, ensure they maintain physical and eye contact as they cross paths.

Remind the performers to vary their movement patterns to avoid predictability. We tend to settle into unconscious patterns. The idea is to become aware of this and either consciously accept the pattern or work to change it—an awareness that is valuable for life and one's creativity.

If they start to fall into a pattern, have them change direction and keep side-coaching to adjust.

Increase the pace until the performers are moving at their fastest speed while still avoiding physical and eye contact.

As they pick up the pace, transition into a run, then gradually slow down until walking at a normal pace again.

WALKING ATTITUDES

Performers are asked to move within the space neutrally—whatever that means to them. Then, they are asked to respond to the suggestions listed below.

Participants should be given time to explore each suggestion. Some attitude suggestions may require more time and deeper exploration, while others may be more immediate. Each will reveal the performers' abilities, imaginations, and willingness to explore and express themselves, as well as the group's prevailing attributes.

It is important to emphasize that there are no right or wrong responses to the movement attitude suggestions; performers should respond in a way that feels appropriate to them.

First Application

Move with strength
Move with fear
Move with love
Move confused
Move paranoid
Move sexy
Move tired
Move happy
Move like a bird
Move sick
Move lean
Move angry
Move drunk
Move like a dancer
Move like an animal
Move overweight
Move like a child
Move old
Move foolish
Move wild
Move joyful

Adapt, modify, or adjust based on needs, goals, and context. At the end of the exploration, have the performers return to an everyday walk.

Walking Attitudes anticipate performance development needs. This activity acts as a casual warm-up, energizing performers and group play, as it encourages the exploration and validation of individual vocabularies and creative and expressive potential.

It is essential to give performers enough time to thoroughly explore each suggestion. Performers should always keep moving.

Second Application

The exploration can be repeated or continued in another session with the following more abstract movement attitude suggestions:

Move straight
Move curved
Move zigzag
Move high
Move cold
Move low
Move wide
Move open
Move stretched
Move solid
Move like air
Move loose
Move closed
Move like water
Move direct

Move hot	Move angular	Move lukewarm
Move indirect	Move slowly	
Move close	Move fast	

Move as close as possible without touching anyone. Try to stay as far away as you can from other performers. Move in quickly and closely. Then return to a normal walking pace.

Reassess your attitudes in relation to your objectives.

After the exploration or session, ask the performers to reflect and share the emotions and thoughts that different attitudes brought up.

What kind of personality comes to mind? What images, ideas, situations, and scenarios pop up based on your experiences, culture, daily life, movies, and other media?

Note: Feel free to repeat, modify, or expand this exploration as needed. For instance, you could revisit the exploration with a focus on animals, machines, or mechanical movements. It could also be developed to include character, scene building, or choreography.

Explorations don't always unfold as you expect. Reactions can be unpredictable, frustrating, yet also revealing. Throughout this book, I have included journal excerpts from my experience working with diverse groups, populations, and conditions across six continents over decades. These insights should offer a glimpse into the dynamics of creating place-based performance and show how my methodology—explorations, explorations, and the process outlined in this book—can work (or not) in real-world situations.

Back in 1993, I was invited to lead workshops and develop a performance for the Sakha National Theatre in central Siberia. This happened shortly after the Soviet Union collapsed, and the Sakha Republic joined the Russian Federation. I was one of the first foreigners to visit the region, which had been closed off due to its strategic military importance and its history of gulags, uranium mining, and diamond and gold deposits. My task was to create a performance based on Sakha's traditional culture as part of a larger effort to revitalize it. During our project, I worked with 27 professional actors from the Sakha, Evenk, and Yukaghir ethnic groups, all of whom are members of the National Theatre's acting company. Their ages ranged from 22 to 71, and each was trained in the Stanislavski method after graduating from conservatories in Vladivostok, Moscow, or St. Petersburg, and performed realistic and propagandistic-style plays. Ethnic minorities had been Russified and were forbidden from embracing or promoting their own cultures for fear that it would threaten Soviet and Russian dominance. The passage below shares my thoughts on the first week of our collaboration, which was an eight-week process leading up to the creation of our original, devised production, *Sardaana*. In the Body and Space section of the book, I will include several excerpts from my Sakha journal to provide readers with a clearer understanding of how a project might develop, the challenges it faces, and the strategies used.

From the Siberian Journal

It was fascinating to observe the reaction to **Walking Attitude**, which mirrored the responses from Slavic Russian performers in St. Petersburg the previous year. They responded to the obvious prompt and stopped there. It was as if they were trained not to exceed the boundaries of what was asked, rather than seeing the prompt as an opportunity to explore and play. It was like they were disconnected, zombies carrying out the request and going no further.

Did they fear expressing themselves differently and risking mistakes? Straying from the group could have political consequences. Or did they simply not know how to play and express themselves imaginatively? Then, there is the proclivity of wanting to discuss each exploration in endless, mind-numbing detail. Was this a habit picked up during the Soviet era, when feelings of disempowerment and isolation led them to exert agency through conversation? Or was it a way to fill their days?

Working through a translator in a large, echoing room made the situation even more difficult, and it felt like everything was working against the project. There were times when I felt completely at a loss due to how little control I had over the situation. I tried to give straightforward instructions for translation, but each time Izolda, my translator, shared my words, there was a brief silence before a loud and lengthy debate broke out. Several people would speak and gesture at once, asking me questions, seeking clarification, and debating with one another. Things would only get worse when the core idea of an exploration sparked reactions before anyone had a chance to reflect.

The more sensory the exploration, the harder it was for them to understand, resulting in lengthy debates. They preferred complexity, perhaps to cope with conformity and the dullness caused by an ineffectual system. Complexity and conflict were part of their daily lives, reflecting the flaws of the Soviet system that defined their reality. The inconveniences and complexities distracted from the government's ineffectiveness, keeping the public preoccupied and focused elsewhere.

When you're focused on just getting by and dealing with the day-to-day, there's no space for reflection or questioning. They were trained to conform, to blindly follow the rules. The real challenge was changing their mindset to be more open to experimentation and improvisation—how to reintroduce play, fun, and creativity into a life that had become so serious and predictable.

In Russian theater, directors often held an almost authoritarian position, which made performers hesitant to trust their instincts as they tried to meet my expectations. They wanted to please me, but first, they had to figure out how. Yuri mentioned that they were eager to discuss their approaches, like how to "Walk correctly."

I encouraged them not to get too caught up in analyzing or pleasing me, but to focus on the explorations themselves. Their expressions showed that they were skeptical, so we had a deeper conversation about what I was trying to achieve. I was proposing a completely different approach to our work, and I knew it would take time—it's a process of trial and error.

FIGURE 1.3 *Walking Attitudes* exploration in the theatre lobby at the National Theatre of Sakha, Siberia

Source: Photo by the author.

Along with the walking explorations, I've included some additional options. It's best to avoid doing multiple walking explorations in one session. Instead, vary them and use the exploration you feel is necessary. Walking is a great starting point for deeper work, and you are in the best position to decide when to do it and how much. Trust yourself. If things do not turn out as planned, don't get discouraged—try something else. Talk to the performers and find an alternative approach. Walking activities are beneficial for preparing for your main work.

WALKING LIGHT

Encourage the performers to move comfortably and naturally in the space. Once they are settled, ask them to develop the following:

- Walk quickly and lightly, as if they are crossing thin ice.
- Walk as if flying a kite.
- Walk as if holding a helium balloon lifting to the sky.
- Walk as if walking on a cloud.
- Walk as if navigating through a dream.
- Walk as if on water.
- Walk as if walking on air.
- Walk as if resembling a spirit.
- Walk as if traveling in a spirit world.

End with walking as an Other than Human Being—a specific animal, insect, plant, or element.

Encourage performers to view their walking weight as a means of expression. This awareness is beneficial in performances that explore different realities, trance states, non-human beings, or mythological stories. All of which will be discussed in the **Place** section of this book.

From the Burkina Faso Journal

It's the first day of a weeklong workshop, from 9 AM to 12 PM, Monday to Friday, for the FITMO festival in Ouagadougou. The workshop is held outside the city center, leading to much confusion about the location and participants. At 9 AM, it was sweltering—the southern edge of the Sahara

Desert—and dust storms were sweeping through. Desertification is occurring as the Sahara gradually pushes southward.

When we initially planned the workshop a few months ago, it was set to take place at the University of Ouagadougou. However, the university has been closed for the past month due to a lack of funding from the central government. President Kampore, often seen as a "Do Nothing" president, is widely regarded as corrupt and unpopular. Such issues are all too common in Africa.

The performers were drawn from several ethnic and tribal groups from across Mali, Burkina Faso, the Ivory Coast, Togo, the Democratic Republic of the Congo, Sierra Leon, Senegal, and Ghana.

Once we gathered, a performer from the Democratic Republic of the Congo spoke about the dire conditions in his country. "Theater is now the only educator in the Congo due to social turmoil. Teachers are absent, and schools are closed. Theater in the Congo is essential for survival."

Every year, the FITMO Festival hosts a series of workshops. For many West Africans, this is their only chance for formal theatre and performance training.

On the first day of my workshop, the organizers surprised me by asking a Swiss dancer who had received a grant to lead a warm-up session. She guided the group through a 20-minute Graham Technique warm-up, which was not relevant to my session and didn't interest the 33 participants. The French government is a major sponsor of the festival and uses it to promote French culture, language, and artists. In the next room, a different workshop was led by a mime who painted her face white. Her presentation sparked our discussion on the cultural body and language, as well as neo-colonialism, setting the tone for our workshop.

Our journey began with **Walking**, exploring what performers notice every day on the streets. It was an explosion of energy, ideas, and creativity. They showcased people with disabilities, quirky individuals, street vendors, beggars, sex workers, a swaggering police officer, a Catholic priest, and traditional activities like selling, smoking, and drumming, which conveyed their lives, a vivid portrait of life in West Africa, which lived in their bodies.

The prompts from **Walking Attitudes** revealed glimpses of traditional dances and rituals. Images of the dead walking, possession, and ghosts appeared when prompted to walk as a spirit or ancestor, accompanied by rhythm, detailed gestures, and facial expressions—a remarkable outpouring of life, imagination, and creativity.

The walking-sick *prompt* evolved into extended scenes of sickness and mourning.

WALKING TENSE

Ask the performers to move as freely as possible, and as they do, assess their bodies and stretch any parts they feel need stretching. It is a self-check.

Then, ask performers to move through every joint in their body—knees, elbows, shoulders, hips, ankles, and so on—until they feel supple, loose, and flexible. Next, have them loosen all their joints simultaneously, so loose that they seem as if they might sink to the ground.

With this awareness, encourage them to walk normally, in a casual, everyday manner.

Next, ask performers to isolate each muscle group, tensing it as they walk naturally. For example, they might tense their shoulder muscles while walking and notice the tension as they move.

Then, instruct them to relax their shoulders, tense their thigh muscles, and continue walking.

Allow time to notice the tension in isolation. Gradually guide the release of tension in one specific area before moving to another. Always maintaining movement.

Additional areas to focus on and release tension may include the hands, face, neck, pelvis, feet, calves, stomach, forearms, and back.

During the exploration, invite performers to observe how tension in each individual body part affects their overall movement, how other body parts react, how their mental state changes, and what kind of human, bird, non-human object, or animal character is implied.

Ask:

When do they move like that?
Consider opposites and balance. If one part of the body is tense, what part is not?
What is the opposite within the body?
What rhythm does the tension create when you move?

Exploration Variation: After the performers have identified and explored the tension in each body area, encourage them to express or repeat a simple, specific line for each tense part.

A possible line could be, "Hi, how are you? I'm doing fine." How does the body's tension impact the voice?

Consider this exploration when developing a character's body, focusing on tension, movement, rhythm, and voice.

LEADING WALKS

Have performers walk as if a specific body part is drawing them forward, guided by an imaginary string attached to that part.

Encourage performers to take time to explore the sensations related to each leading body part.

Encourage the performers to lead in any order with:

Elbow	Eyebrows	Genitals
Forehead	Stomach	Chest
Nose	Knees	Crown of the head
Navel	Eyes	Lower back
Chin	Lips	
Genitals	Toes	

Prompt performers to reflect on each lead by considering the following:

- How does leading with this body part change your view of the world?
- When do you walk like this?
- When have you seen others walk this way?
- What kind of character shows this type of walk?
- What kind of animal is implied when you lead with that body part?
- How do you walk when you lead with that body part?
- What is the rhythm of your walk?
- How does your foot hit the ground?
- What makes your walk stand out?
- Is it lighter or heavier?
- Are you walking on the ball of your foot or your heel?
- When one part of the body takes the lead, another part maintains balance. This shows how the body stays balanced.
- What do you support when that specific body part is dominant?
- Conversely, what is its opposite or contrast in your body?
- Define the lead and its balance. The walk and its rhythm develop from this opposition.

Exploration Alternatives and Evolutions: Run, crawl, dance, skip, or engage in any movement guided by a specific body part, while maintaining the awareness mentioned earlier.

Transform the leading walk into an animal form.

Vary the body parts to demonstrate changes in what is held and the rhythm shifts created by emphasizing each specific body part.

Instruct performers to choose one of the reviewed walks and interact with others.

Ask performers to develop a character voice based on the selected walk. Begin by communicating solely through sound, then gradually add one or two phrases that match the character revealed.

Character Study Continuation: During the session, invite performers to pair up and identify what their partner is initiating—what they carry and their rhythm. Then, they take turns practicing each other's walk by walking alongside one another.

After allowing time for exploration, present the main walk of their partner. Encourage observations and feedback; ask the group to try the walk as it was observed and analyzed.

To continue the study, instruct performers to observe and record three different kinds of walks in their bodies, noting them in various settings—such as a store, on the street, or at a club. They should pay attention to what the individual (or animal, if applicable) is leading with, noting what they are holding and the rhythm emerging from their movement.

Performers will share their three walks with the group during the next session, where others will observe and give feedback.

Ask: *What are they leading with? What are they holding?*

Optional: Ask performers to note one line from someone they observed in public and include it in their presentation.

This activity has proven highly effective across various cultural and university classroom settings. It provides a clear and practical way to connect with the world, transform the body, and foster a psychologically balanced character. It is an excellent tool for character and performance development.

Highly Recommended.

SMILING

As the performers walk by, they make eye contact and exchange friendly smiles. Their pace picks up, and they start running, still smiling. With side coaching, they slow down to approach the nearest person, lock eyes, and share a humorous moment. They keep their cheerful vibe going the whole way.

The performers keep moving and smiling, pausing to make eye contact and playfully make funny faces whenever inspiration strikes.

Encourage the performers to keep moving and smiling, finding new and creative ways to express themselves with their faces and share humorous moments.

Exploration Variations: This activity may involve using the body, stopping in front of another person, establishing eye contact, and doing something humorous that contrasts with facial expressions and isolated body parts.

Note: This exploration is great for breaking down social and creative barriers, easing inhibitions, and improving body expressiveness.

Our Work

For thousands of years, Indigenous peoples have honored and celebrated the connection between bodily cycles, physical awareness, and performance, reflecting a deep understanding of natural rhythms. In many Indigenous cultures, human movement and expression are closely tied to the cycles of nature, with each gesture, dance, and ritual representing a link between human experience and the living world.

Like the Native American medicine wheel, which associates human life stages with the cardinal directions and changing seasons, performers must grasp how their physical expressions fit into larger cycles of meaning. Every culture identifies life transitions with some form of physical recognition, performance, or ritual, whether it is blowing out candles and cutting a cake at a birthday party or exchanging wedding rings. Life transitions are significant in every culture, to varying degrees, from birth ceremonies and baptisms that introduce infants to the elements to coming-of-age ceremonies that parallel biological milestones—such as first communions, bar mitzvahs, and Quinceañeras—with specific performances, regalia, objects, and dances. Diné (Navajo) Kinaaldá ceremony for young women, for example, connects the participant's growing maturity to the ripening of corn and the moon's cycles, which include a run toward dawn to symbolize awakening.

Our path draws inspiration from Aboriginal Australian songlines, which weave through the landscape. These songlines use storytelling and dance to map journeys and teach about the connection between past and present, material and spiritual. They reveal the world as a narrative that shapes our lives—where we are transitory. Performing involves wavering between the particular and the ineffable. This awareness is fading amid anthropocentric egotism, distraction, and technological and capitalist acceleration. Yet, there's an emerging, intuitive yearning for balance. Our work is part of an awakening to find another way to be, drawing on the past to meet present needs and becoming agents of a mindful alternative. The future starts with the power of one: Our goal is to turn consciousness into action.

Viewed as a link between the body and natural forces, our explorations carry deeper meaning. They aren't about mimicry, conformity, or repeating tired notions of our place in the world. Instead, they seek a deeper understanding of our roles within a life cycle beyond time, space, and material existence—both personal and universal, specific and unexplainable. This empowers performers to move beyond inherited, socialized, and cultural norms.

Gaining and expressing performance knowledge is a spiraling, sacred path, with performers serving as a community's avatar, awakening and bringing awareness to deepen understanding and contributing to the evolution of their community and, in turn, to life on earth. Our work is larger than we will ever fully understand, and this is as it should be. From this perspective, performance extends beyond technique, entertainment, or personal expression. It is an acknowledgement that celebrates our contribution to the ongoing cycle of life. Ours is a new form of ritual function, one that is taking shape to meet the needs of our dynamic and complexifying time. It aspires, explores, and experiments to bridge the physical and spiritual worlds, asking and wrestling with questions that may offer some solace, insight, and healing.

Our work sees performance as both a place and a process of evolution—a place that merges the temporal and the infinite, the personal and the universal. Its potential has been marginalized, tempered, and contained, but the vestiges and examples of its form, function, and power remain. It is where everyday actions can take on meaning and understanding, where personal and temporal expression align with the natural rhythms and cycles of the infinite universe, to forward the evolution of planetary alertness. For whom do we perform and why?

From the !Xuu and Khwe Bushmen Journal

As we talked, I realized I had taken many of my basic ideas about performance for granted. When I asked who they were when they performed the story, they looked puzzled and said, "We were ourselves." "But weren't you the old-time (ancestors) people?" I asked, "No, how could we be?" "When you performed the tall grass moving in the wind, you were, what I call, performing." "We were the tall grass and the wind. We can be the grass, the wind, and ourselves."

I tried to figure out how to introduce the idea of performance as pretending and imagining to people who had no concept of it. Maybe my deeply ingrained binary concept of reality was stopping me.

The subject sparked their curiosity, so they asked about it, and I tried to explain the concept of acting. However, they got confused when I described someone pretending to be someone else. They thought I meant soul possession or the way an evil doctor works by pretending to be a good one. Maybe they were right.

From their perspective, everything is connected and part of a larger whole. Since all things are interwoven and belong to the same system, there's no real difference between one thing and another. Identity and sentience are fluid, similar to quantum phenomena. Individuality isn't a primary focus, nor is the need to differentiate oneself from others, as is common in Western culture through pretense or acting. The !Xuu and Khwe Bushmen existed across multiple realities simultaneously, shifting emphasis and adapting fluidly, accessing and expressing what was needed at the moment.

Our discussion about pretending and acting shifted to a talk about metaphor. The idea that one thing can stand for another was new and almost silly to them. The literal way of thinking of the Bushmen didn't surprise me. They were practical people, living in the Kalahari and off the land until recently, without modern tools. The need for metaphor and "acting" as something or someone else comes from the desire to connect with another person, event, or thing. From their perspective, they were already connected.

In the Western context, metaphor, like acting, functions to connect or reveal an idea or feeling by identifying, emphasizing, comparing, and contrasting. The Bushmen showed little interest in such things because everything is self-evident. Things simply are what they are. Everything inherently references what is both itself and something greater simultaneously. Mythology and the world exist within them, not as something separate or distinguished from the self. They had no understanding of metaphor because they had no need for it. Everything is everything. Their human form was a vessel carrying the world. My mind started spinning, forcing me to reconsider all my fundamental assumptions about what I knew and how I perceived the world.

For the !Xuu and Khwe, the concepts of internal and external to the self, are essentially the same. The need for metaphors and acting in Western and other cultures arose when they severed their intimate connection to the earth that they had as hunter-gatherers. Did metaphors then become a way to reconnect with a lost sense of unity? Did acting, like theater, arise when homogeneous groups interacted with others, serving to broaden their understanding of others? Acting lets one become someone else, and through that, understand others. Is the goal of acting and theater to reconnect with others as a way to regain a lost sense of holism?

These questions still rattle me. My experience with the !Xuu and Khwe Bushmen remain a pivotal, fundamentally life-altering experience. One that has enlarged my understanding of performance and the world. I carry my moments with !Xuu and Khwe with me to this day and are now a part of my thinking and feeling.

Exploration

ISOLATING BODY PARTS

Start at the top of the body and guide the performer through each identified body part, isolating each.

Suggested body parts for identification and isolation include:

Scalp, forehead, eyebrows, eyes, nose, mouth, ears, jaw, neck, shoulders, arms (either one at a time or both), hands (either one at a time or both), fingers (individually or as a whole hand), chest, stomach, buttocks, pelvis, genitals, thighs, knees, calves, ankles, foot or feet, toes.

Encourage the performer to isolate each body part and develop awareness through tension release.

Exploring Body Parts

Revisit each part to explore its emotional and expressive capacity. Establish a relationship and dialogue with each body part.

For example, ask the performer to use only their hands to examine the emotion of sadness.

Ideas for exploring emotions and expressions related to body parts include:

Happiness, innocence, hardness, softness, love, pain, pleasure, defiance, fear, strength, weakness, masculinity, femininity, beauty, ugliness, pride, dryness, moisture, contraction, expansion, joy, sorrow, agony, ecstasy, foolishness, silence, vulnerability, openness, closure, calmness, distress, death, life, cruelty, compassion, uncertainty, certainty, sanity, and insanity.

Side coach. Feel free to add or remove emotions and body parts as appropriate. Delve into the expressive and emotional range of each body part, exploring their potential thoroughly or selectively. Keep the exploration engaging and avoid predictability by providing insights and expanding on observations.

Remember that you can revisit and reference this exploration in future sessions.

Highlight how easily or how challenging it is for each body part to express specific emotions or states. Note that some body parts, such as the arms or thighs, may find it more difficult to convey certain emotions.

Exploration Alternative: Discover the connection between body parts and animals. Ask performers to identify and isolate a specific part of an animal's body. This can be done through in-person observation or media. Then, explore how that body part relates to your own body. For instance, your hand might become like the animal's paw, your ears like its floppy ears, or your nose like its snout. How do these body parts move? What are their characteristics, limitations, and abilities?

Exploring the body of an animal, plant, tree, or other non-human being helps you expand your awareness and, in turn, connect with the larger world. Many cultures that focus on hunting and gathering incorporate animals, plants, and other non-human beings into their performances, demonstrating a worldview that recognizes the fluid and interconnected boundaries between humans and non-human beings.

BODY PART SCENARIO

Have performers consider which emotions or expressions are strongest for each body part. If a particular body part doesn't have a strong or personal connection, it can be left out.

The performer is asked to develop a body part scenario by moving from one body part to another, expressing emotion for each.

Then ask performers to establish a pattern of moving from one body part to another.

This pattern should establish itself and be repeatable, becoming a Body Part Scenario.

For example, (1) The forehead expresses fear, leading to (2) the eyebrows, which convey excitement, progressing to (3) the eyes, then chest to express love, transitioning to (4) the hands into a fist, to express determination, followed by joy and relief in the stomach, and so on throughout the body, with each body part inspiring the expressiveness of the next.

Emphasis should not be placed on the logical significance of the sequence. The movement between body parts is driven solely by emotion and intuitive association.

Repeat, listening to the expressiveness of each part and the overall expression of the scenario.

The body serves as the mnemonic, microcosm, and template for our work and for the world.

Encourage an exploration of which body parts best convey each emotion and the limits of what a specific body part can express.

Body part expressions evoke a psychophysical response and lay the groundwork for recognizing or developing a cultural or personal performance vocabulary.

Encourage the performer to allow the emotional connection to flow seamlessly through their body, moving fluidly from one part to another.

Ask the performer to repeat, revise, and elaborate on the Body Part Scenario until it feels natural and instinctive. Encourage the performer to express the story that the sequence of emotions conveys.

Ask:

What is the story?
What is the emotional through-line?
What psychological state and progression does the Body Part Scenario imply?

Remind performers to avoid rationalizing and instead approach the story on an emotional and physical level.

Group presentations of Body Part Scenarios are recommended to conclude and allow performers to learn from one another.

Exploration Alternative: Draw inspiration from animal movements, whether from a single species or a combination, and instruct performers to identify these movements in their own bodies. Once they find the right body part, they explore its expression and develop a repeatable, emotionally engaging scenario.

Note: This exploration uses reverse engineering to develop movement vocabularies as a foundation for performance and expression, blending personal, cultural, and animal inspirations.

BODY PART SOUNDING

Building on the **Body Part Scenario**, this exploration lets you delve deeper.

Repeat the scenario.

A key part of this process is allowing an unconscious, non-rational, and non-intellectual expression to emerge.

The performer's task is to create a vocal sound as an organic response to each emotion or body part's expression. Any vocalization is acceptable as long as it is not recognizable as language. Performers should explore a wide range of sounds inspired by their body's emotions.

Give the performer time to absorb, adjust, and refine body-sound coordination. The Body Part Scenario will evolve from the non-vocalized version.

Make the vocalized Body Part Scenario more storytelling and dramatic. It is a great method, but distinguish between individual body-part emotions and the full-body expression that emerges from the scenario and vocalizations.

Allow the performer the freedom to explore and repeat the vocalized Body Part Scenario multiple times.

Share the vocalized Body Part Scenario with the group so they can observe and comment on its progress, encouraging critical development. A discussion will follow.

- **Use explorations as needed, based on your personal goals and needs.**
- **Trust your judgment and follow your instincts.**
- **Feel free to adjust, repeat, or combine explorations as you see fit.**
- **The order of explorations outlined here is a suggestion, not a strict step-by-step guide. The sequence in this book is meant to give a general idea of the progression.**

SEPARATING SOUND

With discussion and your guidance, performers decide which individual Body Part Scenario they find most inviting and would like to work on collectively.

If there is a large group, split into smaller groups.

Performers learn, focus, and explore the chosen scenario.

It is essential to maintain the integrity of the originating scenario to advance this exploration.

Performers are prompted to elaborate on each vocalization and enhance it with percussive sounds, instrumentation, or other sound effects.

Have instruments or sound-making objects available.

This exploration aims to create a sound score (vocal and instrumental) without intelligible language. Each sound originates from the body part and is an extension of its expressive connection.

The collaborative sharing and exploration of the scenario will yield unexpected results.

Encourage the growth and exploration of the sound score, allowing it to evolve from its initial impulse.

If the original impulse seems lost, ask the originating performer to reiterate it to reaffirm its connection.

Promote the development of a rhythmic scenario inspired by, but not limited to, the original **Body Part Scenario**.

Present the sound scores to the group without movement. Allow time for comments, reflections, and suggestions.

Exploration Evolution: Add clear, repeatable words that are linked and express the Body Part Scenario.

Note: When exploring animal or plant-inspired Body Part Scenarios, consider perspectives, vocabulary, and rhythms. The exploration mixes human, animal, and plant expressions. Embrace the opportunity to co-inhabit.

BODY PART DANCE

Continue working in groups or solo to record each sound score. Groups and individual performers are now asked to revisit the evolved **Body Parts Scenario** sound score, bringing it back to movement.

Performers are encouraged to modify, enhance, and refine how the sounds correspond with the specific body parts to develop a movement sequence or dance-like scenario.

Changes in both sound and body movements are expected and organic to the process.

Urge performers to adjust the sound score or dance, such as repeating a sound to cause repetitive movement of the body part that makes it.

Caution performers to work slowly, breaking the score into sections before returning to the overall dance scenario. This evolutionary process involves listening and working intuitively rather than relying on intellectual reasoning.

Some performers may focus exclusively on the sound score, while others dance or move. Let the process take its course. This exploration can be more challenging for individuals working alone.

Allow time for the exploration to develop and mature. It may take several sessions or become an ongoing, revisited process.

BODY PART DRAMA

This exploration begins by examining the range and expression of a specific body part.

Focusing on one body part taken from the previous explorations, side coach performers through a series of expressive explorations.

Express that body part with grace, pride, fear, gymnastic movements, politeness, strength, comedy, beauty, exhaustion, and add prompts as needed.

After exploring, ask performers to find a partner to interact with and create a dialogue.

Ask them to fully explore the range of foot expressiveness and communication using only that specific body part.

No spoken dialogue is necessary.

To conclude, have performers create a body part drama improvisation using only that body part. Repeat, refine, and organically develop vocal sounds and dialogue.

Exploration Variations: The exploration can extend to other body parts such as the hands, feet, mouth, arms, and facial features. Be creative and don't limit your creative expression.

Note: The Body Part exploration series offers many benefits. It encourages intuitive expression rooted in bodily emotions, promoting holistic sensory and imaginative integration for performance. It helps performers draw on expressive resources and organize feelings from personal, social, and cultural contexts.

Rhythm

Please think of the Circle as our guiding framework for thought and action, with rhythm acting as the life force that connects our bodies to the world. Rhythm is the driving force, the connective tissue, and the inspiration that enables and moves everything.

In this section, we will explore rhythm as a fundamental aspect of body knowing and performance. Experiences with various indigenous and non-indigenous groups have consistently highlighted the importance of re-establishing a sense of rhythm in a performer's body and life. Rhythm awareness awakens and invigorates a different kind of perception—an attunement or synchronization between bodily rhythms and those of our surroundings, including their tones and textures (Abram 1996, 54). I call this awareness rhythm reality.

Imagine the body at the center of a circle, with the circle's edge representing the world's constant motion. Rhythm energizes and connects the body to the circle of life, setting it in motion. The circle symbolizes a concrete idea, while the cycle is an abstract concept—rhythm brings both to life, making them real and tangible.

Our world is built on circles, cycles, and rhythms. This forms the foundation of how we function and is central to our work. Every aspect of the planet—its environment, geology, and biology—is linked to this framework. It's at the core of every human experience, driving individual, cultural, social, political, historical, psychological, economic, seasonal, generational, and millennial events.

Review your day from waking to sleep. You bring your rhythm into the world, which adjusts as you go. Interactions with others, pets, calls, food, news, weather, traffic, appointments, personalities, meals, conversations, feelings, digestion, homecoming, relaxation, work, petting dogs, reading, thinking, TV, social media, shower, and bed all influence your internal and external rhythms. These rhythms come from near and far.

Rhythms surround us everywhere, layered and constantly changing within our bodies. Recognize the polyphonic rhythms of trees, flowers, birds, animals, and insects. Smells reveal odors in the morning and evening, responding to sunlight, darkness, rain, and other weather conditions. Nature has an accumulative, restorative rhythm. Your body is a bundle of rhythms; understanding this helps create harmony within ourselves and others.

Work to stay calm and centered amid shifting rhythms, essential for tapping into the life force that energizes us. Develop this awareness to enhance your work, performers, and community, as it's life's animating force.

Everyone knows their heartbeat—a steady rhythm—and breathing, essential for life. Our hearts and lungs work together in changing rhythms, responding to surroundings, activities, and emotions. Breaths fill the

Earth's atmosphere, while heartbeat drives us, shaping the rhythms of life and civilization.

As humans, we are tied to the daily rhythms of light and dark, which comprise our primary biological clock. These cycles take a bit longer than 24 hours, roughly the same as a lunar day—about 24 hours and 50 minutes. The term "circadian" comes from Latin, meaning "around" and "day." The cycles of the planets affect the light we receive, which in turn sets our body's internal rhythms and influences our sleep, mental state, pain tolerance, body temperature, and hormone levels, such as melatonin production in the pineal gland. This, in turn, impacts cellular, protein, and molecular activity and repair.

But heartbeat, breath, and circadian rhythm are just a small part of something bigger—a complex web of biological, geological, physical, and atmospheric rhythms. We're part of the Earth, participating in rhythmic patterns that stretch far beyond our understanding, encompassing planets, stars, and galaxies.

Earth's two main rhythms are rotation and revolution. Rotation is about Earth spinning on its axis, while revolution is Earth's orbit around the sun. Each solar day, Earth rotates from west to east, creating a cycle of light and heat. This rotation defines reference points: Establishes the equator, latitude, and longitude. The Earth's surface map uses a geographic grid based on these rhythms. Together, rotation and revolution form 24 time zones and cause seasonal changes in day and night length.

Our moon influences tides and ocean currents. Sunspot cycles and Earth's orbital shifts affect climate and life. Ice core and sediment studies reveal cycles like ice ages. Earth's core oscillations can modify energy flow and trigger volcanic-related atmospheric changes. Surface changes like mountains and buildings impact wind, humidity, snow, ice, clouds, and ocean currents—all rhythmically linked.

Our planet is a complex feedback system in which human activity shapes our emotions and health, influencing Earth's life rhythms. These rhythms influence various processes, including women's menstrual cycles, sexual development, and faster puberty and slower aging. Bird migration, regulated by changes in sunlight, exemplifies Earth's rhythms. Sunlight affects the eyes, nerves, and the pituitary gland, which releases hormones that alter reproductive organs and central nervous system activity, thereby triggering migration. Everything influences everything else: Life cycles, biorhythms, and sleep patterns; weeks, months, years, and seasons define, maintain, and contribute to the intricate rhythm of life on Earth.

Human measurement of reality involves extrapolating Earth's rhythms. Humans created seconds, minutes, and hours, asserting control over modern time. Recognizing time as an invention—from celestial movements to clocks and technological impositions—shows humanity's departure from natural rhythms. Tracking how perception and expression of rhythm in life and music evolved reveals the development of humanity and culture.

Our exploration begins with the core rhythms of heartbeat and breath.

ALTERNATE NOSTRIL

This classic yogic practice explores deeper, rhythmic breathing and awareness. It can serve as a prelude to meditation, yoga, or other focused work.

Stand or sit with eyes closed.

Extend your thumb and index finger, resting your thumb on the right nostril. Inhale through the left nostril and hold your breath.

Next, cover the left nostril with your index finger and release your thumb.

Repeat this process in attunement with your breathing.

Now breathe through your right nostril.

Hold your breath.

Place your thumb on the right nostril and exhale through the left nostril.

Continue to repeat and alternate.

This exploration is an excellent way for the group to focus on breathing, rhythm, and self-awareness.

Exploration Alternative: This exploration may start with expelling impurities. The performer fills their lungs, exhales fully by tightening stomach muscles to force all air out, then inhales again and repeats. Use sound to identify expelled impurities.

When the heart is at rest, it pumps 60 to 90 times per minute, creating a rhythmic motion that circulates blood. This rhythm is the foundation of life. When synchronized with breathing, it represents the body's primary, steady, interconnected rhythms.

DIAPHRAGM BREATHING

Part I: Performers stand in a circle, hands on their upper abdomens, breathing naturally.

Side coach: "Inhale and exhale normally to observe your breath. Inhale deeply through the nose to lift your upper abdomen."

As lungs fill, the diaphragm contracts, expanding the stomach. Encourage performers not to force their breath and to visualize filling both the front and back of the lower lungs. Allow plenty of time for deep, easy breathing.

Urge feelings of breath energizing lower limbs and whole body on exhalation. Repeat until it feels natural.

Part II: Ask the performers to breathe in deeply through their noses, filling their lungs to the point of bursting.

Ask them to hold their breath for as long as possible, then breathe out fully and loudly until all the air is expelled.

The performers then imagine releasing every bit of air, making sure their lungs are empty.

After a few regular breaths—in through the nose and out through the mouth—the performers should return to deep inhalations, filling their lungs fully before exhaling completely.

HEARTBEATS

This naturally leads to the next exploration, **Heartbeat**.

This exploration serves as a follow-up to **The Mountain** and **Diaphragm Breathing or** can be practiced independently.

Performers may either be seated or standing.

Begin gently by having the performers inhale through their noses and exhale through their mouths. **Alternate Breathing** can be used or not to initiate breathing awareness.

Ask performers to find a simple rhythm by matching their everyday breathing. Once they achieve a smooth, natural rhythm, have them hold their breath for two full heartbeats—two "lub-lum" sounds.

With their bodies still, the performers will become more aware of their heartbeats. If one performer struggles, suggest putting one hand over their heart to help them feel their heartbeat.

After holding their breath for two heartbeats, instruct them to exhale and return to normal breathing through the nose and out through the mouth.

If they wish, they can exhale softly, then hold their breath for three heartbeats before inhaling. Afterward, exhale and breathe normally, gradually increasing breath-hold time to 10 to 15 heartbeats as they progress.

The exploration can be practiced repeatedly or occasionally, and it can become a regular part of the group's daily warm-up routine.

As performers become more comfortable, they start with fewer heartbeats, like 7 or 10, then gradually increase, e.g., to 10, 13, 17, and 21. It's crucial to allow time to regain normal breathing before increasing the count.

Note: It's a great way to tap into body rhythm, raise awareness of heartbeat, boost lung capacity, and build mind-body discipline.

People experiencing stress tend to breathe shallowly, using only part of their lung capacity, resulting in less oxygen reaching their brains and bodies. Deep breathing methods can boost performers with energy, fostering calmness and heightened awareness.

Explorations like **Heartbeat** sometimes face obstacles. Despite the performer's control, those accustomed to external instructions—such as from the Soviet education system, apartheid-era Zulu, or Ethiopian Orthodoxy—struggle to connect with their bodies. Their thinking is shaped by external forces, leaving them uncertain about their physical feelings. They often seek specific sensations or correct heartbeat patterns. Some worry their heartbeat is inconsistent. This shows how social, political, and cultural influences can embed in the body and mind. Being aware of heartbeat and breathing may lead to deep insights or political awakening.

The human brain is a complex, adaptive system managing organs, chemistry, hormones, emotions, and overall well-being while interacting with constant input from the world. Thought influences our biological, mental, and emotional health, affecting both us and the world. Understanding this interplay is central to our work, helping us learn how our bodies function and thrive. Next, we explore these issues.

INDIVIDUAL AND GROUP MEDITATION

The approach, application, and frequency of mediation practice should be tailored to individual circumstances. It is essential to encourage performers to connect, continue exploring rhythms, and engage in meditative practices on their own.

Ask the performers to find a quiet spot where they will not be interrupted and sit up comfortably with their eyes closed. Lead them through the following sequence during the first few meditation sessions:

- Focus your attention on the sensations of your breath as it flows in and out through your nostrils, gently exhaling through your mouth.
- Do not try to control your breathing but stay aware of your inhalation and exhalation.
- For now, all thoughts other than the sensation of breathing are distractions and should be gradually released from your mind and body.
- Whenever your mind wanders—whether to thoughts, sounds, or other sensations—bring your focus back to your breath.

Some find it helpful to focus on their breath by silently repeating a meaningful word with each inhale and exhale, like "peace" for inhaling and "out" for exhaling. Another idea is to try:

- Count each breath: Say "one" on the inhale and "two" on the exhale, up to ten, then start over at one.
- You can also use sounds like "Om," "ah," or any other sound as a mantra during meditation.

Practice breathing with a word(s)—a mantra—keeping sensations central while the word is a gentle background whisper. Meditate for 15–20 minutes daily at a consistent time—after group yoga is ideal. At home, choose a time when alert.

For these explorations, performers lie down with their heads facing either outward or toward the center of the circle. Outward helps focus on each performer's individual work—heads inward to support the goal of finding a collective breath and building a sense of community. For this next exploration, ask performers to lie with feet flat, knees bent, and back against the ground. This position helps monitor the depth of their breathing.

BODY PART BREATHING

Ask the performers to breathe slowly in through their noses and breathe out through their mouths.

Next, encourage performers to breathe in clear, fresh air—pure mountain air—through their noses, letting it circulate down through their bodies to their feet. There, it swirls around the joints and bones of their feet.

As the performers breathe out, they release any tension in their feet. Then proceed to each body part in the same manner.

Maintain a steady pace and calming tone, using vivid imagery such as fresh, clear mountain air flowing to each part of the body, working up the body.

With each exhale, tension releases from that area. Remember to encourage performers to breathe in slowly through their noses and out through their mouths.

Let your visualization instructions synchronize with the rhythm of your breath. Give performers breaths between each body part before moving on to the next.

A calm and relaxed side coaching voice is key. Suggestions for body part order:

Feet	Stomach	Shoulders
Ankles	Chest	Neck
Calves	Back	Scalp
Knees	Spinal cord	Face
Thighs	Hands	
Pelvis	Arms	

WHITE LIGHT

This exploration can be used in conjunction with a full-body relaxation exploration, such as the **Body Part Breathing** exploration, and as preparation for other visualization explorations covered later.

Eyes shut. Lying down or standing. Performers are asked to visualize a light radiating from their entire body with each inhale through their nose.

At first, this light is small and barely glows. When they exhale, the glow of the pure white light dims slightly.

With each breath, the glow intensifies slightly, but with every exhale, the pure white light dims a bit.

As performers regroup, a glow begins to surround them, intensifying until they can see the glow of others. These others, composed of pure white light, pulse in sync with their breathing—brightening on inhale and dimming on exhale. As the glow increases, they eventually converge, touching each other.

The glow intensifies as the group inhales, then slightly diminishes on exhale, turning their collective breathing into a synchronized pulse of pure white light.

As the glow and successive breaths grow stronger, individual shapes blur together, forming a single large, radiant mass of white light.

Participants are encouraged to visualize the white light rising from the ground and floating upward.

The light then ascends into the sky, passes through clouds, and enters outer space. Here, you can add your own images and details or make the visual theme more meaningful to the group's sensibility.

The white light moves among the stars, consistently pulsating with rhythmic energy. It is eventually attracted to a specific white star and draws closer. The light merges with the star, becoming one with it, blending its brilliant glow and pulsating energy with the star's core.

Give the performers space to enter and inhabit the star—have them sense its energy and the presence of others in the group.

When conducting this exploration with a Greenland Inuit group, star imagery took on an unexpected significance. For the Greenlanders, stars represent the spirits of their ancestors, and the exploration symbolized a reconnection that proved pivotal to our subsequent work.

When you're ready for the performers to leave the star, ask them to gather outside in the darkness of space.

Once they have assembled outside the star, instruct them to return to Earth. Travel back in time, passing planets and other stars, as meteorites streak by until they re-enter Earth's atmosphere.

They may drift through the blue sky, past airplanes, over cities, oceans, and mountains. If you like, you can specify the location, pinpointing the glow's approach to the hometown where the exploration takes place.

I generally suggest performers visualize the building they are in from above as they approach it. The glow enters the building through the roof and eventually settles onto the floor.

As the air moves in and out, the glow begins to fade, slowly diminishing until it becomes just a faint white light.

The overall glow breaks apart into smaller individual ones, and with each new breath, those individual glows shrink away until they are completely invisible.

Then, the glow remains with each person until it is needed again. This inner glow holds onto the memory of their time together and the experiences they shared. It is something they'll carry with them and into our work.

Note: Suggest minimal, straightforward images suited to the context or specific needs, focusing on clarity and ease of understanding. Concrete and tactile images are most effective. Encourage performers not to overthink.

Use a subtle, supportive voice with gentle tones characterized by a steady tempo, smooth rhythm, and clear, concise language. Use simple sentences and precise, concrete images to achieve this.

- **Leading is about giving.**
- **Leadership is an ethic, a gift of oneself.**
- **The core of leadership is not about giving things or even sharing a vision. It's about offering oneself and one's spirit.**
- **The most vital part of a gift is the spirit behind it.**
- **Trusting people to resolve problems increases motivation and yields better solutions.**
- **Effective leadership is given freely and without expectation.**
- **In life, all you really have is what you can give.**

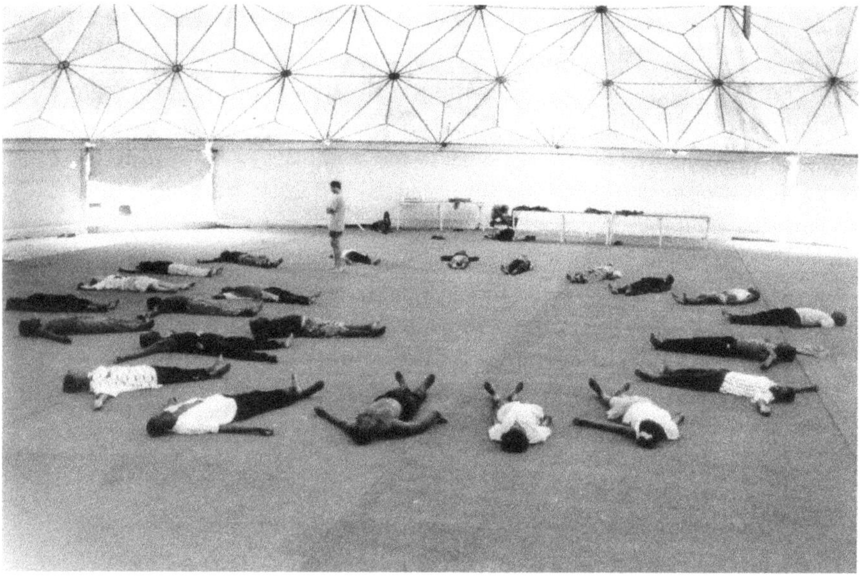

FIGURE 1.4 The author is conducting a visualization exploration in Lusaka, Zambia

Source: Photo Mikael Renlund.

Rhythmic Movement

Through dance, song, chant, and drumming, the heart's rhythms connect with natural world rhythms—animals, climate, spirits, earth. Rhythm acts as a fleeting catalyst, linking and guiding cultures. For the pioneer scholar of religious and shamanic studies Mircea Eliade, rhythm is the world's true revelation.

> Choreographic rhythms have their model outside of the profane life of man; whether they reproduce the movements of the totemic or emblematic animal, or the motions of the stars; whether they themselves constitute rituals (labyrinthine steps, leaps, gestures performed with ceremonial instruments) a dance always imitates an archetypal gesture or commemorates a mythical moment. In a word, it is a repetition, and consequently a reactualization, of illud tempus, "those days."
>
> *(Eliade 1959, 28–29)*

Every indigenous culture I've engaged with features a straightforward rhythmic beat at its musical heart, often complemented by multiple beats. I refer to these as primary beats. They typically form the basis for dance and performance movements. As American ethnomusicologist Alan Lomax observes, "In most musical styles, the performer relies on a single, overarching rhythmic scheme or 'ground plan,' which acts as a reference point for the numerous rhythmic variations that can be developed within it" (Lomax 1978, 49). This simple meter can be double, triple, or a combination of 4/4, 3/4, 6/8, 9/8, 12/8, or similar meters. "Rhythmic relationships link a group together within one overall metrical pattern" (121).

Where do the fundamental rhythms for their dance expressions come from? Was singing and dancing a reaction to the rhythms of their surroundings? As communities developed, did these rhythms evolve into expressions mirroring social and cultural complexity? Nothing arises without cause. Lomax understood that social organization shapes musical complexity. He identified five key levels of social complexity—each progressively more layered and musically sophisticated.

- Extractors: Mainly engaged in gathering, hunting, or fishing. Incipient
- Producers: Early farming societies practicing simple agriculture without animal husbandry before European contact.
- Animal Husbandry: Cultures that raised animals such as pigs, sheep, or cattle before Europeans arrived, often combined with agriculture but without the use of the plow or large-scale irrigation.
- Plow Agriculture: Societies utilizing both plow-based farming and animal husbandry.
- Irrigation: Communities employing agriculture, animal husbandry, plow use, and extensive irrigation systems (122).

The rhythm-focused, conscious body navigating through space is the initial step in expressing body rhythm externally. These following explorations establish a basis for later rhythm, vocal, and movement studies and serve as a reference point for rhythmic connection and expression.

HEARTBEAT MOVEMENT

It is best utilized as a follow-up to the **Heartbeat Exploration.**

Once the performer feels comfortable with the Heartbeat Exploration, ask them to take the rhythm of their heartbeat to different parts of their body, like their fingers, hands, head, or feet.

Initially, encourage performers to connect with their heartbeat rhythm. As they progress, their movements will become less precise and more like an internal response to the rhythm.

After establishing a steady rhythm and exploring it across the body, let the performer move freely in space, driven by their heartbeat. Closing their eyes can help focus.

Encourage performers to switch between different body parts to express their heartbeat. They might start with their hand, then shoulder, then take a few steps inspired by the heartbeat, then move their chest. Suggest trying various expressions.

Whenever the performer feels out of sync with their core rhythm, they should pause to regain it before proceeding.

As the exploration progresses, have performers create a dance-like movement based on their heartbeat's rhythm. You can add variations based on their responses.

Exploration Variation: Ask performers to engage and dance to the heartbeat rhythm. Encourage them to create expressive sounds from their movements.

Exploration Alternative: If space is limited, conduct exploration with all performers seated, eyes closed, or in darkness. This shifts focus from movement to vocal heartbeat. Sticks can be used to create the rhythm, a method successful across cultures. Explanation of **Sticks** follows.

Highly Recommended.

When the same exploration was conducted at Metamorphosis Theatre in St. Petersburg, Russia, it lasted nearly an hour, leading to an improvisational dance and song that mirrored the group's internal rhythms. This process ultimately fostered a shared heartbeat, serving as the basis for all subsequent performance development efforts.

Other explorations focused on developing rhythm and movement awareness, coordination, and expression include the following. These can be used alone or combined with the previous sequence of explorations. They are organized by theme below, without a specific order. Feel free to rearrange, repeat, adjust, and enhance them as needed.

MELTDOWN

Ask the performers to stand with their eyes closed. Visualize a hole in their heel through which sand flows out. As the sand drains, their head gradually collapses and deflates, followed by their neck, shoulders, arms, chest, stomach, and the rest of their body, which crumples and spills onto the floor.

Perform this sequence with a rhythmic cadence, including a verbalized countdown to enhance bodily awareness of rhythm and movement.

Note: This exploration can serve as a warm-up for other floor activities or as a concluding part of a series of standing movements.

CREATION AND MELTDOWN

Have performers relax into the floor at their own pace. Then, ask them to stand, creating a new standing position in sync with your count.

This is initially a slow and measured count of either 100 or 50.

Counting should maintain a regular cadence, as the performers should strive to move in sync with the count, reaching the fully standing position precisely when the designated number is reached.

Repeat the meltdown process with a lower count. For example, if the initial count was 50 or the performers' rise, reduce it to 40, 35, or 25. The performers should be fully melted down by the time this number is reached.

Next, proceed with another creation count of about 15 to bring everyone to standing. Then, perform a meltdown count of 10, followed by creation counts of 7, meltdown counts of 5, creation counts of 3, and finally, a meltdown count of 2. Adjust the counts as needed to fit the situation, always aiming for smooth and seamless transitions between creation and meltdown.

Exploration Alternatives: Adjust the cadence or rhythm of the count, prompting performers to speed up or slow down their rise or fall. For example, you might repeat a number or reverse the count, requiring performers to count backward, such as "35, 34, 33, 32, 33, 34, 35, 36." Vary the tempo by counting quickly, then pausing and resuming with a fast count. Have fun with it.

Note: Feel free to repeat this exploration in your workshop. It works well as a continuity activity, helping performers refocus on their body and rhythm. It's ideal for transitioning between active, floor, or standing work. Get creative—try variations or turn it into a rhythmic dance with drums, music, or instruments.

Highly Recommended.

Another exploration for early rhythm awareness is **Breathing a Lie.** Building on previous rhythm explorations, it explores the psychophysical rhythm of thought, feeling, and action. The goal is to deepen awareness of these internal rhythms as part of holistic existence. The activity is subtle yet effectively illustrates psychophysical and rhythmic dynamics of thought, feeling, honesty, and deception.

BREATHING A LIE

Begin by asking the performer to sit comfortably in a chair.

- Close your eyes and take four deep breaths, inhaling through your nose and exhaling through your mouth.
- Relax as much as possible.
- Then, visualize something you love, like a color, flower, animal, or food.
- Notice how it affects your rhythm, breathing, heartbeat, and atmosphere.
- Focus on one object and silently say, "I love..." Repeat it, feeling how it resonates in your body and mind when you tell the truth.
- What is the rhythmic feeling of that truth?
- Finally, stand up, walk, talk, and move around the room activity.

Allow Time for the activity

- Come back, sit down, and close your eyes.
- Take four deep breaths, inhaling through your nose and exhaling through your mouth. Then, tell yourself a lie.
- Say, "I hate..." (the same thing you just claimed to love).
- Repeat the phrase "I hate..." and observe how your body reacts.
- Pay attention to the rhythmic changes in your mind and body as you lie.

Ask performers to alternate between something they love and a lie, then observe the differences in their bodily, emotional, and thinking rhythms. What happens to their breath and heartbeat? How do they feel?

Exploration Alternative: After completing the above steps, have the performers partner up and exchange a mix of truths and lies. Encourage performrners to notice any rhythmic changes within themselves or their partner. Can the listener distinguish lies from truths? Can the sharer tell the difference between when they are lying and when they are honest?

From the DirectorsLab Chicago Journal

A week-long workshop for directors focused on enhancing their stage direction. I was asked to introduce new perspectives, strategies, and skills. The curriculum is trifocal, comprising (1) presentation of explorations and explorations, (2) explanation, philosophy, and integration, and (3) practical application in real-world settings. My involvement commenced the day after the workshop began.

After introductions and a general discussion to familiarize everyone, the participants' backgrounds, interests, and goals for the week are outlined. Based on our conversations and my impressions, and considering our limited time, I decided to introduce and emphasize **Archetypes**. A more in-depth discussion and exploration of Archetypes are provided below.

Session One (four hours): The performer stood in a circle. Began with **Standing Relaxed**, then moved into **Listening to Feelings** to preface our body-oriented work, slowing the rhythm and setting a more thoughtful, reflective tone for the workshop. How you begin determines all that follows.

Transitioned into **Meltdown**. Once on the floor, shifted into **Diaphragmatic Breathing**, focusing on self-connection, body rhythms, and transforming into White Light.

Break.

We then moved into **Walking**, followed by **Walking Attitudes**. After exploring the attitudes, we introduced **Archetype** characterizations, including the Explorer, Hero, Ruler, Temptress, Lover, Regular Guy or Gal, Jester, Trickster, Innocent, Outlaw, Sage, Caregiver, Ruler, and Creator.

The performers were encouraged to explore archetypes freely, helping to awaken them and set the stage. Although archetypes weren't the workshop's focus, participants reflected on their discoveries and scenarios, showing enthusiasm for future use in their directing.

I outlined our focus on archetypes as deep motifs, emphasizing their role in offering interpretative, structural, psychological, and emotional insights through body exploration. We discussed the archetypal, trans-historical nature of myth, ritual, and Indigenous performance.

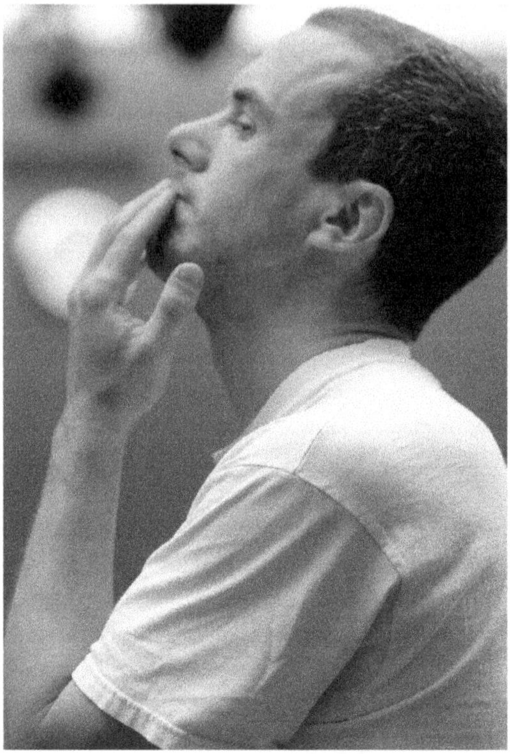

FIGURE 1.5 Archetype exploration, DirectorsLab, Chicago

Source: Photo by the author.

Our next activity creatively and physically shows a person's name and identity through movement, voice, and rhythm, acting as a self-portrait. The group then interprets this, creating its understanding. We explore giving, sharing, trust, creation, and collaboration—the core of collective performance.

Building on earlier work, this lays the foundation for a template supporting future creativity. As we explore body, circle, and rhythm, Names reveals how our names reflect identity and community placement. We are individuals, yet we express ourselves through psychological, social, and cultural influences. Every action shows this. This exploration focuses on identity—specifically our names—shared within a community circle through physical and vocal rhythms to communicate and establish presence, then invite interpretation and acceptance. This awareness is crucial for our ongoing work.

Ideally, this exploration should be conducted after the performers and the group have had about three or four sessions to get acquainted, build trust, establish a collaborative history, and develop a shared understanding.

NAMES

This activity offers performers a valuable way to explore their connections with others, build trust, and develop ensemble collaboration through simple elements. It highlights rhythmic and emotional qualities, emphasizing that every word and sound carries a physical message.

This exploration is particularly beneficial when working with multicultural groups that have difficult-to-pronounce, unusual names.

Form a circle. Each performer takes a turn sharing their name (first or full), paired with a movement. Encourage honesty and spontaneity in their expression.

The way a performer pronounces their name conveys a lot—through vocal emphasis, stress, volume, accent, and cadence, it reflects their self-image and public persona. Their body language and gestures while saying their name also leave a lasting impression.

After everyone has shared their name, the group, with your guidance, repeats the performer's name and investigates its sounds, movements, natural rhythms, and emphasis.

Start by setting an example.

Encourage the group to explore all syllables and vowels of the person's name, paying attention to emotional connections and exaggerating the vocal-physical presentation. Explore vocal variety, intensity, playfulness, resonance, vowels, consonants, feelings, mood, attitude, beauty, expression, and flow, applying the same to physical expression. Consider the performer's personality to shape each name's expression.

Tell the group to focus on integrating the name sounds into the body, exploring where they resonate and connect emotionally. By elongating and emphasizing the sounds rhythmically, they can extend a movement. The goal is to connect emotionally with the name sounds and develop a rhythmic presentation that creates a collective, imaginative snapshot of each performer.

Encourage the group to explore and play until they reach a consensus, emphasizing that there is no right or wrong way to approach or present this activity.

The entire group then learns and rehearses the Names presentation before moving on to the next individual.

After identifying each performer, revisit the group to repeat the names, fostering familiarity.

Using names is an effective way to begin future sessions, helping performers remember and internalize names through vocal, physical, and creative engagement.

This exploration promotes learning names, facilitates group connection, and builds a strong sense of identity. Many performers report that this exploration enhances their connection with others, often recalling colleagues' names years later, especially when performed rhythmically, with exaggeration, and with flair.

Example: In a workshop at the University of Arizona, a performer named Alexander proudly lifted his head, emphasizing the first vowel "Al," clipped with "ex," and softened with "ander." His body moved rhythmically with coordinated expansion and contraction. Inspired by his presentation, the group mimicked the initial vowel with a quick head jerk and exaggerated elongation. At the end of "ex," they produced a sharp, almost sibilant sound, shaking their heads quickly. For the final part, "ander," the group vocally let it fade away and relaxed into nearly slumped postures. Each name became a playful exploration.

Sticks

I have found that using sticks is a great way to externalize a beat. Some performers simply lack musical or rhythmic skills. Many cultures have lost their sense of rhythm due to acculturation, education, or urbanization. Workshop performers in Helsinki showed minimal understanding of rhythm, possibly because their rhythmic awareness had diminished over generations due to urbanization and historical cultural suppression under Swedish and Russian domination. Conversely, participants from the Kwasa Group workshop in Durban, South Africa, maintained a robust and lively sense of rhythm, deeply connected to their dance, song, drumming, and ceremonial customs.

The Greenland Inuit at Tukak' Teatret, having lost their traditional dance and drumming practices, requested that I teach them Alaska Native (Yup'ik and Inupiat) drumming and dance. Many Greenland Inuit endured the harsh effects of early 17th-century Danish colonization and missionary efforts, which banned drumming and spiritual traditions, considering them as devil worship. The introduction of sticks played a vital role in helping them relearn the rhythms of their circumpolar relatives and reimagine and reconstruct Greenlandic dance, singing, chanting, and performance vocabulary.

STICKS

Working with sticks is a collaborative activity where group members gather and prepare their sticks. The process involves going into the woods, selecting their sticks, removing the bark, drying them, and decorating them.

Wooden sticks, measuring between 12 and 24 inches long with a minimum diameter of two inches, are necessary. A hardwood is preferred.

Each performer has two sticks, which can be customized with markings, carvings, and more.

Once the sticks are introduced, encourage performers to experiment with various forms of expression using them—such as lasers, magic wands, rattles, talking sticks, and paintbrushes. Highlight the stick as an extension of their body, serving as a channel to project their self and creativity outward. Think of it as a magic wand that connects with transient or invisible realms.

Create rhythmic scenarios with the sticks.

Suggestions include painting their last home, depicting known people, illustrating landscapes, capturing seasons, portraying dogs, exploring a dreamscape, conducting a symphony, and identifying instruments. Conduct a symphony of the four seasons—spring, summer, fall, winter.

Slow your pace to a gentle rhythm and encourage the performers to use the sticks like flashlights, illuminating their surroundings. Urge performers to connect with the sticks as extensions of themselves, fostering a dance that expresses their feelings about the environment.

Scenarios can be introduced and explored through side coaching. The chosen scenarios may be culturally specific or universal; they are always rhythmic and serve beyond mere illustration. Have fun and work to integrate the sticks into the performer's expressive body.

Note: Sticks have succeeded in various cultural, multicultural, and university settings across Alaska, Europe, Africa, Asia, and the United States.

- **Explorations require patience, persistence, and adjustment.**
- **Our work is incremental and cumulative.**
- **Find a flow that works for you and the group.**
- **Lead by being an example.**
- **Guide gently.**
- **Let the moment and the work speak for itself, and listen.**
- **An open, non-judgmental atmosphere is essential.**
- **Explore with performers.**
- **Share your thoughts and feelings honestly.**
- **Create a safe and welcoming space.**

FIGURE 1.6 *Sticks* during a Ritual Preparation. Metamorphosis Theatre, St Petersburg, Russia (L to R: Sven Holmberg, Morten Ilsoe, and Sergi Popov)

Source: Photo by the author.

Humans naturally adapt their rhythm depending on their environment and company. Whether it's a rambunctious child, someone ill, an elderly person, a party, a political rally, a casino, a funeral, a rock concert, a rave, a bustling city street, or a peaceful night at home, each setting has its own rhythm that we respond to and engage with. Our rhythm either aligns with or stands out from its surrounding context.

MUSIC PHYSICALITY

Provide access to percussion instruments like drums, rattles, tambourines, bells, sticks, and more. Encourage performers to play, make sound, make music, make noise when inspired; quality isn't the focus. Some will join naturally, others may feel pulled or opt out.

Invite the group to move in response to the music.

Continue as long as it is sustainable. Exploration needs time to develop its identity. It may turn into a long-lasting musical and movement free-for-all, where performers move freely and try a variety of instruments. Allow the session to grow and ebb and flow over time. Let it unfold naturally.

Apply this exploration when the group faces a creative block, needs an attitude change, or just wants to relax and have some fun. Join in or just let it flow naturally. It may take an hour or more to play out—allow time. The exploration will find its own ending.

Notes: This exploration's impact is truly freeing. The longer the session, the more comfort, liberty, and expressiveness among performers. These sessions turn into spontaneous "jam sessions" where anything can happen, fostering open communication.

I have conducted this exploration from Stockholm to Zambia and Alaska. As Artistic Director of the Organic Theatre in Chicago, we hosted 11 pm to dawn jam sessions with musicians, dancers, and performers. These events, called "Clam Dance," gained popularity.

Be confident in what you do and how you do it. Inspire others and pass that confidence on to the performers. You set the example and are responsible for maintaining the spirit and quality of the work.

NUMBER MIX-UP

Circle. Each performer gets a number, but they don't know what number the others have been given.

Next, the group comes up with, agrees on, or is given a specific rhythm pattern that is not too long or too short. The beat might be culturally relevant or not.

Establish a rhythm by any combination of foot stomping, hand, face, chest, thighs, head, or arms slapping or clapping. No vocals. Practice the rhythm so it becomes instinctual.

Once the rhythm is set, one performer starts by calling out their number. At the end of the established rhythm, the performer given the number calls out another number.

The established rhythm continues non-stop.

If a performer whose number is called fails to respond by calling out another number within the rhythm cycles, they must leave the circle. When a performer exits the circle, their number is retired.

If a performer calls out a number that is no longer in the circle or was given, they also must leave the circle.

Those who are no longer in the circle no longer participate in the rhythm from outside the circle.

The remaining performers in the circle continue at the established rhythm, calling out another number when their number is mentioned.

Performers might become confused or struggle to remember which numbers are no longer in the circle. This is the point.

The exploration continues until only one performer remains, and that person is declared the winner.

Note: This multitasking exploration involves coordinating rhythm, awareness, and memory, demanding presence, focus, and continuous adjustments to maintain a steady rhythm.

RHYTHMIC MOVEMENT

Performers are dispersed throughout the space.

This exploration centers on finding the body's rhythm. Each performer creates a basic rhythm with vocal sounds.

A basic one-two, one, two, three, or one-two, one-two beat pattern is recommended.

Side coaching. Identify a body part—start with the hand—then move to the shoulders, hips, neck, and head, and ask the performers to move isolated body parts to their steady, unchanging rhythm.

As each body part is moved, emphasize its expressive potential when brought to life by the rhythm.

The exploration should include both large and small body parts, such as fingers, facial features, feet, and toes.

Once the rhythm has been explored throughout the entire body, encourage free-form movement explorations in the space, using the same steady, consistent beat.

Exploration Evolution: Ask the performers to explore a rhythmic call-and-response within their bodies while continuously moving to the beat. For example, when the forehead moves, the foot responds.

Next, infuse the rhythm with emotion to create an emotional dialogue between body parts. For example, an angry right arm interacts with passionate hips, while indifferent shoulders respond, provoking the hungry mouth to chomp with the beat, and so on. Let the rhythm travel and dialogue freely and associatively, without overthinking. It is not about making sense. Have fun with it.

Highly Recommend.

Story Dancing

Yup'ik and Inupiat dance movements express a wide range of feelings and activities like looking, hearing, eating, hunting, joy, age, surprise, talking, and images of birds, seals, mountains, walrus, raven, and northern lights. These dances mimic daily life and develop alongside their hunting and gathering lifestyles. They tell stories through movement, serving as a form of education, reflection, and celebration, while also reinforcing their worldview. Dance is as vital as oral storytelling, both conveying culture and complementing each other—storytelling with words, dance through rhythm and gesture. Dances are communal and participatory, owned by the entire community, encompassing ancestors, the earth, animals, and spirits.

Within this cultural context, the body acts as a performance script, conveying complex messages about human, animal, and spirit world relationships, climate, and survival. While traditional dance isn't purely illustrative or practical, it can abstract feelings and include movements for beauty. Many recent dances depict modern activities like operating a boat motor, using a rifle, playing basketball, or boxing, reflecting both tradition and new influences. Throughout, the organizing rhythm remains constant.

FIGURE 1.7 Story dancing during the performance of *Qayaq*. Tuma Theatre, Fairbanks, Alaska (L to R: Josh Weiser, Wilma Brown, and Mike Lane)

Source: Photo by the author.

MOVEMENT TO MUSIC

The goal of this exploration is to inspire expressive reactions to sound, space, others, and emotions. Its emphasis is on moving beyond stereotypes, clichés, and intellectualized ideas of action, sparking an abstract, surreal imagination, and honestly and personally interpreting the music.

Use recorded or live music.

Instruct performers to listen to the music first, then move in response to the following prompts, using body parts or the entire body expressions.

- Convey a single musical instrument or melody.
- Highlight the main musical line.
- Follow the rhythm line and then create contrast with it.
- Utilize abstract shapes—not human forms—to represent the music.
- Let the music move through the body.

Explore the prompts individually, then communicate with others.

Advise the performers against dancing; that is, no hip-hop, club, jazz, ballet, modern dance, etc.

When engaging with other performers, let the movement occur spontaneously. The main goal is to explore and find a genuine and personal expression. Have fun!

- **Always be honest with yourself and with the performers.**
- **Each moment is unique; you will never pass this way again. Stay present.**
- **Like magic, this group came together against all odds—from countless possibilities in history, time, and space—and here you are.**
- **What can you learn and share?**
- **Every moment is precious and perfect.**
- **Accept without expectations.**
- **You are part of a larger process.**
- **The work is an organic entity—an embodiment of everything and everyone involved.**

MOVING WITH OTHERS

Ask performers to walk comfortably in the space, avoiding others. Then, walk alongside someone.

Move quickly and instinctively adapt to each other's walking rhythm.

Peel off and find someone else, adapt to each other's walk. Do this several times.

Pay close attention to each other's rhythm, stride, style, and demeanor, aiming to create a walk that is collaborative, balanced, and shared.

The goal is to move together as one. Partners should not simply mirror each other; rather, they ought to be inspired by and react to each other's evolving energy and rhythm.

Remain with their previous partner and instruct performers to exaggerate and synchronize their walk, moving as one.

Explore the walk to reveal its situation, character, or objective.

Clarify the extreme, exaggerated walk, making it repeatable and versatile—serious, comic, expressing happiness, sadness, or anger.

Quickly switch to repeat the process with other partners. Do this multiple times.

This focuses and develops quick, adaptive, and intuitive responses.

Exploration Alternative: Suggest different gaits and themed movements, such as walking tall, short, spastic, dizzy, tired, joyful, or walks related to project goals.

Encourage expansion of lightness by moving with ease, playing, and inspiring each other. Start moving as one, then switch to individual walks, then back again.

Progressing to extreme walks, then reverse back to an everyday walk.

MOVEMENT WITH OTHER BEINGS

Select and explore different non-human beings through personal observation, film, video, and audio documentation, focusing on their movement.

Consider domestic and locally available animals and insects, including cats, dogs, squirrels, raccoons, deer, mice, birds, ants, worms, and bees.
Local plants and foliage: Trees, plants, shrubs, grass, vegetables, fruits, etc.
Elements: Rocks, earth, wind, clouds, moon, sun, stars, dawn, dusk, etc.

Ancestors are family members who are no longer alive.

Spirits are entities that have influenced and are central to your culture—such as heroic, religious, political, or mythical figures who have shaped thought and identity.

From a human-centric perspective, moving with a cat might seem childish, while doing so with a rock might appear absurd. It's important to look beyond these biases and hierarchical thinking.

Some beings move like humans; others do not. Each one lives in its own way.

Start slowly and intentionally. Select from the explorations above (or add your own). Study and consider each. Take your time.

Begin with beings that are easier to observe or sense, then move to more abstract entities like wind, dawn, or a river.

Drawing upon observation and research, explore the following:

- What shape, textures, density, and weight define them?
- What similarities do they share with humans?
- Focus on one specific part or aspect of the nonhuman being, explore it, and embody it.

- Continue with other parts and aspects.
- Using this bodily insight, start moving like the being.
- Then, perceive the world as the being does.
- Consider their daily life—waking up, resting, and their sense of time.
- How do days, weeks, months, years, or centuries pass?
- Observe the world from the being's unique perspective.
- What is their rhythm?
- Do they think or feel? If so, what and how?
- What interests them?
- What do they seek and offer?
- What kind of power do they possess?
- Finally, reflect on their relationships with other beings.
- How do they contribute to the circle of existence?

Note: This can be a one-time or a long-term, ongoing exploration.

FIGURE 1.8 Manfred, a Khwe Bushman healer, undergoes a "cooling down," releasing the "heat' from a nightlong healing ritual. His regalia symbolizes and invites the presence of other beings, contributing their sight, hearing, and movement to enhance the healing process. Kalahari Desert

Source: Photo by the author.

Community of Place

The Cosmocentric and the Anthropocentric in an Age of Complexity

Performance is not merely embellishment. It is an intrinsic ontological condition—the fundamental technology through which humans engage with a place. Indigenous performance systems serve as evidence and models of humanity's original way of knowing: The cosmocentric understanding that human awareness resides within, not above, a living whole. What modernity dismisses as animism was, in fact, a complex acknowledgment of distributed agency among all elements of a place. The *yua* of the Yup'ik and the spirits recognized by the !Xuu Bushmen are not primitive projections but empirical acknowledgments of relationality as the core structure of reality.

For 300,000 years, humans developed through direct sensory dialogue with terrestrial systems. Performance arose as the neural technology connecting separate entities—human, animal, plant, geological, atmospheric, ephemeral—into a unified community. This was not just a metaphor. The Yup'ik term *agayuliyararput* combines dance and prayer because the distinction itself is an anthropocentric fiction. What indigenous systems understood—and what quantum physics, systems theory, and ecology now confirm—is that separability is an illusion. Marcel Mauss clearly expressed the principle: "Each one is the whole, and the whole is formed through it." Performance was the practical method for embodying this non-dual awareness.

The community of place is not merely a location. It encompasses the entire ecology of presences—incarnate, ancestral, geological, seasonal, spiritual—organized through rhythmic structures that function as information systems. Rhythm is not an aesthetic choice but an ontological necessity: The temporal framework through which multiplicity attains coherence without hierarchy. Rhythm is the manifestation of energy that moves, transforms, and gives shape to the universe. Indigenous performance systems understood this and maintained place-specific rhythms not just as cultural practices but as survival technologies, ways of maintaining balance within systems too complex for analytical understanding. The forms were not symbolic but operational—they did not represent a relationship with totality; they enacted and participated in it.

Yup'ik elder Harold Napolean describes how *Yuuyaraq* (the way of being human) is defined as the way the Yup'ik live in harmony with this spirit realm and its inhabitants. "To the Yup'ik, the land, rivers, sky, seas, and all creatures within are spirits and thus sacred. They are born into both the physical world—marked by the Bering Sea, Yukon, and Kuskokwim rivers—and a spiritual realm. Their arts, tools, weapons, kayaks, umiaks, songs, dances, customs, traditions, thoughts, and actions all carry the influence of the spirit world beings" (Napoleon, Madsen, and Alaska Native Knowledge Network 1996, 5).

Contrast this with the anthropocentric shift. By framing human awareness as an observer rather than a participant, and separating the subject from the object, modernity broke apart the community of place into resources waiting to be extracted. Written language began the move toward abstraction; each new technological medium—print, photography, digital networks, and now AI—has accelerated the retreat from direct sensory experience into systems of representation. The effects are not just cultural but existential. We now live in a world of surveillance, datafication, virtual realities, and algorithmic control—a complete reversal of the cosmocentric model. While place-based indigenous systems shared agency among all inhabitants of a place, today's power structures concentrate it within technological and economic oligarchies, claiming control over planetary systems.

The outcome is predictable pathology. Depression, addiction, political polarization, conspiracy thinking, and authoritarian resurgence are not separate problems but symptoms of a cosmic displacement. A species that evolved for 300 millennia within a community of place now finds itself disconnected from that matrix, replacing participation with simulation, relationships with control, and presence with narrative. Mass media spreads apocalyptic fantasies, nostalgia, superhero myths, and metaverse escapism because facing our real situation directly is too painful: We have made ourselves strangers in our home.

Yet, crises create opportunities. Place-based systems endure not as museum relics but as practical alternatives to anthropocentric collapse. What they provide is not a romanticized return but a profound reorientation. The challenge facing our modern condition is the same as that of our hunter-gatherer ancestors: Surviving within complex and fragile systems. Climate destabilization, mass extinction, resource depletion, and technological displacement—these create a new hunter-gatherer reality that requires similar resourcefulness, vigilance, sensitivity, and recognition of interdependence.

The shift from an anthropocentric to a cosmocentric awareness is not a luxury of theory but a necessity for survival. Malidoma Somé clearly states the obligation: "We owe the cosmic order because we are individually and communally responsible for its maintenance" (Somé 1993, 14). This is not mysticism—it is systems thinking. Every action produces consequences that ripple through networks too vast for precise calculation. Place-based performance technologies were explicitly developed to navigate this reality, to sustain balance within larger totalities beyond individual understanding. They achieved this not by dominating but by participation in cycles beyond human timeframes.

Performance, when properly understood, provides a means to grasp complexity. It is not about performance as entertainment or mere representation, but about performance as the active recognition that we are always already part of communities and places that extend well beyond human boundaries. The !Xuu Bushmen's trance dance did not symbolize transcendence; it was a technique for achieving a temporary dissolution of the boundary between subject and object, allowing direct experience of totality. What modernity needs is not copying these forms but understanding their purpose: They were ways of cultivating ecological awareness before ecology became a formal concept.

We stand at a crossroads. The traditional anthropocentric systems are clearly breaking down—economically, politically, environmentally, psychologically. What will replace them remains uncertain. Indigenous knowledge offers the way: Not a return to a prelapsarian past, but rather the recognition that awareness and community of place are inherently linked. Performance can spark this awareness and provide tools for reestablishing sensory engagement with the living totality—something we never truly left behind but merely convinced ourselves we had transcended.

This is not about adopting ritual forms or appropriating other cultures. It is about understanding that performance—serving as consciousness technology, a method for coordinating dispersed agencies, and a means of maintaining rhythmic coherence amid complexity—is precisely what an age of planetary crisis demands. The wisdom lies not in specific practices but in the cosmocentric epistemology itself: The recognition that we are integral parts of systems far greater than ourselves, and that our survival depends not on control but on participation, not on domination but on maintaining relationships with all other inhabitants of place.

The ongoing revolution is about decentralizing. After centuries of elevating humans to the pinnacle of reality, we must recognize our actual position: One species among millions, one form of awareness among countless others, heavily dependent on systems we did not create and cannot control. This acceptance is not defeat—it is liberation into reality as it truly is. Place-based performance systems demonstrate what awareness looks like when it acknowledges its embeddedness, recognizes that the "I" is always already "we," and understands that place is not just a backdrop but the protagonist.

We are, whether we recognize it or not, becoming planetary hunter-gatherers navigating unprecedented complexity. The question is whether we will develop the necessary consciousness technologies or continue in anthropocentric fantasy until collapse makes the choice irrelevant. Place-based awareness offers not definitive answers but orientation—not solutions but ways of being that have lasted far longer than industrial modernity. Performance, understood as enactment rather than mere representation, as participation rather than observation, and as a method for sustaining a community of place in an age of fragmentation, may prove to be the technology we most urgently need.

Making Sounds

The following series of explorations focuses on the sounds and rhythms produced by the body.

TOUCHING SOUND

Partners.

The first partner lies down relaxed with their eyes closed, while the second partner kneels or sits nearby.

Using a gentle touch with two or three fingers, the second partner slowly and softly makes contact with different parts of the body. The first partner releases a sound directed at the touched area, such as the stomach.

The goal is for the first partner to "vibrate" the body part being touched. This touch helps focus the first partner's breath, sound, and energy.

The sound can be steady, drone-like, or rhythmic.

If the vocalizing partner has trouble directing their sound to a specific part of the body, the second partner can gently touch and shift attention to another area.

Suggested areas to explore include the shin bone, forearm, forehead, nose, collarbone, solar plexus, various ribs, knees, toes, and thighs.

Note: This activity effectively demonstrates how sound is created in various parts of the body and helps to open the body as a resonator.

Highly Recommended.

BODY OF SOUND

This exploration can be done while standing, sitting, or lying down, alone or in a group.

Eyes closed to better focus.

Guide performers to breathe in through their noses and out through their mouths.

Establish a slow, steady breathing rhythm before moving forward.

Provide gentle prompts encouraging performers to breathe into each body part, starting with their feet.

Visualize a white breath entering, traveling down to the feet, wrapping around muscles and joints, then gently massaging the feet, toes, and heels.

On exhalation, visualize releasing tension and give the exhale a color.

Invite performers to produce a sound on each exhale for specific body parts, using any sound that matches the tension and release, and to associate each exhale with a color reflecting the tension released.

If making a sound is difficult, suggest simple options like "oh" or "ahh," emphasizing that each body part has its own unique sound, tone, and color.

Guide performers to inhale into each part, starting at the ankles and moving up to the calves, knees, then the top of the head, encouraging a distinct sound or varying the intensity if using a single sound.

Maintain deep nasal breathing, allowing the breath to flow smoothly through each area of the body.

Continue this process by inhaling, visualizing, and directing breath to a specific area of the body, then releasing it with a sound and a color.

Highly Recommended.

MANTRA

You can conduct the activity either individually or as a group.

Performers are seated.

Discuss and agree on a short phrase for everyone to explore.

If working in a group, form a circle and guide performers through a quick relaxation or breathing exploration, or build on another calming activity. Dim lighting and eyes closed to enhance focus.

Side Coach: *Repeat the short phrase for several minutes, maintaining the same tone and rhythm as you hear and say it—become familiar with its vowels, consonants, and cadence. Imagine the phrase as a landscape you traverse, noticing any emotions, thoughts, or images it evokes.*

Let the sound resonate throughout your body, paying attention to how your breath, chest, throat, vocal cords, mouth, tongue, lips, facial muscles, and skull resonators produce it.

Observe which sounds influence specific body parts and which bones resonate. Then, begin exploring creatively by repeating, extending, distorting, shortening, reversing, breaking, and rearranging the sounds, using sonic elements as puzzle pieces to assemble and reorganize. Have fun with the process.

Note: Give enough time for this exploration to develop—usually around 15 minutes for performers to start freeing themselves and diving deeper. Often, the process takes on a life of its own and can extend to an hour or longer. The sound release may become emotionally and personally intense.

Highly Recommended.

From the Siberian Journal

We moved straight from yoga into the **Mantra** exploration. Because of the group's cultural sensitivity about voice, I was interested in how this activity would go. I asked for a Sakha word with some interesting sounds. A few options were suggested, and we decided on *UuChuGae* (phonetic), which means "good" in Sakha. This was a great choice because of its three distinct syllables, and its positive meaning contrasted nicely with the Sakha people's usual pessimism.

The activity was done while sitting down. After a minute of repeating the word, they immediately started exploring the sounds—repeating, rearranging, stretching, changing rhythm and emotion, singing, and playing with them in every way—treating the word only as a sound resource, ignoring its meaning.

Based on my earlier experiences this week, I joined in to reassure them and give an example. They tend to overthink, feel insecure, and believe they must do everything "correctly." I wanted to make sure they wouldn't get stuck in a limited, literal understanding of the exploration.

My worries were unfounded. Doing the exploration again let me explore and understand Sakha culture through vocal and bodily expression.

Things progressed quickly. The exploration lasted over 45 minutes—it was hard to get them to stop. They were elated and, by all measures, it was a big success, showing a lot of variety, cadence shifts, rhythm, repetition, singing, wailing, celebration, and full engagement with the activity.

The exploration built a bridge between the work—yoga, meditation, and overall process—and their culture. It connected many ideas and offered an unexpected way to express and release personal and shared stress and tension.

Took a break.

We reconvened in the theatre's meeting room to watch a video of some of my previous work. The group had asked to see it. However, the theatre doesn't have a VCR or a TV—both were stolen. One of the actors brought theirs. The timing to view my work was perfect.

As expected, they responded strongly to the Metamorphosis Group (St. Petersburg) and Kwasa (Zulu) pieces. Many actors were concerned that our work together would be all dance and movement. Some were worried they would have to perform shirtless, as the Zulu performers did. I reassured them that theirs is a strong song and vocal tradition, and the work would likely follow that path.

Several especially liked the Zulu work—they said they had never seen anything like it before. They discussed and commented on the video. They realized they had little exposure to other theatre work.

BABBLE

Partners seated and face-to-face.

The performers use babble–nonsensical sound expressions of their choosing–that reflect how they are feeling physically and emotionally.

Partners communicate through babbling, creating a babble dialogue.

Side coaching. Suggest a situation characterized by a unique way of speaking that frames the improvisation.

Adjust the following suggested list as needed, based on cultural circumstances and goals.

They are mathematicians who describe their day, surroundings, and each other using numbers.

Infants communicating via baby talk about the future, routines, pooping, eating, and family.

Martians speaking a made-up language about their travels, the strangeness of Earth, and their home planet.

Foreigners use an exotic, unknown language to talk about their day and problems.

Two birds communicate only in bird talk about flying, eggs, nesting, worms, food, and a frightening encounter.

They are foolish and discuss life, its unexpected course, the meaning of life, the apocalypse, and frustrations.

Victims of severe injuries from accidents or war communicate through grunts, coughs, and cries. They describe their pain, reflect on how they became injured, their past health, and contemplate death.

They are close friends who whisper so quietly that even their partners cannot hear. They share embarrassing moments, achievements, secrets, and face fears.

Note: A popular exploration among various groups. Reduces inhibitions in language expression for both beginners and multilingual performers. It provides a vocal mask, allowing performers to freely express thoughts and feelings that might otherwise stay unspoken. Many performers found it psychologically freeing and insightful.

Highly Recommended.

SOUND OF OTHER BEINGS

This exploration builds on human vocal expressiveness and **Movement with Other Beings**.

Referencing observations from "Movement with Other Beings," ask performers to select five non-human entities—one from each category: Animal, Plant, Element, Ancestor, and Spirit.

Reflecting on the movement of each selected being, encourage performers to explore a vocalization.

Performers might start with what is obvious, but they should aim to go beyond it to achieve a deeper, more intuitive understanding and expression of, for example, a rock, a cloud, or a sunrise.

The vocalization should convey not just the essence of the being but also its function, perspective, and role in the world.

Once performers feel confident with the sound of another being, encourage them to combine and develop it into an action, a gesture, and ultimately a movement.

Find a partner and they converse with other beings.

Yes, rocks can move and speak. They do so differently from humans. Many Indigenous groups refer to rocks as elders, the oldest ancestors.

Consider that rock in your yard. It has existed since the planet's beginning, changing form and location over millions of years. They are the bones of the earth, holding history, a journey, and the stories of what they've endured. They can educate us and remind us of a perspective we have lost. We share the same minerals, making us, in a sense, relatives.

By giving other beings voice and movement, we embody and understand them, allowing their expressions to live through us—this is how they share what they know. Yes, rocks communicate in their own ways. A diamond on a wedding ring, a headstone, the granite pillars of a government building, Stonehenge, Easter Island, a Zen Garden, mountains, and riverbeds all have unique voices, roles, and places.

Our work thematically explores interactions with other beings, enriching awareness of the bond between humans and non-human entities. These beings are part of our world but have been excluded from our awareness to our detriment.

Exploration Evolution: Once the vocalization is identified and explored, ask performers to develop a repeatable sound and/or movement pattern. This pattern might evolve into a rhythm, a chant, a sound poem, or a mix of these elements, serving as a coded way to evoke a non-human entity during a performance.

The final theme of our work is **Balance**. The **Circle** functions as both a diagram and an operational motif. **Rhythm** acts as the articulator and animator of the circle. Now, **Balance** completes and enhances as it shapes the objective, vision, and purpose of our work.

Balance

Balance signifies a state of equilibrium between conflicting forces—a fundamental principle shared across many cultures as vital for human well-being and environmental sustainability. Instead of leaning toward the extremes of domination or disorder, balance offers a moderate approach that leverages the strengths of opposing forces while mitigating their risks. This timeless insight is especially critical now, as humanity confronts the outcomes of deep imbalance in our interactions with each other and the Earth.

The idea of balance is ingrained in humans from early on. The Cretan myth of Daedalus and Icarus illustrates a core Western warning against extremes: Flying too low causes sea spray to weigh down your wings, while flying too high melts the wax with the sun. The only way to achieve freedom and survival is through moderation—maintaining balance.

Ancient Greek philosophy articulated this wisdom with the idea of the Golden Mean. Socrates explained that virtue is found in choosing the mean and avoiding the extremes on either side. He illustrated this with examples: Exclusive devotion to gymnastics creates hardness and ferocity, whereas focusing only on music results in softness and weakness. Achieving a balance between these pursuits fosters harmony, beauty, and goodness. Aristotle expanded on this concept, illustrating how virtues are balanced states between extremes—courage, for instance, lies between cowardice and recklessness, and generosity between stinginess and wastefulness. This idea was so integral to Greek thought that the phrase "Nothing in excess" was inscribed at Delphi's sacred temple.

Eastern philosophy independently arrived at similar insights. Buddhism's Middle Way emphasizes that enlightenment exists between indulgence and asceticism. Confucius's Doctrine of the Mean states that excess is equivalent to deficiency. These teachings recognize that extremes, even virtuous ones, can cause suffering and imbalance.

Indigenous, place-based worldviews express balance most fully, seeing it not as an abstract idea but as a lived experience. Alaska Native perspectives highlight the synthesis of information gained from engaging with both the natural and spiritual worlds to live in harmony with natural laws. This view of balance as reciprocity describes a dynamic relationship where humans give and receive in sustainable cycles.

Place-based performance embodies this holistic approach through action, rhythm, and sound, representing a way of being that integrates community, place, and spirit. This nuanced understanding of interconnected systems reveals what Western thought has often overlooked: Balance requires active participation rather than static equilibrium.

Healthy ecosystems exemplify these principles by maintaining a dynamic balance in which genetic, species, and ecosystem diversity remain stable while gradually adapting through natural succession. This balance supports all types of life and acts as a blueprint for sustainable human systems.

Human civilization has disturbed this balance. Our human-centered perspective has led to complex social and cultural structures that promote overpopulation, exploration, and the commercialization of the Earth. This has created a relentless, self-sustaining system that accumulates imbalance, endangering the entire system of planetary life.

This ecological disruption reflects the cultural trauma faced by both Western and non-Western populations. Indigenous communities find themselves out of balance, torn between traditional intelligence and Western modernity. Traditional values—focusing on long-term planning, preservation rather than growth, community bonds over individual pursuits, and stability over change—seem disconnected in a world that values instant gratification.

Cultural displacement manifests in high rates of suicide, alcoholism, and drug addiction, illnesses, spousal and child abuse, social upheaval, and cultural change. The denigration and loss of indigenous, place-based cultures represent not just human tragedy but planetary impoverishment, as these communities serve as the planet's elders, holding essential wisdom about sustainable living.

Western culture faces its own crisis of imbalance. Disconnected from our bodies, the natural world, and community, we experience existential alienation, seeking empty comforts in materialism while basic needs for belonging and purpose go unmet. The signs are clear: Wars, refugees, pandemics, mass migrations, global hunger, unchecked development, pollution, deforestation, species extinction, social and economic inequities, greed and corruption, consumerism, crowded cities, and the rise of government, corporate, and technological authoritarianism. These symptoms and the desperate efforts to manage imbalance reflect the extent of the Earth's environmental imbalance.

The solution lies in achieving synthesis. Western culture's greatest strength has always been its adaptability and transformation. Now, facing planetary limits and internal crisis, the West must turn inward, simultaneously acknowledging the end of its era, taking inventory, recognizing limits and victories, admitting wrongs, adapting, and reinventing. Our moment requires mindfulness.

Balance begins with re-examining and enacting fundamental changes in how we relate to our bodies, communities, and the planet. These relationships form interconnected circles that embody both beauty and mystery. Changing our way of being—our responsibilities and engagement with the earth—is the path to changing the world. The principle of balance, recognized across cultures and throughout history, offers our most explicit guidance for navigating current crises. In a world torn between extremes, the middle path remains the way of wisdom, sustainability, and hope.

Balance begins with the body, which changes thought, perspective, others, community, and the world. Become the change envisioned.

Reminders

- **Engage fully in each exploration, maintaining focus and completing it thoroughly.**
- **While some tasks are simple, do not dismiss their straightforwardness; instead, embrace what simplicity teaches.**
- **Every step serves as a foundation for the next.**
- **Rushing betrays and undermines our work and offering.**
- **Cheat the work and cheat yourself.**
- **Dedicate your full effort and aim for excellence.**
- **Well done is its own reward.**
- **Our work is bigger than any one of us.**
- **Our work is a dynamic, living entity.**

The next explorations focus on balance awareness, breaking physical barriers, and developing trust among performers. They are cumulative, deepening understanding, preparing for performance while fostering a connection with others and the broader world.

STAND-UP

Performers are asked to partner up and sit on the ground back-to-back.

Side Coach: *Interlock your arms, bend your knees, and place your feet flat on the ground close to your buttocks.*

The goal of the exploration is to stand up in a single, smooth movement as a unified team. After successfully standing, sit down smoothly with your arms linked.

Repeat the exploration to improve the fluidity of standing and sitting.

Next, unlink your arms and stand or sit while leaning into each other. Once you master this part of the exploration, try sitting and then standing as quickly as possible.

Exploration Evolution: This exploration may unfold in stages with groups of 4, 6, or 8 performers. The core principles and processes stay consistent. Performers link arms and stand closely, maintaining contact to accommodate differences in weight, height, and balance as they attempt to stand.

FIGURE 1.9 *Balance and Roll Around* during a workshop. Lul Theatre, Addis Ababa, Ethiopia (L to R: Nigussie Balie, Meaza Takele, Nuyredin Nesro, Menahel Teshome, and Michael Million)

Source: Photo by the author.

BALANCE AND ROLL AROUND

Continuation of **Stand-Up** or offered separately.

Performers select partners who are similar in weight and height. Note that they will be in close bodily contact with their partners, and that gender preference may also be a consideration in the selection process.

Partners decide who goes first.

The first partner provides stable but relaxed stability, adjusting their weight and balance to accommodate the second partner who leans gently but firmly into them.

Maintaining solid contact between bodies is essential.

The second partner slowly begins to roll around the first (stable) partner. They roll with around relaxed, but not as dead weight, mainly using their shoulders and chest while leaning and moving around the first partner.

The stable partner adjusts their balance, footing, and shifting weight to support the second partner as they move or roll around their body. It is an intuitive and bodily give and take between the partners.

Encourage slow, gentle rolling movements. Highlight the importance of partners developing an awareness of their body's natural sense of balance.

Performers rolling may do so with their eyes closed; they need not limit which parts of their bodies they use to lean into and roll with.

Switch and repeat.

Exploration Evolution: Create groups of four who then lean into one another. Ask performers to identify one performer who will roll around the others as they adjust to the rolling partner.

Advise the performers to work on making their balance adjustments as smooth and unified as possible.

Exploration Evolution: The group moves as a single unit while one performer rolls. They must adjust their balance to accommodate the rolling individual and stay mindful of the group's constantly changing weight distribution, thereby maintaining an intuitive, shared sense of balance.

Highly Recommended.

Happiness is not a matter of intensity but of balance, order, rhythm, and harmony.

Thomas Merton

Balance and centered awareness explorations aim to reawaken and develop an instinctive ability to respond and stay balanced with another body, thereby recognizing balance within oneself. Once embodied, this sense of balance extends outward, cultivating an awareness of the importance of maintaining equilibrium.

BALANCE AND CENTER AWARENESS

Partners. One partner is asked to stand comfortably with their eyes closed, focusing on maintaining a neutral standing balance.

With gentle pressure, the other partner gives the standing partner a gentle push, then waits as their partner's balance responds.

The gentle push should target various body parts, including the shoulders, chest, back, forehead, thighs, knees, sternum, hips, etc.

The pushing partner needs to allow the standing partner time to sense, register, and adjust their balance.

The balancing partner is encouraged to verbally describe how their center has shifted and how it returns to its original position of neutral balance.

Partners then switch roles.

Note: This exploration fosters awareness of how centers move and how an adjustment impacts their entire body. Encourage performers to appreciate how their natural balance mechanism restores their center and stability.

Exploration Alternative: The pushing partner applies greater force to destabilize the standing partner and may push more frequently. The aim is to enhance sensitivity and technical skill in helping the performer regain their balance. Advise performers to avoid speeding up too rapidly or applying excessive force.

From the Siberian Journal

10 AM. Started with some yoga stretches. Moved into the **Meltdown-Creation**, adjusting the cadence of the count and backtracking. Did some other standing stretches. Focused on hand stretches.

I asked each participant to choose a partner for the **Balance and Roll Around**—a mistake—and I should have started with the Stand-Up to keep it simple, then progressed to the next exploration. The **Roll Around** was consistently too fast and somewhat out of control. The exploration requires greater awareness and physical sensitivity for one's partner.

They seemed to operate from tension, as if they were hauling bags of potatoes. Not much listening occurred to allow the exploration to explain itself. The exploration felt rushed and forced, so I decided to step back.

Went to the **Stand-Up**. They aligned with the exploration's physical objective. However, many chose to push or muscle their way up rather than share and support each other equally. Peter (my translator) and I demonstrated and explained how the back needs to be sensitized. I mentioned that it is a large part of the body, often neglected, yet it contributes greatly to our being and expression. I gave the example of the Sakha shaman costume, decorated with numerous back ornaments, which was given significant importance. The back needs protection, but it also functions as a shaman's main means of communication with the spirit world. This connected with them.

I continued working with groups of three, four, six, and eight people, having them combine **Balance and Roll Around** to **Stand-Up**. I did this to encourage physical interaction among them. This was their first time engaging in this kind of activity. Some participants lost interest and sat down on the side, while others stayed involved. It is becoming clear in this workshop who should be part of the performance group and who should not. The ensemble is effectively selecting itself. Basically, those who attended the yoga training sessions are the ones staying interested—they seem eager and open to trying something new.

Introduced **Floating**. Once again, people rushed into it without fully understanding the process or being aware of their actions. The men appear to be driven by tension—resistance rather than a sense of flow. They can come across as rough, which is off-putting, but the women in the group endure it. This roughness might be a result of their own lives being filled with tension—pushing and pulling, whether at the store or in most areas of life. The current economic and political uncertainty and tension might also play a role.

Some performers seemed exhausted, so we called a ten-minute break. I do not think this group has put in a physical workout like this in years, if ever. Many are in poor shape, and several smoke and get out of breath easily.

STAND OFF

A continuation of **Balance and Center Awareness**. Presented as a sequel to that exploration or separately.

Partners stand face to face. Similarly sized partners, gender, and physical comfort should be discussed before beginning the exploration. Decide who begins.

One partner softly pushes or acts as if to unbalance the other.

If the pushed partner moves one or both feet, the pushing partner earns a point.

The pushed partner earns a point if the pushing partner moves one or both feet or loses balance while pushing.

Increase the pushing intensity as agreed by the partners.

If one partner falls, stumbles, grabs, or wraps their arms around the other, they score two points. If neither partner loses balance, no points are awarded. Determine how many points are needed to win.

Side Coach: *You can use tactics like faking, dodging, or feinting, using sound and psychology to make your partner lose their balance. As you and your partner become more comfortable with the rules and scoring, the exchange can grow more intense and physical.*

Exploration Alternative: Conduct the game in slow motion.

ONE LEG BALANCE

Partners. The size and gender difference do not matter.

Each partner balances on one leg and must lean on their partner for support. Try with and without interlocking arms.

The partners then try to move across the space, each using one leg and supporting one another. They need to develop a travel method that suits their weight, height, and balance dynamics.

After crossing the space, the partners are instructed to return using their other leg while continuing to support each other.

Exploration Alternative: Performers travel back-to-back using only one leg.

Lao Tzu on Ying Yang

We may never know where life originates or where death leads; nevertheless, I will attempt to share some insights. The utmost yin is strict and cold; the utmost yang is intensely hot. The strict and cold emanate from heaven, while the intensely hot derives from the Earth. The two merge and penetrate each other to create their union, from which all things emerge. Their actions alternate through the four seasons and day and night. Life has its place of emergence; death has its place of return. This opposition between beginnings and endings forms an endless circle whose conclusion remains unknown to all.

Kristofer Schipper, The Taoist Body

FLOATING

Offered separately or as a continuation of **Stand Up** or as a part of a series that includes **Stand Up**, **Balance,** and **Roll Around**.

Partners should be of similar height and weight. They stand back-to-back.

One partner is designated as the lifter, placing their arms beneath the other partner's armpits and looping their arms around the partner's. Together, they lock elbows. The lifter then bends their knees to position the small of their back under or below their partner's buttocks.

The lifting partner then raises their partner onto their back. Once the partner is secure, they unlink their arms. The partner on the back relaxes on the lifter's back and "floats."

Next, the lifting partner bends at their knees, gently giving their partner a ride that rocks them back and forth. It's essential to ensure the lifted partner feels safe and secure during the ride. The initial position should be balanced, allowing the lifted partner to relax. As the lifting partner provides stability, the lifted partner floats and balances.

Side Coach: *As the lifting partner, you can sway side to side, bending and straightening your knees, to give your partner a smooth ride. Your goal is to build trust and help them unwind and enjoy the stretch of the ride.*

Imagine yourself as the ocean waves, creating a gentle rhythm of lifts and descents, using knee lifts to mimic a raft floating on the water.

Encourage the lifted partner to surrender and trust the process. Invite them to relax their heads and picture themselves drifting on the sea, with a gentle ocean breeze caressing them.

Some individuals struggle to relax and let go. Utilize this exploration to focus on balancing, trusting, and releasing.

Highly Recommended.

The body connects different worlds and embodies all times, ancestors, and spaces in heaven and earth. It can unify, equalize, harmonize, and control the cosmos, restoring the world to its rightful balance.

From Chinese Theatre (282)

FIGURE 1.10 *Floating* exploration, Lul Theatre workshop, Addis Ababa, Ethiopia (Bottom and Top, Lealem Berhanu and Menahel Teshome)

Source: Photo by the author.

ALTERNATIVE MOVES

Define the perimeter of the space using chalk, tape, or another marking.

Performers are asked to stand scattered and frozen in the space at the start. Their bodies and arrangement balance the space.

Begin. When a performer feels inspired to move, they initiate actions that upset the balance of the space. They use movement, gestures, levels, sounds, and any other means to create disturbance, then pause and pose.

The group freezes for a five-count, absorbing the disruption and imbalance in the space.

After five counts, another performer responds to the first performer's movement to rebalance the space.

The exploration continues, with a performer either disturbing the space's balance or rebalancing it.

As the performers become more comfortable with the exploration, reduce the five-count freeze, then eliminate it.

Performers respond spontaneously and constantly either unbalance or rebalance the space.

Encourage performers to:

1 Improve their focus.
2 Listen and sense with their whole bodies.
3 Learn to identify and rebalance spatial imbalances.
4 Be aware of the presence and movements of other bodies.
5 Develop their instinct to read a space and react spontaneously.
6 Develop an intuitive sense of balance and the performer's role in creating it.

This exploration enhances the development of ensemble and improvisational skills by fostering active listening, physical reactions, and collaboration with others.

MOVE AND CENTER

Performers move easily and freely in the space, remaining aware of their center of gravity and how their balance changes.

Performers are then asked to respond to these prompts.

Move with curves, whatever that means to you.

Give time for each movement prompt exploration.

Prompt suggestions (add or adjust to needed):

Move straight
With height (on toes and off)
Low
Smooth
Jerkily
Sideways
Unpredictable

Stay aware of your center and how your balance shifts.

While moving through the various prompts, occasionally ask performers to freeze so they can assess their center of balance.

Side Coach: *On three counts, make three jerky movements until you flatten on the floor.*

How did your center of balance move?

Now rise to standing with three counts or movements, then lower yourself until all upward and downward movements to the floor become smooth and flowing. Stay conscious of your shifting center of balance.

Now move quickly through the space, navigating around others while staying centered and balanced, weaving with as much curve as possible, yet still able to weave. Always stay aware of your adjusting center of balance.

Freeze. What is your center of balance?

Now move chaotically, deliberately disrupting your center of balance. How far can you push your center of balance?

Return to your usual walk and movement in the space, focusing on your center.

Move with the maximum amount of space between you and others, then with the smallest possible distance without touching.

Be aware of how your center of balance shifts relative to others.

Note: Advise performers to continue this awareness in their everyday lives.

FIGURE 1.11 *Move and Center* Exploration, Metamorphosis Theater, St. Petersburg, Russia (L to R: Sergi Popov and Elena Stepahova)

Source: Photo by the author.

Moving

The upcoming explorations aim to deepen our understanding of how the body naturally seeks balance. The performer's craft involves clarifying and expressing the action-reaction within their body and, ultimately, within their collective performance and community. Our focus is on balance as the main goal, striving to make it an instinctual and constant awareness.

MAKE A MOVE

Invite the performers to relax their right arm and let it move freely.

When prompted by your signal, instruct performers to freeze the gesture and hold it steady.

Side Coach: *Do not interfere with the gesture; instead, notice the feelings and thoughts it triggers and expresses.*

Recognize the natural impressions that emerge from your body's expression. Move your right arm again, holding it steady without altering its position, and feel the connection from your hand through your arm and shoulder to your face and eye muscles.

Make a slight adjustment to the arm. Consider what effect this small change creates. Each minor adjustment alters both the change itself and the overall body expression, with every small tweak adding to a more complex whole.

Every movement forms part of a unified expression and is part of a broader balance, which is instinctively guided by its connection to the entire body.

Movement and adjustment are also linked to your emotional and mental states. As you move and refine that movement, you make mental and emotional adjustments. This process connects to all parts of the system that continually adapt, interact, and seek an evolving balance among emotional, physical, and mental factors. Your body and life are in a constant state of processing, reconciling, and balancing.

Every change in body language or attitude results from a complex interaction of environmental, physical, emotional, biological, intellectual, and spiritual influences. Just as the world seeks balance and stability for health, well-being, and survival, so do we. Many subtle and not-so-subtle shifts happen constantly, often without our conscious awareness. The performer's role is to develop a heightened wakefulness of themselves and their surroundings. This is awareness of how interconnected the world's dynamics are can be cultivated, trained, and refined. It involves aligning the body's inner dialogue with an understanding of the world's dialogue, making them one. The performer's sacred role is to be fearless, aware, open, vulnerable, generous, and honest, reflecting all of this to the community.

Performers, to varying degrees, possess innate sensitivity and awareness. Some are born into a lineage or heritage, while others have a unique spark of light or are driven by a calling to serve. For some, it is a place in the world; for others, it feels like a curse they cannot escape. In different cultures and eras, these individuals were seen as gifts essential to community flourishing—such as shamans, healers, mediums, prophets, and diviners—known by many names but sharing the same sacred purpose. They move between the known and unknown, making the invisible forces that surround and inhabit the world visible. Their role is to maintain balance within their community, requiring heightened awareness. Their actions, words, and ways foster awareness and healing, thus restoring balance. Functionally, they reinforce and reaffirm the community's order and continuity within the cosmos.

Today's performers still hold a special place—they transform and shapeshift, appearing possessed or attuned to the unseen realms, venturing into dark, mysterious parts of the collective unconscious, existing between the material and the ephemeral worlds. Respected but often regarded with suspicion, they command both reverence and skepticism. Like shamans across cultures, they are revered role models and catalysts of awareness, set apart as vectors and shapers of energy. The flow of life and moves through them, channeling vitality to awaken and energize their communities. Their expression captures a profound understanding of the complexity of the present, past, future, and all possibilities of the world. The performer, transcending the limits of time and space, demonstrates a sense of presence and connection to something beyond the self and the moment. They serve as a sacred conduit, oscillating between human and other-than-human, between the finite and the infinite, the tangible and the ineffable.

From the Korean Journal

Much of the unexpected has come to light during my time as a visiting professor at the Korean National University of the Arts. I was invited to teach "Traditional Theatre," with my role being to find modern expressions of the traditional. Sometimes, it takes an outsider to point out what an insider cannot, especially in a culture that emphasizes conformity.

In developing my approach to cross-cultural performance pedagogy and using traditional vocabularies to express a contemporary context, I identified a "cultural diviner process." This methodology uses performance explorations and explorations as diagnostic tools to understand cultural patterns and unlock creative and expressive potential.

The explorations serve as methods to engage with and analyze the culture itself, revealing sacred roots beneath it and fundamental human needs. My limited experience with Mudangs, the Korean shamans, sparked this insight. Now, these connections seem obvious and limitless. Each culture responds differently to creation; the common element is the "body" of that culture—its underlying assumptions, resistances, and latent potentials. This process reflects traditional divination practices: The diviner interviews participants, knowing they already have the answers but need guidance in reading, interpreting, and acting on them to restore balance.

Workshops focus on skill-building, raising awareness, and providing training. The performance development stage expands on this with a more interactive and comprehensive approach. Student interactions and feedback influence and steer the process, similar to the diviners I've collaborated with in Africa, who listen and observe to interpret patterns. This is a cultural exploration aimed at discovering and establishing a shared understanding, with performance serving as the tangible shared action. A collective effort is both affirming and healing.

Each culture's response to creative challenges reveals its fundamental views on authority, creativity, community, and personal expression.

Korean students demonstrated that the key to artistic breakthroughs isn't about abandoning cultural identity, but about exploring new methods to express themselves within familiar traditions. The aim was to help students discover their unique creative voices while respecting both their heritage and innovation. Listen carefully, observe keenly, and understand physically. All the knowledge you need is available. You are a conduit.

Working in Korea offered an ideal environment to explore these tensions. Each exploration revealed unseen cultural influences and opened new avenues for expression. This iterative process taught me how to adapt to a new culture and preserve traditional art forms amidst change, while Korean students explore the freedom to create and express their individuality.

Implications for Cross-Cultural Performance Work

The cultural diviner process now plays a key role in my approach to cross-cultural artistic projects. It indicates,

- Performance workshops serve as advanced tools for cultural diagnosis.
- Every culture possesses its own unique creative potential, which emerges through specific inquiries and explorations.
- For genuine cultural exchange to occur, it is best to work within existing cultural frameworks rather than imposing external models.
- The combination of cultural identity and creative exploration leads to the most meaningful artistic discoveries.
- Performance functions both as a means of cultural preservation and a catalyst for change.

MAKING MY MOVE

A continuation of **Making a Move** and may be applied separately.

Ask the performers to raise their arms in a simple gesture.

Side Coach: *Extend your right hand in front of you, palm facing left, not because you want to, but because I ask you to. You're willing to follow along without knowing where this will lead. Welcome to the opposite of improvisation: Earlier, you chose a gesture, and now you are doing something that's been directed. Accept this gesture without analyzing its meaning intellectually or emotionally. Embrace it and let it speak for itself without judgment or qualification.*

Ask the performers to explore what the gesture evokes. They can either comment aloud or note it for their own reference.

Allow time for reflection. They may move to a neutral position, then shift to the assigned position to notice the difference.

Finally, ask the performers to respond to a second included movement and experience what it conveys.

Side Coach: *Something is given to you from the outside that differs from the free movement you previously made, but if you fully embrace it, it becomes the same. It becomes yours, and you become it. If you can experience this, it will clarify the entire question of texts, authorship, and direction. Performers and all artists recognize that freedom arises when the outside is integrated within, creating a perfect blend that turns into something unique and new.*

You balance the external with the internal. It doesn't matter what it is or where it comes from. It is now part of a balance, just as you are part of an ongoing social and cultural harmony through time and space.

Ask the performers to revisit and transition smoothly between the neutral position and the two given movements. Observe the connection. Let sensitivity and body awareness guide even the smallest muscle interactions. The work remains subtle.

Ask which emotions the gesture evokes.

- *What images does it evoke?*
- *How does the gesture affect your thoughts?*
- *What do you think you are expressing?*
- *Is that expression clear?*
- *How can it be made clearer?*
- *How can it be made ambiguous or hidden?*
- *How should you respond, and what is your next move?*

Invite performers to sense the movement as they transition from the first to the second movement and then into a neutral state.

- *What happens during this sequence of movements?*
- *Meaning starts to emerge.*
- *What changes occur?*

Consider how an Action and a Reaction together create meaning and maintain an intuitive balance. An Action is a conscious initiation, while a Reaction responds to that action, establishing significance and equilibrium. A single movement is generally neutral or limited, but when two movements occur in succession, they converse and begin to generate meaning.

- *What meaning can we find in the relationship between the two movements?*

Side Coach: *Now, add three more movements to reach a total of five. Perform all five, ensuring each movement provokes an emotional reaction, expresses something, and creates meaning.*

This exploration demonstrates how performance traditions centered on form, function, and communication—often through rich movement vocabularies inspired by their surroundings—may lack expressiveness on their own. However, when integrated, they develop into a coherent and meaningful expressive system.

To capture a form from the environment is to extract it from the world, abstracting and organizing it into a system of expression and meaning.

Note: Form-based performance is a lively expression of place and holistic connection.

In many traditional performance cultures, formal expression is essential. Originally grounded in a specific location and era, the form carries and conveys the spirits and ancestors that preceded it, enabling them to perform in modern contexts.

The form reaffirms and explores cultural roots, existing before and shaping its content.

Western culture often depicts this through stereotypes and archetypes. Character types like the cowboy with a white hat symbolize and strengthen culturally ingrained figures and meanings, linking the past with the present and the individual with their community.

These forms last because they're essential and bring a shared meaning system to life. Expressing them serves as a reminder, a reassurance, and an affirmation of cultural continuity and significance, as well as a model and a point of reference.

MOVEMENT FLOW

Groups of four to six performers.

One performer initiates a movement, which is then shared and imitated by the group.

Next, another performer organically builds on the original movement, transforming it into their own expression.

Encourage freedom of interpretation while maintaining consistency and a natural connection with the group and the movement that inspired it.

The movement should flow as if it's circulating around and through the group, resembling amoeba-like motion.

Side Coach: *The movement circulates through the ensemble until it becomes a smooth, instinctive flow that shifts between action and reaction, establishing an inherent group balance.*

The speed of this movement accelerates, establishing a precise rhythm for each evolution. The group becomes a rhythmic unit of smooth continuity.

Exploration Variation: Naturally develop the movements by incorporating non-verbal sounds.

Exploration Variation: The performers navigate different settings like a city street, park, or museum, responding to each space's atmosphere. For example, when someone enters a room or city park, performers react with nonverbal movements or sound.

Next is the performer's balance.
Following that is the community's balance.
Finally, there's the balance of the world.

These next explorations emphasize body-based characterizations influenced by performers' everyday experiences. They offer a framework for analyzing and assessing culturally specific shapes, movements, and situations from a psychophysical standpoint.

BODY SHAPES

Each performer takes their turn standing in front of the group and assumes a body shape.

Each shape must have one specific focal point, such as an upraised right shoulder, a stiff left leg, a pointing finger, or a furrowed forehead, that the rest of the shape references.

Inspired by the shape, each performer then interprets it—not literally, but abstractly.

Side Coach: *After assuming the shape, listen to it carefully, letting it transform you both psychologically and physically, and triggering responses such as personal memories, images, and associated emotions.*

When you're ready, move in response to the shape—use the shape as a reference for your germination movement.

Walk, sit, participate in various activities, and when you're prepared, make sounds.

This continues with each performer presenting a body shape, while the others explore that shape in turn, each in their own way.

After exploring each shape, performers choose three shapes to investigate further.

Side Coach: *Create a phrase or sentence that feels natural to your body. Build a character from that shape. Explore the rhythm and system of how the character moves and balances in their world.*

If you feel you're losing the essence of their body shape character, pause, return to the original body shape and its focal point, reconnect with their inspiration, and then continue.

The body shape offers a perspective, reflecting attitude and emotion, as well as a way of being, seeing, and engaging with the world.

Exploration Development: As each performer returns to work with their preferred body shape, repeat the process described above for review. Then, within the context of their body shape character exploration, encourage the performers to interact with others.

For example, they might greet each other, search for something, move objects around, sit down to eat, send someone on an errand, or assign a task reflecting their character. Let the performers practice their characters in action or initiate a scene.

Allow the dialogue and actions to unfold naturally. Think quickly to develop the scenes.

Later, performers can share their interactions with the group and seek feedback and reactions. This process helps refine characters, scenarios, and performance ideas.

Exploration Alternative: Performers take on the form of another entity, like a tree, plant, animal, or architectural piece. This can be done individually or collaboratively.

Note: This exploration allows performers to directly observe how they influence their emotions and act as conduits for their environment. It is essential for developing an effective approach and expanding their performance vocabulary. Additionally, it promotes the integration of traditional and popular cultural elements with the distinctive expressions of individual personalities and cultures.

This exploration supplements other research that focuses on an external-to-internal, psychophysical approach to character development.

BLOB

Groups of 5 to 10 performers form a cluster, where everyone is connected, and they begin to sway until a silent agreement is reached on where and how to move.

Pay attention to the group's balance and rhythm. When the balance shifts and the group move, be alert to any imbalance and work to restore it.

Then, the group moves in whatever direction the communal mind chooses.

Experiment with different movements, including swaying, jerking, hopping, staggering, dancing, rhythmically shaking, shuddering, and adjusting the group's balance at low, medium, and high levels.

Exploration Alternative: Sounds are added to accompany movement.

STOP-START

Part I: Performers navigate through the space, encouraged to change directions while remaining attentive and responsive to each other.

When one performer chooses to stop, the others follow, whether suddenly or gradually, with the group responding accordingly.

Once everyone has stopped, anyone wishing to move again may do so independently. The group can remain stationary for a while—they are not required to move simply because others are moving, nor must they stop because everyone else has stopped.

They continuously alternate between stopping and moving, staying attuned to each other's actions, fostering collective awareness and a shared movement rhythm over time.

Part II: Building on the established stop-start sensitivity, prompt the performers to begin again by moving, stopping, starting, and, when inspired, jumping. Keep going until the group moves, stops, starts, and jumps as one unit.

Part III: Building on the previous point, various movements can be included. For example, they might shuffle to the left, turn around, nod their heads, and more.

Part IV: Another way to expand the exploration is by adding language, a single word or phrase. For example, a sequence might include performers stopping, turning around, jumping, saying, "You are wonderful," and then starting again. This sequence can change each time. The difficulty of the sequence depends on the group's ability to find a balance, keep rhythm, and stay aware.

TRIANGLES

An improvisational exploration.

Three performers are asked to imagine a romantic relationship among themselves, with two of them designated as the "couple."

The third performer plays the role of the "home wrecker," whose goal is to try to draw one of the "coupled" partners away. The "lover" within the couple attempts to navigate the situation, feeling conflicted about leaving and unhappy when staying. Meanwhile, the "abandoned" partner works to restore balance to the relationship.

Ask how interpersonal balance shifts and how the group's center of balance changes over time. Also, encourage performers to observe their positions relative to the triangle's overall balance.

Non-verbal sounds can be incorporated into the scene. The dynamics of the relationship may shift after the improvisation concludes or as it unfolds. Include scenarios involving same-sex relationships and diverse racial or cultural backgrounds to deepen performers' emotional understanding and expand their range.

Create new triangles with unique challenges for performers at your discretion.

Exploration Variation: Performed outdoors, whether on the street, in the woods, or in a park, this activity involves partners encountering a third, non-human entity they can connect with, forming a triangle. As partners shift their attention from one non-human element to another, they take time to reflect and explore how the third element influences them emotionally, intellectually, and spiritually. For instance, interacting with a boulder feels very different from engaging with an automobile, a sculpture, a fire hydrant, or a sign. The purpose of this exploration is to observe and understand how adjustments are made, emphasizing subtlety and nuance in the process.

Exploration Variation: As the exploration unfolds, introduce language—short phrases or single words work best; otherwise, many performers may switch to verbal communication. After establishing language, prompt the performers to notice how language influences their actions and reactions.

THE FLY

Ask the performers to identify and fix an imaginary point in space. The focus point is a fly positioned just beyond arm's length from the performer. The distance to the fly stays constant.

Side Coach: *Begin by focusing on a fly, allowing it to move freely in space—raising it high, lowering it, or circling your body. Then follow the fly with your head and body, mirroring its movements as it zigzags, hovers, approaches, and varies its rhythm.*

The fly gets closer to you, landing on your arm, nose, forehead, shoe, etc.

The fly invades your clothing—pants, underwear, shoes, etc. Be specific and respond to the fly's movements.

The fly escapes from your clothes and chases after you, transforming into an angry attack fly! The fly dives and strikes as you try to stop it.

You reach your limit and confront the fly, focusing and waiting for the right moment before quickly smashing it with both hands.

Note: Discuss how rhythm was vital to the process, and how balance helped to organize and refine the sequence as it naturally evolved. Also, discuss how humor, character, and drama arise from achieving balance and then pushing beyond it.

From the Siberian Journal

Sample Session

Isolating Body Parts: Tension-Release. Working through the body—starting from the floor, lying down, and then moving in a circle. The group stood up to join in the **Creation & Meltdown**. I led the exploration several times, beginning with a 50-count, then 30, 20, 10, 5, 2, and finally, 1. As we worked together, they moved in unison with the warm-up explorations around the circle. Each person created a stretch or movement with a sound, and everyone else in the circle mimicked it. The next performer in the circle then built on this, enhancing the movement or sound and transforming it into something new. The exploration proceeded smoothly, culminating in Maria naturally falling into the dance steps and melody of a traditional "Oosoki" Circle Dance inspired by the summer solstice.

Although they collaborated for years, they never developed into a cohesive team, as they lacked a shared vision or common goals. The group also lacked interpersonal imagination, creativity, and physical connections—elements missing from their training and understanding of theater. This approach conflicted with the communal spirit rooted in their cultural heritage.

Circle Definitions: From the tallest to the shortest, and from the oldest to the youngest (though the women might have added a bit of exaggeration), birthday months, and from the darkest to the lightest clothing colors. This exploration was simple but fun for them—creating a lighthearted, new way to see each other. There was plenty of laughter and a playful atmosphere. I think this also introduces a fresh way to bring fun and playfulness into work.

Given the challenges I faced earlier in the week, I decided to fully explain each exploration, integrating it into the workshop rhythm. Before each exploration, we pause to gather the group and provide an explanation.

This approach, by contrast, offers minimal explanation and focuses on preserving the rhythm and flow, similar to practices in Western or English-speaking settings. In Western contexts, participants tend to exhibit higher self-confidence, are more willing to take creative risks, are open to exploring the unknown, and often display a greater sense of playfulness.

I asked each performer to select a partner for the **Balance and Roll Around**—mistakenly. I should have begun with the back-to-back **Stand Up** to keep it simple before moving to the next exploration. The Roll Around was often too quick and somewhat chaotic. This activity demands greater awareness and physical sensitivity toward one's partner. Participants seemed to approach it with tension, as if they were asked to haul bags of potatoes. There was little engagement, preventing the exploration from conveying its purpose. Because the activity was so hurried and stressful, I decided to revisit the approach.

They responded to the straightforward, physical goals of the exploration; however, many muscled their way through instead of sharing and supporting each other. Peter and I demonstrated and explained how the back needs to be sensitized. I emphasized that the back is a vital part of the body, often overlooked, yet it plays a crucial role in our being and expression. I used the example of the Sakha shaman costume, adorned with numerous ornaments, to illustrate its significance. The back needs protection, but it also serves as a conduit to the spirit world.

Explorations

As we advance in our work, you'll observe that the main themes—**Circle, Rhythm,** and **Balance**—begin to merge. This blending is intentional. Over time, these themes will intersect more deeply, developing a unified approach that culminates in the performance event. Each theme is essential on its own, but its real strength lies in its combination to create a complete, harmonious whole—lived and performed.

Circle—the diagram and symbol
Rhythm—the energy and articulator
Balance—the objective and vision

Below, we will continue to examine our themes of Circle, Rhythm, and Balance, and expand on how they interrelate. As we move forward with these **Body** explorations, they will increasingly anticipate and intersect with the **Space** and **Place** phases. This is how it should be when working toward a cosmocentric way of being in and with the world. The main emphasis of the **Body** phase is on awareness and expression, while social and cultural issues will be addressed in the upcoming **Space** phase. The **Place** phase will concentrate on performance, including its development, creation, rehearsal, and execution.

Review the explorations below and utilize them as needed. Ensure they are diverse, engaging, and challenging. Incorporate variety to maintain interest, unpredictability to stimulate critical thinking, imagination, and fresh perspectives, and include jarring elements to provoke response perspectives. Combining explorations encourages new and surprising connections, ways of thinking, seeing, and being. Please review the sample sessions throughout this volume to better understand different ways of moving forward with our work. Ultimately, the sequence depends on your specific context, group, and goals. Trust your instincts, let go of expectations, and follow what naturally unfolds. Always stay attentive to what the work is communicating.

Naturally, all of this can be more challenging in practice. Just like in life, the journey is rarely direct. Explorations might lead you down unforeseen paths. Don't get discouraged—view them as part of the natural process and concentrate on the positives. Each activity and exploration will introduce you to something new. You're part of this journey. If anything is unclear, don't hesitate to discuss it with the group. Feel free to share your thoughts, emotions, and experiences with others—and yourself.

Your Work

Maintain a journal to plan and reflect. Stepping back might reveal insights. Pay attention—avoid laziness and complaints. Remember, there are no problems, only issues.

When organizing sessions, explore options. Follow your plan but stay flexible.

Review, reflect, and tweak your approach regularly. It's a natural process with no single correct method. The path will become clear. Balance thinking with feeling.

Trust your instincts; doubt can be destructive. If you choose wrongly, face it—your intelligence will guide you. Consult your group and adapt. Leadership shapes performance.

Set clear goals for each session, connecting them to your broader project.

Keeping a journal is like an ongoing conversation with yourself. Use it to process thoughts, work through ideas physically or vocally.

You're part of a growth process that demands focus, generosity, and commitment. Half-measures won't suffice.

There are no shortcuts. This challenging work requires dedication and stamina. Keep in good shape; you can't cheat the work.

Follow your own approach and discover what works best for you.

Each project is one-of-a-kind, and you only get one chance.

Make it count by giving your all.

Embrace the uncertainties and joys along the way.

Stay attentive but calm and receptive.

Your behavior and mindset influence the outcome.

The effort you invest determines what you gain.

If unsure, just believe—positive thinking spreads and can inspire.

Don't focus too much on the details. If something doesn't go as expected during your session—whether it's a change in location, time, weather, or if you decide to alter your approach—have a backup plan and make a well-informed decision. This is part of the process—and life. Trust yourself and your team. Be bold, take a leap of faith, and believe.

Plan ahead, but remain adaptable and ready for last-minute changes. I've frequently spent considerable time preparing a session only to abandon it and choose a different approach because the circumstances or energy in the room didn't align.

At the National Theatre School in Krakow, a casual comment triggered a revelation that prompted an immediate change. During a workshop at the National Theatre of Kenya, I had to adapt everything at the last minute. I was told there would be 25 actors, and I prepared for that. On the day of the workshop, nearly 100 people arrived; it was raining, and the room had roof leaks. At Addis Ababa University, I was informed the workshop would have "no more than thirty actors"; instead, nearly 200 attended. While leading a workshop at Karamu House Theatre in Cleveland, one performer in the all-black, intercity company started speaking in tongues (glossolalia) during a visualization exploration. Having experienced speaking in tongues at black Pentecostal churches, I instinctively placed my hands on the performer, synchronized my breathing, and invited her to return. She did. The group was as stunned as I was, prompting us to explore Black Pentecostal Church culture as a resource we had overlooked. That event inspired our work and became a key reference that shaped our performance.

I see the world as a continuous, unfolding story that we all participate in. Each event and detail is interconnected, forming a larger narrative. Your task is to become a reader of this world, where everything is accessible to you. The world is always communicating, flowing through and around you. Every moment carries a voice. Stay present in the world, and the world will be present in you.

And yes, the global narrative can be overwhelming. The important thing is to stay aware, calm, and adaptable. Act with kindness and generosity. Remember, you have a choice—either accept or decline. I have no expectations. Just like nature, everything is inherently perfect and occurs as it should.

Body and Spaces

CROSSING THE FLOOR

Performers are lined up at one end of the room or space.

Side coach and provide images to inspire as they cross the floor.

Examples of images:
Cross the floor as if walking on shards of glass.
Thick, damp grass.
Cow dung.
A minefield.
Dead bodies.
Fire.
Slippery ice.
Water.
Sand.
Jumping from ice floe to ice floe.
Stone.
Uneven rocks.
Wet paint.
Grease.

Performers move across the floor freely, interpreting your image prompts either literally, abstractly, or metaphorically.

CLIMBING THE MOUNTAIN

Performers gather on one side of a space and are asked to imagine the floor or ground as a sheer rock cliff. They find themselves at the bottom, looking up. They "climb" by crawling along the ground or floor.

Their goal is to ascend the mountain and help others in the group do the same. This exploration requires teamwork.

The group imagines they are climbing a rock face, gripping small crevices and cracks between the rocks—this is quite "dangerous," so they need to be cautious and support each other. Depending on their imagination, they might even fall from the rock face.

Encourage performers to either develop a plan or just begin climbing—supporting one another and climbing over and around each other to reach the top, the other side of the room, or space.

LOG-ROLLING

Performers lie side by side on their stomachs, forming a line of "logs.'

The performer at one end then rolls over the others, moving to the end of the line.

Once at the end, they take their place at the opposite end.

As the first performer begins rolling and progresses down the line, the next performer starts rolling, and the process continues until everyone has had a turn.

Encourage gentle, slow, and steady rolling to prevent injuries.

Note: This activity is excellent for breaking down physical barriers and understanding how performers of various weights and sizes collaborate. And it's enjoyable, always good for many laughs and smiles.

From the Siberian Journal

We took a break and reassembled in the fourth-floor ballet studio. Work in the ballet studio began with **Crossing the Floor.** The group lined up on one side of the room and crossed.

Thin Ice: Something they could relate to—some performances were very effective, and the variety was engaging to watch—showing many different emotional ranges and techniques used to cross the ice. Kuprion was the only one who crossed quickly, then sat at the other end and waited for the others—uncertain if this meant he was bored, unchallenged, or simply indifferent to the exploration. Standouts included Kiril, Misha (whom I referred to as Misha 1 because there are two Mishas in the group), Elena, and Anna.

In response to Desert (which they only know abstractly), there were some lively and diverse reactions—some quickly engaged in creative extrapolations, such as putting their shirts over their heads, crawling, and walking on their shoes across the floor. Walking on Broken Glass was difficult to watch. Several took minutes and seemed to be in significant discomfort as they crossed, demonstrating their ability to portray pain convincingly.

Regarding Grass: Maria commented that it was a shame to walk on the grass, implying that grass is meant for cows to graze during winter, not for human footsteps.

Concerning Walking on the Moon: They experimented with anti-gravity walking.

About Flying a Kite: It took some time to explain about the kite through a translator, but they finally understood—it is unclear how many had flown a kite before, given the limited leisure time in Russia these days.

In response to Walking on the Lines of the Floor, many participants walked in a line to maintain their balance, demonstrating their dedication and technical skill. They know how to turn a simple prompt into a creative situation. They are beginning to engage in more imaginative free play with the work and share their ideas. A few lost their balance and fell humorously, showing they are starting to play; Maria, in particular, was quite the comedian.

Regarding Dream Walk, the most abstract of the walks—it took a few minutes (learning to give them more time) for the activity to fully develop. It began with literal movements and interactions, gradually becoming softer and more ephemeral as they engaged with each other.

In response to Nightmare: An organic transition into a horror story and an energy shift—filled with much terror and mock attacks on one another.

I thought we had finished the exploration. I mentioned that the nightmare section reminded me of shopping in Yakutsk. They misunderstood and thought I suggested they should do "shopping." Suddenly, there was a burst of energy, long lines, arguments, and lots of vocalizing—characters started to appear, shop attendant-type characters venting their frustrations, filling the room as smaller scenes began to unfold. They fell into it, and it was hard to stop them. It was incredible—it struck a real nerve. Food shopping is a big part of their lives, and I should have realized they would associate it with nightmares.

Next, **Climbing the Mountain:** Several groups started their ascent up the "mountain," some helping others while some crawled over each other—almost everyone was fully immersed in the activity, demonstrating great focus and dedication to the goal—highlighting their Stanislavski training. Some tripped and "died," while others turned the exploration into dramatic stories—they thrive on challenges and simple problems to solve—curious and recurring traits.

They needed to cross the floor, so we did the **Log Roll**. It was exciting, and they really enjoyed it. Even though they are an acting company with many members who have worked at the theater for 20 to 40 years, I sensed that, for many, it was their first time engaging in such a physical activity for fun. It was a true revelation for both the company and me. We discussed this afterward, and they agreed with my observation.

BIG FOOT

Prepare in advance.

Whether by wearing larger shoes or making shoes from local materials such as cloth, boxes, cardboard, paper bags, tree bark, banana leaves, or grass, the key part of the exploration is creating shoes. These shoes should be oversized, like clown shoes. Performers are free to paint and decorate them as they wish. The oversized shoes help performers understand scale in their movements, creating a sense of awkwardness that encourages more expressive, exaggerated upper-body movements.

This exploration also helps performers break out of their usual walking patterns, prompting them to navigate space and interact with others in new ways.

Invite the performers to walk and explore everyday activities or attitudes (see **Walking** for inspiration) while wearing the oversized shoes, letting the **Big Foot** inspire "big" movement throughout their bodies.

Side Coach: *How do the shoes impact their balance and rhythm? Your perception of reality and one another?*

Evolve a character inspired by the shoes, walks, and movements you are experiencing. Go to where it takes you.

Exploration Progression: Encourage performers to create a voice, dialogue, or lines for their Big Foot character.

Create an improvisation scenario for two Bigfoot characters to perform or ask them to come up with a situation based on their character explorations.

Exploration Alternative: I also conducted a similar activity in which performers were asked to insert bologna or another gel-like substance into their shoes. While it differs greatly in squish factor, it results in a humorous experience.

Performers should bring in and wear old tennis shoes, if possible. Overall, this activity is great for enhancing improvisational skills and laying a foundation for character development and performance.

GRIDS

Performers move the space as if walking on an imaginary grid of straight lines.

Moving precisely and without premeditation, the performer turns at a 45-degree angle toward another straight line at any time they choose.

Side Coach: *Walk and stay on the grid. If another performer occupies their grid line, redirect to a different imaginary line.*

Speed up your grid walking and focus on making your turns more accurate and frequent.

The grid is shrinking. Walks and turns must become closer, quicker, and more precise. Keep eye contact with each person you encounter as you move across the grid.

Note: Many thanks to Anne Bogart and Tina Landau for inspiring this exploration and for their inspirational Viewpoints work.

Exploration Variations: Compare walking in rural versus urban areas. Highlight how the leisurely, open countryside contrasts with the crowded, bustling city life. This concept was effectively demonstrated during a workshop at the Estonian National Theatre School in a small rural village. It clearly showed how people's bodies and rhythms adjust and transform, and how environments are intertwined with and transmit these rhythms.

POLES

Each performer receives a bamboo or dowl pole. Ideally, each pole measures one meter or yard in length, with a diameter of ¾ inch to 1 inch.

Performers are then instructed to follow a sequence of actions:

Side Coach

- *Examine the pole closely and observe its features.*
- *Hold the pole in front of you and evaluate its weight, size, texture, and how it feels in your hands.*
- *Experiment with balancing the pole in different ways, using various parts of your body to discover new techniques.*
- *Move with the pole, following its rhythm.*
- *Use the pole to defend yourself and to attack.*
- *How does the pole influence your movement?*
- *Create shapes with the pole, such as birds, animals, trees, plants, and monsters.*
- *Play with others using your pole.*
- *Create rhythms and sounds with the pole.*
- *Develop a rhythm with the pole that incorporates your name.*
- *Make the pole a part of your identity.*
- *Form a circle with the poles.*
- *Pass the poles around the circle, starting slowly and then increasing speed.*
- *Each person takes a turn saying their name with the pole, and others imitate.*
- *Create a group environment with the poles, like a pole fortress, forest, or ocean.*
- *Make a collective creation with the poles, moving and producing sounds together.*

The above provides an example and a rough outline of the sequence of events. Feel free to add or remove elements as necessary. Allocate time for performers to explore and experiment. Offer coaching and reflect on the ideas that arise from the group. Ensure there is sufficient space for performers to express themselves freely.

Note: This exploration is inspired by Lolita Lesheim (my ex-wife). Thanks, Lola!

Highly Recommended.

Lao Tzu transformed his body. His left eye became the sun; his right eye, the moon; his head became Mount K'un-lun; his beard, the planets and constellations; his bones, dragons; his flesh, four-footed creatures; his intestines, snakes; his stomach, the sea; his fingers, the Five Peaks; his hair, trees and grasses; his heart, the Flowery Dais; as for his two kidneys, they were united and became one, the Real and True Father and Mother.

Kristopher Schipper, The Taoist Body

FALLING DOWN

This exploration can be used as a standalone activity, as a warm-up, or as a transition to imagination warm-ups.

Encourage performers to walk freely, changing directions and speeds as they feel inspired, moving faster and slower, and generally feeling relaxed and open. This free movement acts as a warm-up for exploration.

Part I

Performers are instructed to stop, safely fall to the ground within their comfort zone, then get up and continue walking. They repeat the falling process multiple times, each time falling differently.

Side Coach: *Choose another performer in the group and form a partnership.*

Keep moving freely around the space, following each other at equal distances.

Fall again. Stand up and walk, then fall again, this time mimicking how the person you're following falls. Imitate their fall's feel and style.

You will notice a change in how you fall, creating a shared falling pattern.

Repeat the follow-and-mimic actions several times to build consensus and develop a rhythm.

Now, find another performer, follow them, and repeat the process.

Part II

Form two groups of four performers each and move freely and together across the space.

Develop group sensitivity, awareness, and a shared feeling of connection.

When inspired, one performer loudly calls out to the group that they are falling (e.g., "Thomas is falling!"). Then the group members rush to help that person as they start to fall and support them to land gently on the floor.

Continue this process with each performer.

Let the pace of falling increase as the group gets more comfortable.

Pay attention to how you help keep balance and maintain the group's cohesion. Allow a group rhythm to develop naturally.

Part III

Depending on energy levels, the group may choose whether to move on to Part III. Members name other performers, and once a name is called, that performer starts to fall. The group then promptly supports them, aiding a gentle landing. As the exploration proceeds, performers will learn to self-regulate, adjusting their falls according to the group's ability to catch them.

Exploration Variation: Expand the exploration to involve the entire group. Begin with free movement, then have a performer call out that they're falling with others nearby rushing to assist them. Gradually increase the speed so multiple falls happen at once. Focus on creating and maintaining a steady, self-regulating pace that everyone can sustain.

If a performer chooses not to participate in an exploration, that's okay. However, they must remain supportive and respectful of others' work.

Performers might not understand an aspect of the work, which is perfectly fine. Some things fall into place, and others do not. All you can do is stay open and let the work explain itself. The performer is not obligated to agree with, participate in, or understand everything—each person in their own way and at their own pace.

Our work starts with understanding and calls for honesty and a willingness to engage in dialogue. It only asks that you stay open.

Our work is not separate from everyday life. Find ways to connect the work to life. Our work and life are intertwined.

Focus and Mindfulness

Our explorations of **Circle, Rhythm,** and **Balance** are wide-ranging and adapt to goals, contexts, and group dynamics. **Walking,** similar to other practices, can be revisited, deepened, and explored in various ways to enhance our work and move toward performance.

Our work is a lifelong journey of growth and discovery. It involves gradual exploration that boosts performer confidence and unlocks the many resources and skills we possess. The performer already holds the necessary knowledge; our role is to access this potential and sharpen their abilities. In doing so, we reaffirm and forward the value of our work.

Below, we approach our work from another perspective, encouraging performers to analyze and enhance the nuanced interactions between body and mind.

WALKING MEDITATION

Performers are asked to walk slowly and mindfully.

While walking, they remain completely silent, avoiding any communication or eye contact.

Eye contact must be avoided at all costs.

Begin with 20-minute walking meditations, and if appropriate, extend the session to 40 minutes, an hour, or more.

Walking Meditation isn't just about walking; it can also include standing still, sitting, and observing. The goal is to remain mindful, release any concerns, and simply be present in the moment.

Encourage performers to let their thoughts, images, and feelings come and go. Like **Listening to Feelings**, they should not attempt to control their thoughts but instead allow them to surface and observe them. Allow things to unfold naturally.

This exploration can take place indoors, outdoors, or a combination of both—in the woods, a field, or any quiet, undisturbed area.

As the group becomes familiar with the exploration and repeats it, try moving to a semi-public space with ambient sounds.

Your role is to observe. Pay attention to the performers, their movements, the shifts in rhythm, and how the group deepens its feelings as the experience unfolds.

This exploration greatly enhances self-awareness and encourages a higher level of collective coexistence, not through social or verbal interaction, but through presence and mindfulness.

At the end of the exploration, ask the performers to form a circle without making eye contact. Then, on your command, have them raise their heads simultaneously to see one another and the circle.

Note: Performers often need time to adjust to this exploration process. We live in an era accustomed to moving quickly, and it can take time to slow down, especially in urban settings where our interactions with machines and technology set the rhythm of our daily lives.

Encourage performers to be patient. It usually takes 15–20 minutes to relax and let the exploration unfold. Once that occurs, it is truly remarkable, with many people saying it's "comforting." This exploration has a subtle yet lasting effect.

Remind performers to stay mindful of their actions throughout the exploration and the rest of their day. Ask them to reflect on their emotions and the discoveries they have made. Please encourage them to apply what they have learned in their daily lives, working to become more mindful and present.

Highly Recommended.

THE JOURNEY

Part I: Request performers to locate a spot on the workspace perimeter.

Side Coach: *Choose a spot or mark on the wall opposite your position, or, if outside, a tree, building, or similar object.*

Concentrate on the spot and move toward it as slowly as possible. Ensure your movement is quiet and smooth, avoiding any vocal sounds.

Move toward the spot as if it is the most important thing in the world. The slow pace helps you notice the details of how your body parts move and how your body shifts its balance.

If you cross the space and another performer blocks your path to the spot, stay focused—pause if needed to let the performer pass—and then proceed toward the spot.

This exploration integrates movement, fluidity, and mind into a centered meditation. The gradual journey toward a simple, singular goal serves as a metaphor, illuminating what you prioritize in life.

Your journey represents patience, endurance, and the development of awareness in life.

When you reach your destination, come as close as you can to your spot, face it, and then close your eyes. Breathe in through your nose and out through your mouth, focusing on the essence of your life.

If some performers are moving too quickly, suggest they slow down. After three or four performers arrive at their spots, have them turn (with their eyes still closed) and then open their eyes. Tell these performers to identify the first spot they see on the opposite side of the space and approach it as slowly and smoothly as they can to their new destination.

Part II: Encourage performers to vocalize as they move to their spot.

Side Coach: *Begin with quiet breathing sounds or hums, then gradually increase your vocalizations (if you choose) to express growing anticipation and emotional development as they approach your spot. When you reach the spot, create a "song" or vocalize to celebrate your success.*

Part III: Some variation in travel speeds and distances is expected.

When the performer reaches their location and has "sung" to celebrate completing their objective, ask them to close their eyes and keep singing. Once most performers have arrived at their spots, ask them to turn to the center.

Side Coach: *Turn into the center of the space and let your "song" softly fade away.*

Open your eyes and identify another performer across the space as your next target in your journey. Move as slowly as possible toward that performer, and as you do, gradually develop a vocalization that expresses your feelings about your approach.

Expect some performers to move more slowly, and that is fine.

Two, three, or more performers might identify and approach each other—this is acceptable. Alternatively, you can whisper to specific performers to indicate who they should move toward, which is especially useful in larger groups.

Part IV: When the performers gather, encourage them to keep singing or vocalizing in celebration of reaching their destination.

Side Coach: *Once together, close your eyes, listen to one another, and create a vocal jam. Allow the jam to develop and change naturally. If a dance or movement emerges, follow where it leads.*

If there's interest, the exploration could develop into a collaborative vocal, rhythmic, and movement jam session for the entire group.

Note: This exploration significantly boosts focus, concentration, and body control. It acts as a metaphor and symbol for life's journey and the awareness needed to stay fully present. Moreover, it sharpens the performer's technical skills by encouraging patience and concentration of the mind, body, and emotions—maintaining mindfulness and focus in action, then shifting that focus completely.

Highly Recommended.

From the Siberian Journal

My goal was to focus on concentration and more meditative work because I felt the group needed to connect with their inner selves. But as soon as I explained **The Journey**, several performers immediately wanted to know what a "spot" was.

"Why do we have to go slowly?" "What's the point?" "Why can't we just get to it?"

These questions sparked a heated discussion that disrupted the work's rhythm. It took some time, but eventually, they understood that the Spot was a metaphor. The exploration itself is simple. What was confusing was why they were making it so complicated.

I believe this was just the Soviet way of doing things—always looking for a complex solution when a simple one was right in front of them. Complications were how they occupied their time and lived their lives. They hesitated to accept the simplicity of things, fearing that if it was too easy, it was suspicious.

We also examined their lives and how the Soviet era had shaped their perceptions and interactions with the world, making them both suspicious and obedient.

Most of the group eventually engaged with the exploration and focused on their designated spots. Some participants found it hard to concentrate, reporting that their eyes felt strained or that they drifted off. Others walked to their spots with their eyes closed, possibly due to a language barrier, even though we had reviewed the instructions multiple times. A few hurried to their spots and then waited for the next instruction, indicating discomfort with the process and their bodies—they preferred simple, direct tasks and were conditioned to obey, complete, and wait for further guidance. In the Soviet Union and at the Sakha National Theatre, the director held absolute authority—reflecting the autocratic Soviet system. Performers were not permitted to think, only obey.

It was fascinating to see the differences in how cultures interpret the exploration—I should not expect what I think should happen but accept what happens. By accepting, I gain insight. Each individual reveals a bit more about themselves, as does the group, culture, and context. Be patient and listen, observe, and accept, not expect.

The Body phase involves combining various explorations based on your understanding of the context, performers, needs, and project goals. These explorations should blend vocal, rhythmic, musical, and character work with activities like imagination-building, storytelling, skill-sharing, discussions, physical training, and ensemble development. It is crucial to keep challenging and teaching, while making sessions engaging and enjoyable to maintain interest.

Aim for unpredictability—alternating between personal, imaginative, and meditative explorations, physical, group, and scenario-building activities. Variety encourages performers to adapt, reorient their focus, and build resilience. This versatility equips them to respond intuitively to challenges by drawing on a diverse range of skills and abilities. Developing strong, capable, and confident response skills is a key goal of our **Body** work and will be crucial as we move into the **Space** and **Place** phases.

Creating an atmosphere of participation and inclusion, combined with challenge and surprises, engages and sustains our work. Strive to keep the process transparent, open, and shared. This is your responsibility. You are responsible for maintaining and advancing the work (encourage, but do not push too hard). Inclusion, challenge, and being part of something greater than yourself are deeply satisfying, contagious, and exciting. They make hard work feel effortless and its achievement rewarding.

Next, we add more playful explorations and explorations to make your work sessions more varied.

Play

The boundary between play, ritual, and sport has always been flexible—what Huizinga describes as a "fluid contrast" where seriousness and play continually interchange (8). In ancient societies, this divide was almost nonexistent: Athletic contests doubled as sacred festivals, games served as a form of divination, and every ceremonial act was performed in the "spirit of pure play truly understood" (5). The Greek agon contests, the potlatch ceremonies of Alaska's Athabaskan, Tlingit, and Haida peoples, as well as judicial trials by fire and combat—all were simultaneously contests, rituals, and games, creating temporary worlds set apart from everyday life. Play, ritual, and sports are integral to societal structure. In our practice, we can rediscover that primal unity, using games not just as warm-ups but as tools for wakefulness—vehicles for rhythm, tension, alternation, and ecstatic focus that define both ritual and deep play. These games are not trivial; they are techniques for accessing Huizinga's "realm of the beautiful and the sacred," where performers operate "above the level of the serious" (19). Each game forms a unique magic circle with its own rules—agreed upon yet binding—and an ordered perfection amid everyday chaos.

TAG

Tag is one of the oldest and most popular games. Its origins could be traced back to a period when humans needed to develop skills such as chasing and hunting, or were themselves being pursued.

In tag, one player is designated as "it" and tries to chase the others. When they tag someone, that player becomes "it." The main objective is to catch others or avoid being caught.

This game is an excellent way to energize a group and increase blood circulation. It's fun for people of all ages and backgrounds and can be effortlessly integrated into a variety of workshop activities.

Exploration Variation: The person who is "it" announces who they are when tagged. For example, after being tagged, they might call themselves a predator like a Lion, Polar Bear, Bee, or Wolf and label those being chased as an Impala, Seal, Rabbit, or similar. This can also expand to include culturally specific figures, such as a malevolent spirit, a mythological figure, or a historically significant opponent like a slave trader or colonizing soldier, or even Freddy Krueger. A great way to identify and explore the predators of the historical, cultural, or pop cultural imagination.

Here are a few different versions of the classic game of tag.

ALTERNATIVE TAGS

Similar to Tag with the amendment, the rules are that a person can only be tagged on the head or another specific body part.

Additional rule changes might include, for example, that the chased player must have their feet on the floor to be tagged out. This variation alters the player's movement to a mix of running and jumping.

Another option is a tag where the chaser is the permanent "it." In this case, the chaser must tag the other players on a part of their body that limits their movement. The tagged person must then hold the part of their body that was touched until someone else releases them. The chaser's goal is to slow down and hinder all the other players' movements.

Freeze Tag is another option. Here, the pursuer tags others, who then freeze in place. They can only unfreeze when touched by another player. The goal is to freeze all the performers.

AMOEBA TAG

Here's how it works: One person starts as "it." When they tag someone else, they don't become "it" themselves—instead, they link up, arm in arm, to form a chain of people, like an amoeba.

These two then tag other people, and the process repeats: The three of them link up and keep chasing the others. If a fourth person gets tagged, the amoeba splits in two, forming two smaller groups of two people that track down the others.

This game continues with more tagging, linking, and splitting until the entire group forms one big amoeba.

GO-TAG

Everyone squats in a single line, facing each other in alternating directions.

At one end of the line is a runner, or the mouse, and at the other end is a chaser, or the cat. They are set in motion with the cat's goal of catching and tagging the mouse, while the mouse tries to avoid getting tagged. Both players can run in either direction along the line.

The chaser can pursue the mouse by tapping the back of any player and swapping in. The tapped person instantly becomes the cat. The game focuses not just on chasing and tagging but also on teamwork and strategy.

The game can also be played with the cat and mouse swapping in and out, replacing players in line. This makes the game extremely fast-paced and demands quick thinking.

Every performer needs to keep track of who is the cat and who is the mouse. For clarity and better gameplay, the cat and mouse should shout "cat" or "mouse" as they tag and replace others in line.

You can vary the rules by making the cat only run in one direction, such as counterclockwise, which adds more excitement and strategy to the chase.

Encourage players to develop strategies and maneuvers instead of relying solely on speed and strength. When a mouse is tagged, that player leaves the game. The cat becomes the mouse, and someone at either end of the line becomes the new cat. The game continues until only two players remain: A cat and a mouse. All the other players have been eliminated.

FIREMAN LIFT RELAY

Players are divided into two teams and form a single line, standing across from each other and facing one another.

At the ends of each line, you will find two players who start the relay. Each is a member of the opposing team, with the goal of moving their team across the space.

The relay begins with your signal. The two end players sprint to the first person in line (their fellow team member), then use a fireman's lift to lift and carry them back to the opposite line.

Here's how the Firemen's Carry works: The carrier lifts the other person onto their right or left shoulder, whichever feels more comfortable. The weight of the person carried is balanced on the carrier's shoulders. The carrier reaches through the other person's legs with one arm and grasps their dangling wrist with that arm, holding it near the back of the leg. For instance, if the carrier has the carried on their right shoulder, their right arm is extended through the other person's legs and grasping their right wrist. The carrier's left arm is free—their free hand for climbing down a ladder. The process is reversed for carrying on the opposite shoulder.

The person being carried is placed in the opposite line, then runs back across the space to bring the next team member into the line. The person who was carried then carries the next teammate back across the space, and this continues until every team member has been carried across. The first team to successfully carry all their members wins.

This next activity is one I have frequently used with great success in many different settings. It's a fun strategy game that enhances body awareness and movement skills while fostering community. It's an excellent way to end a workshop session and can be repeated. Performers often request it to hone their skills and apply new game strategies. I learned this game from actor Jason Alexander (famous for his role on *Seinfeld*) when we were students at Boston University. He was then known as Jay Greenspan. Thank you, Jay!

SCORPION

Performers stand apart and spread out across the space.

Game Rules Overview

A scorpion is selected. Their goal is to tag or "sting" everyone else. The others try to avoid getting "stung."

When stung, a performer must scream to alert others to the scorpion's location.

The stunned player then opens their eyes, leaves the game, and moves to the side, staying silent. Any movement distracts the other players.

It is essential that the scorpion make a loud and clear "ssssist" stinging sound when they sting. This alerts the other players and keeps the game moving with interesting results.

After the scorpion has stung everyone except the last player, that person also becomes a scorpion. Then, the two try to sting each other. It's a battle to the death.

To start, all players close their eyes, and you move around talking and repeating the rules to distract from who will be chosen as the scorpion. With a tap on the shoulder, that player becomes the scorpion, and the game begins.

Players and the scorpion must always stay moving, even just slightly, and they cannot hide. Those who open their eyes to sneak a peek are usually easy to spot (their body language and movement give them away) and are asked to leave the game.

Offer side coaching when necessary, encouraging players to keep moving and avoid peeking. Dim the lights for a better effect.

To play, participants must constantly read the space and the presence of others with their bodies, balancing body awareness with the rules of the game.

The game can be captivating to watch and enjoyable to play. Strategies emerge, performers' bodies become alert, adjusting speeds, playing levels, darting and responding to sounds, changes in air currents, and movement of others.

Until the concept is well understood, it's recommended to choose a more assertive person as the scorpion for the first few rounds.

Game Alternatives: For larger groups or when time is limited, consider selecting two scorpions. If two performers are remaining who are not scorpions, all four will receive a scorpion sting and face off in a duel to the "death."

Interstitials

Here are some playful imagination activities that can be paired with other offerings or as interstitials between explorations.

"X" ERCISE

Invite the performers to move freely within the space.

When inspired, they bring their hands and arms across their torso, forming an "X" shape. This "X" then bursts outward from their bodies, extending their arms and legs wide to release their energy into the space.

Then performers return to walking and repeat whenever they feel inspired to project their "X" energy into the space.

CAPTURING SPACE

Performers are encouraged to expand their bodies to occupy as much space as possible using different body parts. They should move through the space, actively engaging their hands, legs, mouths, and other body parts to grasp and capture it. This includes utilizing their shoulders, arms, lower body, knees, buttocks, eyes, ears, and nostrils, ultimately involving their entire body.

GIVING AND TAKING $

Ask the performers to move around the space as if they are giving money/energy to each other, envisioning an endless flow of money originating from their center.

They should ensure everyone in the group receives a generous amount, reflecting their feelings of generosity.

The exploration can also involve giving money to an imaginary, appreciative audience.

Next, instruct the performers to take imaginary money/energy from everyone and from the surrounding space.

Each performer then deposits their money at the center of the circle, creating a large pile. Finally, encourage them to strut, dance, and prance joyfully, feeling wealthy and happy.

DISORIENTATION SPINS

Apply this exploration randomly and unpredictably, whenever and wherever you think it's necessary, to disrupt and energize the work—especially during the initial body phase of our project.

Ask performers to spin in either direction for a set amount of time; I suggest starting with one minute. You can adjust the duration as needed.

They spin with their arms extended and eyes open and may collapse to the ground or remain on their feet once they stop.

The goal is to become comfortable with randomness and the unexpected, so performers can develop a response to a world that is spinning out of control, become disoriented without warning, and improve their ability to adapt, adjust, and thrive.

Note: If some performers respond adversely to the spinning, they may be excused or attempt to regain their balance by concentrating on their thumb.

Psychophysical

As you might have guessed by now, the psychophysical approach to performance is central to our work and is the overarching perspective of this book. Our work views physical action and psychological intention as interconnected parts of a single creative process. This methodology, heavily influenced by Grotowski's lab (Osiński 1986, 70–89) and Phillip Zarrilli's work, and grounded in traditions ranging from Stanislavski's system to Eastern performance methods, highlights that the body itself is a vessel for memory, emotion, and meaning-making. The performer doesn't just use the body to display pre-existing internal states; instead, the body creates and uncovers these states through specific physical actions (Stanislavski 1936, 167). In this way, a gesture isn't simply an expression of feeling but a gateway to genuine experience—the body guides the mind into new realms of expression and meaning. Through disciplined training that combines vocal explorations, movement work, and what Grotowski called "organic reactions" (Osiński 1986, 51), performers build the ability to access what could be termed psychophysical memory: Corporeal knowing that goes beyond intellectual understanding.

This approach dissolves the Cartesian division between mind and body, recognizing instead a continuous connection where physical impulses and psychological truths arise together. When performers perform explorations that turn a facial expression into a full-body state, or let static body positions suggest character and action, they practice a form of consciousness technology that respects the body's intelligence. In this way, the psychophysical method becomes more than just a technique for making believable characters. It is a philosophical stance on performance—one that sees the body as the primary tool through which we understand and express our deepest truths.

These explorations focus on removing physical barriers and improving intuitive, bodily interaction to enable us to "think" with our bodies.

FREE CONTACT

This advanced exploration is intended for mature performers who have fostered a sense of comfort and trust.

This exploration requires a dark or dimmed environment, where performers are positioned within the space and can contact each other through any of the senses, using sound but no understandable language.

Ask performers to observe the thoughts, feelings, and memories that may emerge through free contact interactions.

Emphasize gentle, respectful, and safe touching, exploration, and communication, with performers free to engage, pursue, refuse, avoid, or end contact at any time.

Encourage playful interaction and let communication guide the experience. If a performer wishes to avoid contact, they must clearly state their desire.

Advise performers to focus less on thinking and more on feeling in an uninhibited manner. Allow the exploration to evolve into deeper levels of interaction and communication beyond initial curiosity. This activity enhances awareness of personal, interpersonal, and non-verbal communication.

Note: Any issues related to the exploration should be discussed collectively.

WHAT IS GOING ON?

Performers are encouraged to move freely and naturally within the space, experimenting with unusual and creative ways to use their bodies.

They are asked to adopt different positions—low, high, on their toes, squatting, or spinning—and then freeze. Approach performers randomly and ask what their body shape in the freeze position suggests they are doing, e.g., digging a hole, dancing ballet, begging for money, walking on the moon, or picking flowers.

Only ask a few performers at a time, then continue with the collective move and freeze. The goal is to be spontaneous and intuitive, not to give performers time to think about their shape. If they can't answer quickly, move on to another performer with the same question: "What's going on?"

Repeat this process several times, and have performers respond quickly, with little time to think, letting their bodies suggest an action, thought, feeling, situation, or occupation through body memory.

Exploration Variations: Ask performers to freeze as before, then suggest a setting. For example, a factory, an Olympic track and field event, a cocktail party, or a horror movie. Given that context, ask them what they are doing, who they are, and to identify the origin of the body memory they are referencing.

The exploration may evolve from their frozen positions into action. Characters, relationships, and situations can develop from here.

Note: Refer to the diverse and innovative vignette techniques devised by Augusto Boal for animating a body shape, developing character, and crafting scenarios. Many thanks to Augusto Boal for his pioneering and important contributions! We stand on his shoulders.

WHOLE BODY COMMUNICATION

Performers are asked to study their faces in a mirror—preferably a full-length one, but a handheld mirror can suffice. The aim is to explore various facial expressions. Participants can do this as a group or individually at their own pace.

Side Coach: *Start by observing all the ways the face communicates, both overtly and subtly.*

Examine and consider:	*The avoiding eyes*	*Breathing fire*
The wink	*The strong jaw*	*Smelling something hot*
The raised eyebrow	*Wide-faced surprise*	*Breathing smoke*
The smile	*Closed-face discontent*	*The angry face*
The frown	*The relaxed open mouth*	*The bad smell*
The crinkled forehead	*The tense open mouth*	*The sweet smell*
The smirk	*The tongue hanging out*	*The crossed eyes*
The sneer	*The pointed tongue*	*The jutting chin*
The fixed eyes	*The flared nostrils*	*And on and on...*

What does each expression evoke emotionally? What memory does it draw upon? What is the origin of the expression?

After identifying and exploring various facial expressions, focus on one specific expression.

Let that facial expression evolve into a full-body gesture. For instance, your entire body becomes a "bad smell" or a "relaxed open mouth."

Exaggerate each expression in a cartoon-like manner, emphasizing the entire body, and then bring it into movement.

Explore and perform each expression individually, pausing to freeze it as a "moment" or "snapshot" when they finish.

Then, explore the expression's voice. What does the expression sound like? Talk like?

What kind of character, concerns, and worldview do they have?

Then, interact with others as your body character expression.

Exploration Evolutions: Ask performers to create a scenario by blending specific facial expressions with their body movements. They can develop a storyboard-like sequence, pausing at each gesture to narrate a story and ensure the scenario flows smoothly and continuously.

The exploration can naturally develop character voices and dialogue from specific expressions or sequences. Individual expressions or dialogues might be shared with the group.

Note: This is a great examination of facial and body expressions, showcasing a method for developing characters from internal to external expression. It enhances performance and creativity and serves as excellent preparation for mask work, which we will explore in the Space phase of our project.

Trust

GROUP LIFT

Each performer lies on their back on the floor, and the group takes turns lifting them.

Members are evenly spaced around the participant—typically six to eight people are needed for safe lifting.

The group touches the performer to establish contact and build trust.

They synchronize their breathing while positioning their index and middle fingers underneath the performer.

On a three-count, they lift the performer to their chest, ensuring balance before moving them a short distance. They then gently and gracefully lower the performer, maintaining balance throughout.

The activity is performed in silence, with each person taking turns being lifted and carried. The focus is on intuitive group coordination and balance; trust and concentration are strengthened.

For larger groups, multiple lifting teams may be used.

It's important that the activity is conducted with safety in mind.

Exploration Alternative: Incorporate imagery into the activity. The group might imagine carrying, for example, a fallen hero, a wounded person, a mummy, a human sacrifice, or a large fish.

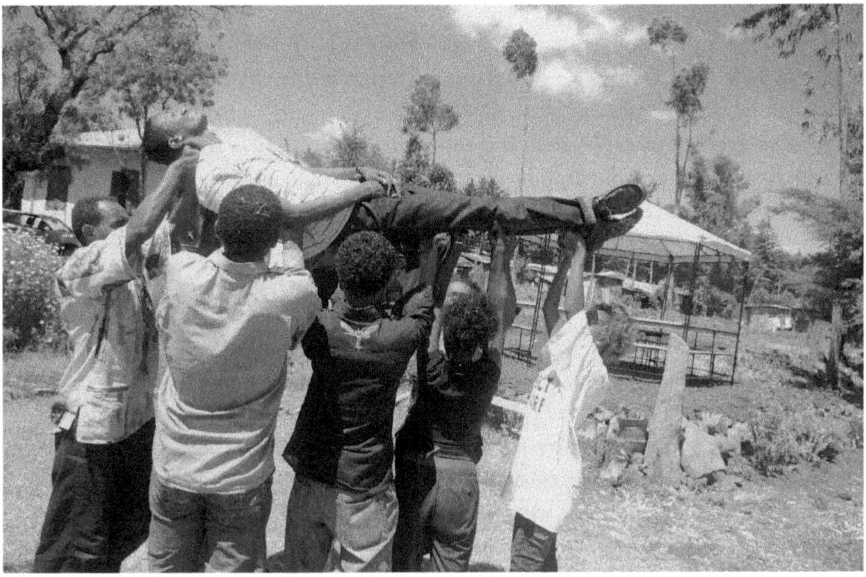

FIGURE 1.12 Nibret Baychekin being lifted in the *Group Lift*, Addis Ababa, Ethiopia, November 2009

Source: Photo by the author.

FLYING

A performer is asked to volunteer. Start with one actor standing at least four feet above the ground on a ladder or another platform.

Five to eight performers are needed, and they line up in parallel, facing each other directly beneath the "flying" performer.

The catchers are responsible for catching the "flying" performer. They lift and lock their arms with the individuals across from them, forming a line of locked arms that will form a net when the flyer falls.

Catchers should be advised not to be rigid in their legs or arms, but to give as the flyer falls into them.

The flyer may choose to fall face-forward or face-backward into the catchers. Flyers may leave their eyes open or close them.

When ready, the flyer nods, and the group begins a collective three-count breathing sequence in which the flyer participates. Breathe in and then count one, two, three together; on three, the flyer falls into the waiting arms of the catchers.

The group rotates, with each performer taking turns flying.

As with all activities, no performer is compelled or required to participate if they have any reservations.

The flyer should NOT sit in the catcher group. The flyer should be relaxed but maintain enough rigidity in their body to fall straight and along the entire length of the catchers.

Generally, place stronger catchers at the upper and middle parts of the catchers' line.

Depending on the group, a second round of exploration may go higher. However, it should not exceed two meters or six feet.

This activity can provide an incredible feeling of freedom and help overcome fears, while also strengthening trust and teamwork within the group.

Warning: This can be a very risky exploration, and if not conducted with full understanding by all participants, it could lead to serious injury. Take extra time to demonstrate and clarify the exploration for everyone, especially when working across cultural and language differences.

FIGURE 1.13 Michael Million, *Flying*, during a workshop at Lul Theatre, Addis Ababa, Ethiopia

Source: Photo by the author.

DISEMBODIED HEAD

Partners. One person lies on their back, eyes closed, remaining relaxed. The other sits or kneels at their head, whichever is more comfortable.

The goal is for the person lying down to relax and surrender their head to the other. The sitting partner gently cradles the other's head, lifting it carefully and moving it slowly to the sides. Focus on smooth, gentle movements.

As trust builds between the partners, the person lying on their back begins to relax their head more. The sitting partner can then gradually increase the difficulty of their movements, such as moving the head more quickly from side to side, lifting it higher, or turning it farther.

When the partner is lying on their back, they will naturally resist or anticipate the head movements. It is essential to move the head without a set pattern and to allow the partner lying down to fully relax and release their head.

Once one partner finishes, they switch. The exploration lasts between 10 and 25 minutes with each partner. Afterward, each person should slowly stand and walk around to fully feel the weight of their head–like a disembodied head.

From the Siberian Journal

Sample Session

It was chilly after the break, so everyone wore coats to stay warm. To boost their energy and warm up, we played a game of **Ameoba Tag.**

Next, we moved into **Mass Movement.** They had trouble understanding that one person moves first and then others follow, creating a mass. Honestly, I hadn't thought it through myself, which might have been part of the problem.

In **Balance the Room,** two groups insisted on balancing with everyone holding hands, which worked fine.

Next, we played **Pig in the Middle,** which was a huge hit. They insisted on playing it multiple times, wanting to do it right—whatever that means—rather than feeling it out.

They mentioned feeling uneasy about their own emotions and that their feelings take a backseat to understanding the physical and mental aspects of the exploration. This self-awareness surprised me.

I introduced the **Group Lift.** They enjoyed the activity—the feeling of freedom and collective physical effort drew them in. They insisted on doing it again.

We ended the morning session with **Diaphragmatic Breathing** and then moved into **Touching Sound.** For some, breathing remains very shallow. Is it tension? Is it connected to how they perceive reality? Breathing shallowly to avoid feeling, to escape the pain and difficulty of their existence? Could it be a symptom of the city's fast pace or the anxiety over so much change and the total upheaval of life as they once knew?

Aggression

The connection between aggression and performance practice goes deeper than metaphor—it extends into the neurobiological foundations of human ritual behavior itself. From the Paleolithic hunt to the Greek sacrificial altar, human communities have consistently directed aggressive impulses through structured ritual forms that both contain and transform violent energies into social cohesion (Burkert 1983, 19–21). This paradox is central to performance work: The very forces that threaten to destroy community—aggression, violence, domination—become, through ritualized repetition and collective participation, the mechanisms by which community is created and sustained.

Neuroscientific studies support what ancient practitioners sensed: Rhythmic, repetitive rituals synchronize limbic activity among participants, creating shared states of arousal that promote collective action and help control dangerous, aggressive impulses (D'Aquili and Laughlin 1996, 134–135). Our work, like a ritual space, offers a safe environment where performers can confront and engage aggressive energies without resorting to actual violence. Through activities examining power, resistance, victimization, and dominance, performers engage with what Burkert calls the core human condition: The unavoidable tension between the urge to kill and the need for restraint (20).

These next explorations are not just psychological explorations but involve biological imperatives—aggressive behaviors that helped humans survive through cooperative hunting, territorial displays inherited from primates, and submission rituals that prevent violence within the species. When performers explore aggression in the studio, they engage with an ancient practice: Ritualizing violence in a way that lets human groups experience and release aggressive tensions without harming social bonds. Although this process can be uncomfortable and raises important ethical questions about power, consent, and violence, it is crucial because repressed aggression doesn't vanish but seeks less conscious and controlled outlets.

Our work acknowledges and encourages performers to experience aggression firsthand, to feel its energy within their bodies, and to observe how playful competition can quickly evolve into real hostility. Through embodied practice, they learn what ritual has always taught: That the journey through violence does not lead to its complete removal but to its transformation into community, meaning, and regenerative strength.

FINGER FOLLOW

Part I

Ask partners to face each other.

One partner uses their index finger, while the other stays still, relying only on their eyes to follow their partner's finger. The moving partner is encouraged to explore different levels, move around, and adjust the distance from the stationary partner.

Part II

The stationary, observant partner now moves their eyes and head to follow the index finger.

Part III

The stationary, observant partner uses their entire body to follow the moving partner's index finger. Moving through space, exploring different distances, and using levels are encouraged.

Exploration Variation: All performers follow your finger. First, they watch with just their eyes, then with their head and eyes, and finally with their whole body.

RESISTANCE

One performer holds onto another, using passive resistance and passive aggression to keep their partner in place.

The partner being held struggles to break free, screaming, fighting, and wiggling in a similar passive-resistant and passive-aggressive manner.

Next, the holder is told to recognize that their partner is a suspect the police need to detain. The held partner should also adjust their response to match the holder's new attitude.

The holder then realizes the partner is having a seizure and must keep them restrained for their own safety. The held partner responds accordingly.

The holder then understands that the partner might kill them if released. The held partner reacts to this shift in the holder's attitude.

Create various scenarios involving holding and being held, switching roles and partners, and repeat.

Note: This exploration introduces physical contact and explores elements such as aggressiveness, victimization, and violence. It is also beneficial for stage combat preparation, emphasizing control and physical sensitivity.

PIG IN THE MIDDLE

Three performers participate. One stands in the center, while the other two toss a ball back and forth, aiming to prevent the central performer from catching it.

The throwers may employ tactics like deception, taunting, and teasing to make the central performer uncomfortable.

Physical contact is not allowed. If the central performer catches the ball, they swap places with the thrower. The game continues with a new "pig in the middle."

Note: The exploration emphasizes humanity's tendency to harm others through playful or teasing actions. It encourages the formation of emotional and psychological bonds with performers for future development. Performers are encouraged to explore and express any emotion, even if it feels uncomfortable. A common challenge in many minority communities is victimization, a delicate topic often sidestepped. Exploring victimization allows performers to experience its effects firsthand and to better understand its historical and cultural background.

PULL APART

Select two pairs of performers to create partner groups.

One group should position itself to be as inseparable as possible.

The other group, when given a signal from you, tries to separate the first group as effectively as they can within 30 seconds or one minute. You will keep track of the time.

Only pulling is allowed—no tickling, punching, scratching, or other violent actions.

Switch the roles so the partners attempting to separate the other group do so. Continue this process until a final championship round, where the two most inseparable groups compete in a showdown.

CIRCLE PULLS

A circle is drawn on the ground to mark the group's boundary as they stand. The circle can be made with chalk, in the sand, or by another method.

The group is asked to hold hands. At your signal, the performers try to pull as many people as possible into the center of the circle. The pull lasts for a set amount of time (e.g., 5 or 10 seconds). Once a performer's feet cross inside the circle, they are out of the game.

After the first round—or pull—a second, tighter circle is drawn, and the performers are asked to hold hands again. Then, at your signal, the second pull begins.

As the remaining group becomes smaller, a smaller inner circle is drawn. The game progresses in a similar manner until a champion emerges.

Note: This exploration targets specific physical goals. The activity helps break down physical limitations and encourages the expression of aggression and competition.

LIFE BOAT

An area on the floor is marked with chalk to represent a "Lifeboat."

Ideally, mats or carpeting should be used to create a lifeboat and the area around it. This allows for more aggression and increased safety.

All performers crowd onto the "Lifeboat." Someone must be thrown overboard every one or two minutes; otherwise, the entire "Lifeboat" sinks.

The goal is to be the last person remaining on the lifeboat.

Performers are allowed to push and shove each other off.

You act as a referee and timekeeper, counting down to when the next person must be thrown off.

Performers who enter the "sea"—the area outside the lifeboat—with at least half their bodies are eliminated and become sharks. You decide when they have entered the sea.

Sharks may grab any performer still on the lifeboat and pull them into the sea.

Exploration Alternative: The lifeboat sinks more gradually, taking about two or three minutes each time. You keep track of the time and may give a countdown. Similarly, the exploration continues until only one person remains on the lifeboat. This slower pace allows strategies to develop.

Note: Proposing a life-or-death scenario evokes strong emotional responses. When such scenarios are explored repeatedly and over extended periods, they often result in more aggressive behaviors and higher levels of physical violence. Weaker participants, frequently women, are more vulnerable to being targeted, which leads them to form bonds and collaborate to challenge the stronger individuals. These innate group dynamics serve as a foundation for future dialogue and development. Additionally, alliances based on gender tend to form naturally. Occasionally, performers deliberately throw themselves into the sea.

From the Zambia Journal

To address inter-tribal tensions, some performers were deliberately assigned to perform the tribal dance of a rival tribe. Initially, this caused some grumbling and thoughtless "correcting" from those whose tribal dance was being attempted; however, a sense of shared learning and cooperation eventually emerged. In this setting, I leveraged my outsider status to ask them to do things that insiders might not typically consider or request.

Dancing the dances of other tribes helped identify historical and tribal prejudices, break down barriers, expand performers' performance vocabulary, and foster understanding and respect for different tribal traditions. This exchange cultivated a mindset and approach that affected both the work and the performance: Traditional displays evolved into a language open to everyone, not limited by tribal identity. The group ultimately saw this open cultural exchange as a resource, strength, and important part of their collective identity.

MASTER-SLAVE

Three performers participate in this activity, with two acting as partners and the third serving as the scorekeeper. The focus is on exploring master-slave dynamics, where performers select who will take on the roles of "master," "slave," and "scorekeeper," with all roles rotating in turn. Participants can also decide or you can specify the type of points awarded and when.

Here's the process: The "master" holds a newspaper or slapstick, while the "slave" must follow every command and wish of the "master." The scorekeeper uses a pad and pen to track points.

The "master" then commands the room or space to be rearranged—objects moved, the manner of moving them, and the attitude during the process. The "slave" receives other minor commands and must accept them cheerfully.

The "slave" can employ any strategy to tease or earn points from the "master." They might make faces, be overly polite, talk back, mumble, or cause trouble. Each successful insult earns the "slave" points, while being caught results in point deductions. The "master" can punish the "slave" and earns points for actions obeyed.

A time or point limit can be established to determine the winner. Roles then switch, and the activity repeats.

Exploration Alternative: Two groups are formed: "masters" and "slaves." The need for scorekeepers and scorekeeping is removed.

As mentioned earlier, the masters give commands to the slaves, and the "slaves" also act against their "masters." Depending on how mature the group is, the stakes of the exploration might be raised to include public humiliation and the ordering of the "slaves" by the masters in public settings.

Note: This exploration frequently involves historical, colonial, and political undertones. It can be used as a method to reveal ingrained images of self and culture, highlighting difficult issues for the group to reflect on and possibly transform into a performance. In Zambia, this exploration quickly became a metaphor for the political, racial, sexual, and social oppression faced by one group from another. The Polish National Theatre School in Krakow highlighted gender disparities and sexual abuse. At the Riksteatern in Stockholm, it identified unspoken intercultural tensions and prejudices between Swedes and members of immigrant and Sami communities. At the University of Pondicherry in India, tensions between Hindus and Muslims, Tamils and non-Tamils were brought to the foreground.

Highlighting complex and underlying issues unites the group, deepens our exploration of self and others, and fosters trust.

As we progress into the **Space** and **Place** phases of our work, we will delve deeper into Aggression and all of our Body explorations.

Performing

The following explorations prepare us for the transition from the **Body** phase of our work to the **Space** phase.

BEING WATCHED

Partners.

One performer sits silently, without expression, encouragement, or reaction, while the other stands to perform, doing whatever they like—being watched. There is no language, only movement and sound.

The only instruction is to perform.

The performer may react to being watched in any way they feel; often, they will initially contort their face, dance, sing, move, and gesture to express their feelings about being observed. The passivity of the "others" seems to provoke them even more.

At first, those being watched will go through their "bag of performing tricks," things they have done before to entertain and engage others. They start by performing actions and adopting characters and expressions that have worked for them in the past—tried-and-true routines. These familiar performance bits are things we all have and pull out from time to time. Much of it will be childlike, popular shtick, and personal expressions that have worked for them before. Over the years, behaviors are shaped by parents, relatives, teachers, coaches, observation, imitation, and admiration. We adapt and integrate gestures, attitudes, behaviors, perspectives, and voices from others, molding them to fit our needs. In doing so, we often lose our authentic selves.

You'll notice the performer starting to run out of energy, perhaps even looking at you with a desperate search for an end. This usually happens around 15 minutes into the exploration and marks a breakthrough, when individuals move beyond their usual performance language and become genuinely expressive. It is a subtle, almost therapeutic process.

Allow the exploration to continue for at least 20 minutes per person, and if possible, 30 to 40 minutes. I have let the exploration run for as much as an hour with each performer, and the results are remarkable—layers of performance built over a lifetime peel away, revealing a perceptibly different person, free of artifice, fully present and authentic.

Decide when to switch.

Each partner is watched, in turn, by the other.

This exploration highlights several key points: (1) Humans act both for themselves and others. (2) When they don't receive a response, they often overcompensate, pushing their efforts further to get a reply. (3) There is a tendency to "perform" for an audience, aiming to meet expectations—whether real or perceived—and using expressions that are culturally and socially familiar and have proven effective before. (4) The burden of performance we all carry acts as clutter, blocking authentic responses and true expressiveness. (5) It is acceptable for performers to rely on their tricks—this exploration aims to raise awareness of when and how they do so consciously.

This exploration reveals the extent of the accumulated performance vocabularies we all carry and use unconsciously, and how this accumulation is both a resource and an obstacle to achieving authenticity in performance and life.

Exploration Variation: Same as above, including language. This enhancement broadens the performer's expressive range. Usually, adding language necessitates extending the exploration time to attain a genuine level of authenticity.

Note: Overwhelming performers have noted how revelatory, emotionally, and creatively freeing the exploration was for them.

Highly Recommended.

FIGURE 1.14 Menahel Teshome enacting *Being Watched*, during a workshop with Lul Theatre, Addis Ababa, Ethiopia

Source: Photo by the author.

PATTERNS

The exploration builds on and continues **Being Watched**.

After each performer has "watched" the other, they respond by performing a movement pattern inspired by the other performer, drawing on memory and impressions.

Once the response movement pattern is established, the performer repeats it and then shows it to their partner.

After both have created and repeated their respective patterns, each one develops another repeatable movement pattern. They then combine the two patterns and evolve them so they move through space, adding sound. The combined patterns should trigger connective movements, which should be encouraged.

Suggest that the performers evolve the patterns into a "journey" through movement or dance.

The other partner may respond by adding sounds or singing to support the performing partner's movement patterns.

Pause and invite the partners to reflect on their discoveries. Have them share what they observe, what moves them, and their interpretations with the performer. Encourage them to repeat the exploration and switch roles.

After completing the same sequence, guide the partners to merge their movements, dance, song, and sounds to craft and explore the story of their journey.

This story might feature a hero, an antihero, and a bard (the storyteller) who can narrate or comment using words, images, movements, or sounds. These elements can be tailored to the context and to the workshop and project goals.

COMPOSITIONS

Part I

Multiple groups of three to six performers are arranged in the space.

A leader is selected to guide their group through slow, simple movements. The aim is to keep everyone involved. Repeat the movement, adding more depth and detail.

When leaders feel it is time, they pass the leadership to another performer silently through a gesture or glance. The new leader begins with the last movement and transforms it into their own. The rest follow.

All performers get a chance to lead at least once.

Then, encourage performers to move and explore the space, maintaining the same principles of moving together and passing leadership.

Part II

As the group's confidence grows, they might increase their distance during play while remaining responsive to their rotating leader and maintaining movement, tempo, and rhythm. They may encounter other groups; if so, encourage them to cross and mingle, passing through and around one another while maintaining the integrity of their own group.

Then encourage the groups to intersect, become aware, and respond to the movements of the other group.

Meanwhile, each group continues to pass on its leadership and move with a group mindfulness.

As the interaction progresses, encourage the groups to engage in movement dialogue—talking with their bodies and movements.

Culminate the exploration by having groups move in unison to form a unitary group movement and identity. Leadership is shared and passed fluidly between performers.

Exploration Variation: Incorporate music or rhythm and encourage performers to interact with it.

Note: An excellent exploration for developing movement-based performance.

Highly Recommended.

PUSH, PULL, SPIN

Part I: Performers are asked to move naturally in the space with their everyday walk.

Then, they are asked to change direction multiple times, turn around, and stay aware when their movement patterns become predictable.

Ask the performers to move at half normal speed—not slow motion, but about half as fast as their everyday walk. Avoid touching each other.

Ask the performers to walk in a curved way—whatever that means to them.

Side Coach: *Move in a curved manner and occasionally bump into others. Maintain half speed to prevent injuries. Let the bumping happen naturally and* *blend into the curved movement. When a bump occurs, go along with it, letting it lead where it will.*

As you move at half speed in a curved way, occasionally bumping into someone else, and now incorporate Spin.

Continue moving in a curved manner. If someone bumps into you, spin away or respond to the bump. You now have four movement elements to draw from: Half-Speed, Curved, Bump, and Spin.

Now another movement element: Push.

Now, move in the space, occasionally push another performer, who may, in turn, spin, curve away, or bump into someone else.

Maintain half-speed, react, and flow freely to every Push or Bump.

The next element: Pull.

You may now pull other performers when an opportunity comes along.

Play and have fun with these six movement elements: Half-Speed, Curve, Bump, Spin, Push, Pull.

Maintain half-speed and remember the elements you have at their disposal. This is you movement language.

Part II: Once the group has learned and become familiar with the movement vocabulary listed above, it is time to introduce music. I usually select instrumental music by Steve Reich, Philip Glass, or similar artists. The key aspect of the music is its slow, even pace.

Move with the music and continue as before.

Next, introduce music that aligns with your project's objectives. R&B, Hip-Hop, Classical orchestrations, movie soundtracks, etc., it all works.

Ask the performers to integrate it into their movements, allowing it to guide them.

Now, work on creating scenes. For example, "Modern Dance." Inspired by that title, with its image and concept in mind, use their vocabulary—Half-Speed, Curve, Bump, Spin, Push, and Pull—to interpret it.

Use music or sound that compliments the scene.

Other titles might include "Urban Violence," "Bad Acid Trip," "Love at First Sight," "Wild Party," "Drinking Problem," and more. Create titles that align with your overall performance goals or reflect cultural and community themes.

Another option is to create a story with sequential scenes. These scenes can be either parallel or linked to your main goals.

At a workshop hosted by the University of the Arts in Helsinki, the theme was "Death and Rebirth." The first scene was titled "A Normal Ordinary Day," followed by "Sudden Death." The sequence continued with "Face of Death," "The Spirit Leaves," "The Spirit Plays," "Time for Re-Birth," "Birth," and finally "The Dance of Life Returned." All scenes were performed using the Half-Speed, Curve, Bump, Spin, Push, and Pull vocabulary.

This exploration goes beyond just free creative expression, sensitivity, and group collaboration. It also establishes a specific vocabulary that guides and shapes how we express ourselves, which will be increasingly important as our work progresses.

Notes: Adjust movement vocabulary elements according to preferences, goals, context, available space, and situation. If gym mats, sprung floors, or grass are accessible, consider including "falling" as part of the movement. For performances requiring rapid transitions, focus on elements like "stillness" and "jerk." To incorporate cultural relevance, utilize live or recorded cultural or traditional music to inspire and motivate the storytelling aspect of the exploration. Additionally, a specific myth, legend, or cultural story can serve as a narrative reference to guide Part II. Have fun!

Highly Recommended.

THREE MOVEMENTS

Part I: Partners are asked to create and present three movements each, for a total of six movements.

These movements can be random, unusual, cultural, or otherwise interesting, and each partner should be able to perform the other's movements.

The preparation time is limited to five minutes. The partners then share their movements with the group, without providing any comments or judgments. These six movements serve as their vocabulary.

Part II: Develop these movements into a scenario.

Side Coach: *Let the movements suggest a scenario organically. Movements may be repeated as many times as necessary; however, each partner must use each of the original six movements at least once within the scenario. The movements may change in intensity and rhythm.*

Parameters: One partner must occupy a space; the other partner enters, then exits, with either or both partners leaving at the end. The scenario can also include "connective" movements to transition smoothly between actions.

Movements can repeat and occur in any order that fits the scenario. The scenario doesn't need to be logical. Each partner must perform each of the six movements at least once. There is no sound or language at this stage.

The scenarios are presented before the larger group.

Part II: In the next step, include sound or dialogue.

Partners may revise their initial three-movement scenario but must keep the original movements unchanged.

Present these scenarios to the group again and open for discussion.

Performers work, explore, and develop the scenarios further, using the original iteration as a foundation for a performance or as creative sketches.

Note: This low-risk, small-scale approach is ideal for initiating performance development across various groups. Additionally, it effectively prepares performers for future creative projects and explorations their creative skills.

This exploration acts as a precursor to developing a **Ritual Preparation**, which will be addressed in the **Space** phase of this book.

Highly Recommended.

From the Tanzania Journal

From a directing and performance development workshop at the University of Dar es Salaam.

During the second week of our three-week workshop, we focused on identifying a current problem to explore through performance. The group decided to concentrate on AIDS, a significant issue impacting many students.

- Performers identified words, ideas, images, and feelings associated with AIDS.
- Pambano—fight, struggle
- Mshtuko—shock
- Kiburi—arrogance
- Kibyongo—deformity
- Angamizi—catastrophe
- Mshangao—surprise
- Mnyako—dryness (lifelessness, dying)

The **Three Movement** exploration was limited to these self-imposed ideas and elements, which helped focus the work and prepared for scenario building and performance.

The partner groups then refined their movement scenarios by adjusting, editing (learning from and being inspired by others' presentations), adding sound, emotional nuance, and language.

Performers presented their work to the larger group.

Some examples from the session:

A scenario about a relationship—man and woman meeting, contracting, dealing, and dying: A short work that condenses time, space, and relationships to their essentials.

A scenario about the effects and nature of AIDS, how it impacts a family, and the medical care it needs.

A scenario that emphasizes what needs to be done, presented in a style that alternates between the comic, ironic, tragic, and larger-than-life.

A scene where a performer begs for help, shaking abstractly while AIDS takes control, passes through, and gradually destroys a body and spirit. The other performer circles, avoids, and judges the other's sexual promiscuity.

A scene in which one performer watches her sister die. As she narrated the story, she stroked her sister's face, brushed her hair, and sang a traditional grieving song.

Another scenario unfolded in a doctor's office, featuring a realistic narrative style as a doctor and patient talk, chant, sing, and then mourn in their native language.

Other scenes were equally heartrending—snapshots ranging from realistic to grotesque and symbolic. One depicted the cutting off of a penis, others zombie-like walking, and a tribal mating dance that dissolves into a slow dance of sickness.

The scenarios and movements formed a rich resource from which we created a workshop performance, which was publicly presented two weeks later.

Into Spaces

This marks the conclusion of the Body phase of our work as we transition into the Space phase, which aims to deepen our understanding of how the body interacts with different spaces. These spaces—social, cultural, political, economic, educational, and religious—intersect and influence our behaviors, emotions, and thoughts. Human beings have always been encoded; this encoding has helped organize and sustain human civilization throughout history. From the territorial markings of our primate ancestors to the sacred circles used in Greek sacrifices, from the architectural layout of the polis to the digital landscapes we now inhabit, humans have continually created, challenged, and negotiated spatial orders (Burkert 1983, 5). In today's complex, layered world, we navigate through multiple overlapping spaces, constantly negotiating and moving through our lives.

Our Space phase explores how the Body exists within spaces shaped by inheritance, imposition, and choice. We inherit spatial codes from our ancestors, such as cultural personal distances, gendered public and private divisions, and sacred versus profane geographies embedded in rituals. Spaces are also imposed by structures of power, including surveillance architectures, urban zoning, and invisible boundaries that define who belongs where.

We also select and craft spaces—the liminal moments of performance itself, transforming a warehouse into a sacred space, and forming communal circles to distinguish sacred ground from the profane. This awareness and reflection help us seek the wholeness of the circle, synchronize rhythms, and find balance. Ritual, arts, artists, performance, and performers serve to reveal the spatial codes that structure our lives, to establish temporary autonomous zones where alternative spatial relationships can emerge, and to cultivate spatial mind—a somatic awareness of how spaces shape us and how we, in turn, reshape the spaces we occupy. If the Body phase encouraged us to focus on our instrument, the Space phase will guide us to understand the contexts that give that instrument meaning, constraints, and potential.

SPACE

The Space section examines the multiple overlapping realms humans inhabit—physical, social, cultural, spiritual, technological, and psychological. It distinguishes "space" (abstract potential) from "place" (meaningful, organized system), arguing that contemporary life overwhelms us with competing spaces while offering little coherence. Drawing on cross-cultural cosmologies from Hindu, African, Mayan, and Greek traditions, the section presents narrative as humanity's essential tool for organizing spaces into meaningful wholes. Through explorations exploring circles, personal and cultural vocabularies, archetypes, masks, and ritual preparation, performers learn to identify, embody, and navigate spatial complexity—becoming guides who illuminate connections across diverse spatial realities in an increasingly fragmented world.

Introduction

Our hunter-gatherer ancestors understood that their physical limitations made them vulnerable in an unpredictable environment. To survive and evolve, humans relied on cooperative agreements, leading to the development of social and cultural structures. Today, our civilization flourishes because of our ability to create, adapt, and manage systems of meaning—building and organizing spaces, atomized units that surround, shape, and give meaning to our lives.

These spaces are in constant movement, continuously evolving, merging, coming into existence, and fading away. This sacred dance of creation appears across cultures worldwide, demonstrating how the universe unfolds

DOI: 10.4324/9781003686422-3

through divine motion rather than fixed laws. In Hindu cosmology, Nataraja's cosmic *tandava* symbolizes the harmony of creation and destruction. African myths depict creator gods like *Amma* balancing elemental forces. The Mayan universe is expressed through choreographed movements that sustain harmony among celestial, earthly, and underworld realms. Greek tradition views Apollo, "The Dancer," as a symbol of cosmic order, while Buddhist ritual dances represent the middle path. Modern spaces continue and embody the spirits and deities of the past, perpetually dancing and holding within them the way of the universe—creation and destruction, seeking balance.

But before we get too far ahead of ourselves, let us examine two fundamental terms: "space" and "place." Often used interchangeably, these concepts require precise distinction. Traditionally, Space signifies physical presence or the potential for human activity—abstract and open. Place, by contrast, refers to specific events or myths that maintain definitive positions in both time and space (Augé 1995, 82). An empty house represents space filled with potential; once occupied and personalized, it becomes a place.

Conventional thinking portrays space as open, free, exposed, and vulnerable—like a desert or tundra without established paths or signposts. This perspective sees space as unencumbered potential—a blank canvas awaiting human meaning—remaining "free" because no human frameworks have been imposed upon it.

This idea reveals an unconscious human-centered bias. When we use geographical, political, or economic frameworks, space is interpreted and integrated into a human-defined place, showing how human-focused views often overlook or ignore the sovereignty of non-human or place-based systems.

Consider the desert: By naming, mapping, and constructing roads through it, we align it with human systems, diminishing its sense of openness and freedom. When humans establish borders across "empty" areas, these spaces become entangled with political, ethnic, cultural, and economic frameworks.

Yet, the desert possesses its own patterned systems exclusive of humans—its own circles, balances, and rhythms that exist independently. As human civilization progressed from hunter-gatherer societies to urbanization and globalization, the act of marking and systematizing has often reflected anthropocentric hierarchical superiority and territorial control. This arrogance is the root of planetary disequilibrium.

Any "empty" space—deserts, tundra, oceans—possesses its own sovereignty. Science defines the spaces of the physical world; religion focuses on the spiritual; politics and economics on social relations; and governments and law define and regulate spaces. This act of claiming, naming, and organizing exemplifies colonization—the act of human possession to control, exclude, or exploit space. What should be a respectful dialogue among humans and other-than-human beings has been reduced to a one-sided monologue. Yet, the world requires dialogue. Each entity communicates in its own way and on its own terms. We grow, learn, and survive through dialogue.

What seems empty is actually full. Look again. Take a moment and observe the space around you—the gaps, the in-between, the spaces occupied by materiality. You might easily say most of your environment is open and empty—a decidedly anthropocentric materialist view. Yet, it is filled with aliveness, teeming with energy from the past—every thought, word, love, hurt, feeling, ghost, and deity lives in the empty. Empty is not so empty, rather it is alive with history, ancestors, spirits, and feelings—mysteries beyond our reckoning. The breath of dinosaurs, Julius Caesar, every plant, animal, criminal, hero, victim, and saint all recirculate through the oxygen in our closed atmosphere. There is a good chance you have breathed the breath of Napoleon, King Tut, George Washington, and Marilyn Monroe at least once in your life. Add to it the newest ghostly inhabitants that today move the world: Bluetooth, the internet, sound, ultraviolet, and satellite waves, to name but a few. Materialists view only what is physically present and measurable, often dismissing much of our environment as empty and insignificant—seeing and appreciating only a part of the world's abundance.

Place embodies the organization and rhythmic alignment of various spaces. Our work arranges and clarifies, making real, for a brief, fleeting moment, the intricate web of coexisting spaces to illuminate the wavering moment between the seen and the unseen world.

When I move to a new house, it begins as "space"—without furniture and belongings, it is potential, undefined, unorganized, and unassigned meaning. By bringing material artifacts from various spaces together—furniture, personal artifacts, family photos, a pet, grandma's coffee table, and books arranged into an architectural space—you create an ad hoc, cohesive system. By arranging multiple entities/spaces to reflect your history, identity, and needs, you make it your place. Such is the diagram of our work: To identify, signify, and arrange spaces into an erstwhile place.

Narratives

Narrative is the act of rendering spaces to create a place.

Cultures consist of narratives formed through images, symbols, myths, gestures, rituals, regalia, food, and behaviors. Walking through my suburban

Dallas neighborhood, I find myself immersed in expressions of Western cultural perspective—a lived narrative made manifest.

Cars observe speed limits, following social norms and legal enforcement. A neighbor waves hello, reinforcing community bonds. Homes occupy legally defined, socially recognized plots built using mathematical principles, architectural knowledge, and cultural heritage. These structures embody values: Spacious front lawns indicate a pastoral yearning and national mythology; driveways signal connectivity to broader networks and participation in mobile culture; my home's mid-century modern architecture communicates a historical perspective and life philosophy; the brick façade connects us to the earth, authenticity, and proximity to Mexico, where these bricks originated.

My neighbor's house features immovable shutters, classic columns, arched eaves, and an unoccupied porch—elements that showcase a blended history, myths, status, and memories. I interpret clouds and blue sky through scientific, spiritual, or aesthetic frameworks. Walking my street means traversing an environment dense with narrative spaces I engage with, affirm, and help create.

Each of us shapes our cultural narrative through thoughts, speech, and actions. Every thought and action simultaneously reflects and alters our spatial systems.

Humans are storytelling and space-organizing creatures—capacities that have driven our evolutionary success—narrative functions not merely as a vessel, but as a dynamic and necessary extension of human initiative. We have been successful, it can be argued, because we organize and constantly reorganize spaces into narratives as needed. We are comparable to a computer, and a computer is like us, functioning through software and hardware: Space = bits and data, narrative = processor and program. We create the tools, and these tools, in turn, shape us. Historically, tools have included everything from a club, a bowl, and a knife to writing, the wheel, printing, electricity, automobiles, computers, smartphones, and AI. This latest tool aims to extend human capabilities and improve our survival in ways beyond imagination. Throughout the eons, one unwavering human survival tool has remained: Narrative.

Narrative functions as home: An integrating operating system, venue, and medium where different spaces converge. It acts simultaneously as action and location, vector and container, specifically designed to organize and define spaces into coherent wholes. Our work is to organize space, create narrative, and animate the spaces in between.

Space Age

Today's world teems with numerous overlapping, interconnected, yet sometimes disjointed spaces. This complexity evokes anxiety, uncertainty, and the thrill of possibility—transforming reality entirely. The remarkable proliferation of spaces in recent decades stems from interconnected forces: Globalization, rising authoritarianism, social media, artificial intelligence, religious and political extremism, corporate influence, consumerism, the effects of COVID-19, the climate crisis, and deepening inequality.

These spaces compete for our attention, seeking integration into our lives while offering endless options. We find ourselves swimming—perhaps drowning—in the abundance of space, with little time to understand, integrate, and transform so many spaces into coherent places.

Responses to this spatial diversity have been intense and often desperate. Religious and political fundamentalism represents pushback against space-mixing, striving for clear, familiar order systems that offer comfort and security. We witness fixed boundaries and rigid categories being breached everywhere.

Meanwhile, growing multiculturalism and interdisciplinary approaches herald fluid, flexible, non-hierarchical spatial "flows" that challenge traditional concepts of space and place. Geopolitical dynamics, global media, economic shifts, travel, internet connectivity, and social media have rendered geographic mapping and national identity nearly obsolete. This explains why corporations like Google, Microsoft, Apple, and Facebook have become culturally influential—they facilitate new narratives, mediate innumerable spaces, and map our emergent, collective place—the world.

This historical moment demands innovative cartographic and narrative representation. In our chaotic world, we need fresh maps and stories. An evolving, inclusive, participatory, information-driven narrative emphasizing accessibility and interconnectivity is taking shape, creating a fundamentally new kind of world.

The performer identifies, interprets, and navigates these varied spaces at once material and conceptual, fixed and mutable, individual and collective, particular and universal. This is our new – old reality. Adapt or die. Acting as an avatar and surrogate, the performer is a guide, showing the connections, contradictions, and complexities of existing across diverse spatial realities. Exploration and reimagining will create fresh narratives. All that has preceded is reexamined in light of the task at hand. Deconstructing and reconstructing are processes of integrating for now and futures unknown and unimagined. We are in a moment of transition, an axial shift to a new consciousness and way of being—the Cosmocentric way.

Performance is our work. A functional technology, a haptic map revealing the spinning and turning, and pop-up spaces that surround. The performer is a reconfiguration of the psycho-paleolithic hunter-gatherer, navigating, embodying, and expressing fluid landscapes. Their role is to illuminate the unseen, foster coherence and balance, and serve their community. It is all about thriving, sustaining, and contributing to that big, magical, and overwhelming something we are all a part of but will never really, fully understand.

Traditional Indigenous ways of the past are fading, reviving, and transforming. This is the way of the world—neither good nor bad—it simply is because that is what it should be. Many Indigenous ways have vanished, becoming part of earth's memory. Yet, they are still there. Their wisdom transformed lives, still in the patterns of creation, in our biology, in the world all around, in every sunrise and sunset, in the cycles of the moon, and in the vast inhabited spaces in between everything. This is our moment, one of recalling, reimagining, and recreating it all.

Cycles

Our exploration of spaces begins with circles and cycles. That space of life and death we all inhabit is a recurring cycle. In the realm of nations and power, we've been there before and will return. That space of relationships, friendships, aspirations, war, jealousy, fear, poverty, sickness, discovery, growth, decline, victory, disappointment, gratitude, love, education, greed, care, comfort, and more—you name it—we've all experienced it in one way or another before.

All of it, psychophysical manifestations, imaginings, and entities, inherited, adjusted, and passed on. Spaces are big things when you are in them, small things when you are out. The names, the faces, the fashion, the circumstances, and details are different—Ancient Rome, Victorian England, Ming Dynasty, the Aztec kingdom, the Swinging 60s—all variations on eternal, universal themes. Nothing is created or destroyed—it just moves and changes. Creation is always, must always, be creating. We are part of the universe's ongoing process, a work in progress driven by momentum and the sum of everything that has come before. We are reanimations of animations. Everything that was, is, and will be lives in this moment.

We are all, along with every plant, animal, microbe, mountain, river, and rotation of the earth, one small part—a space—spinning in a vast, incredible, ongoing, cyclical, and ever-reconfiguring narrative of becoming. The spaces we inhabit are cyclical and transforming. You are, and every part of your life—and this book—is an expression of cyclical momentum.

The question is how to (re)engage with the cycles of life, the cycles of all the spaces we inhabit? Such a vast and overwhelming abstraction, seemingly tangible but evaporating in our grasp. How and where to begin?

The Body phase of our work was decidedly more functional, specific, and constrained by the frame and in relationship to the body, exploring its physicality, functionality, perspective, and myriad abilities. The Space phase builds on this by moving into more conceptual and abstract considerations. At times speculative and poetic.

In the work that follows, we will focus on space and spaces as units, energy packets, and entities. As attractors and holders that give form to feelings. They give meaning to the world and carry life forward. Like atomic particles, they cycle, vibrate, and have a rhythm, seeking balance. Just as the universe reflects the particular, the particular reveals the universe. In this phase, we will attempt to identify the biological, personal, social, cultural, and historical spaces we inhabit, understanding them as continual cycles, appreciating their often-unnoticed impact on our lives. The aim is to engage, dialogue, and report back—such is the way and value of the artist.

We begin with the everyday and what is in front of us: Acknowledging and examining the spaces we occupy, starting with the mundane, evident, and routine. The simplest elements often prove the most overlooked and undervalued, and are openings to vast, far-flung, and unexpected vistas. Observing, listening, and sitting with the obvious is simple, humble, and crucial. A method, an ethos, a way. We move through landscapes of space daily without the attention they deserve. We return to what we know—or what we think we know—the circle. The circle is the physical revelation of the cycle. It all begins with what we see.

CIRCLE DEFINITIONS

In a circle, instruct the performers to arrange themselves quickly and instinctively without talking or overthinking. You can also set a time limit of 30 seconds to a minute.

Adapt based on the needs, context, skill level, and goals.

Ask the performers to arrange themselves in a circle.

- By height
- Colors they are wearing (from bright to dark)
- From oldest to youngest
- Birthdays (January to December) or astrological signs
- From the least hair on the head to the most hair
- Strongest to weakest
- Prettiest to ugliest
- Lightest to heaviest
- Lightest to darkest
- Tallest to shortest
- Macho to feminine

If appropriate and as the group becomes more comfortable, steer the exploration toward a playful and insightful activity. This can strengthen the key themes of self-reflection, honesty, and acceptance. Adapt it to your context and needs. Potential next steps could include:

- Smartest to dumbest
- Largest feet to smallest
- Most sane to least sane
- Calmest to most anxious
- Richest to poorest
- Most talented to least talented
- Humble to arrogant
- Honest to dishonest
- Anxious to most relaxed
- Generous to stingy
- Liberal to conservative
- Sexual to frigid
- Spiritual to secular
- Most to least political
- Most emotional to least emotional
- Arrogant to humble
- Talkative to shy

Note: This exploration will cause discomfort and should be followed by open discussion. It can spark sensitive issues related to race, appearance, identity, and preconceptions. Participating in these conversations can foster trust within the group and enhance communication. It is crucial to recognize that the group, like our work, is a dynamic and complex entity.

From the Stockholm Journal

Shikasta, a diverse group sponsored by the National Theatre of Sweden, included refugees and immigrants. There were some Swedes, and several Sami, Finns, Africans from several countries, Kurds, Palestinians, Poles, and an Iranian. It was a group formed with good intentions that reflected Sweden's liberal-democratic openness and acceptance. Many had escaped terrible circumstances and were grateful for work, stability, and a new life that most had only dreamed of.

However, during our first session, I sensed something beneath the surface, something unsaid, so at the end of the day, I asked questions—obvious questions that an outsider can voice and be forgiven if they offended. Questions led to a broader discussion about the group's ethnic, social, and political differences—topics that had previously gone unspoken despite their shared experiences.

Members acknowledged their prejudices and assumptions, which lived on the edges of their perceptions of one another. Issues related to skin color, intelligence and education, gender, sexual orientation, lifestyle, religion, and social and economic status, even the food they ate, were openly discussed. Although they had worked and known each other for some time, there was surprise. People looked at one another with eyes wide and thoughtful stillness. Honesty burns brightly.

We all make quick judgments, categorizing others based on visible traits and fitting them into culturally inherited stories. We have all, at some point in our lives, thought ourselves better than another.

We hide, self-censor to fit in, and mask until it becomes a reality. Silent habits and perceptions are normalized, seldom acknowledged, and less often examined. The role of the performer is to reveal. Our workshop found its theme. To explore the narrative, intuitively and collectively woven, to be accepted and to flourish. We only had two weeks together. We agreed to identify and interrogate the individual and collective spaces inhabited, and how they navigated, reshaped, and expressed themselves.

WHY DID YOU DO THAT?

A performer is invited to step into a circle and engage in any activity or behavior for 30 to 40 seconds to establish their role.

Then, a second performer joins and asks, "What are you doing?"

The first performer responds verbally, giving any answer except the actual activity they are engaged in.

Example: The first performer skips and replies, "I'm flying a kite."

The second performer joins the circle and begins miming kite-flying.

After establishing the kite movements, the first performer asks, "What are you doing?"

The second performer replies, "Walking my dog."

In response, the first performer adopts the action suggested, saying, "Walking a dog."

Performers may repeat an action or response already given.

The back-and-forth continues until a performer cannot immediately answer when asked, "What are you doing?"

If a performer asks, "What are you doing?" and responds with their current action or repeats an action or repeats a response already given, or if a performer fails to keep up with the quick exchange of thoughts and actions, they exit the circle. Then a new performer enters to replace them and continues with, "What are you doing?"

The activity is ongoing and non-competitive, so there is no winner—only play.

This exploration requires intense focus and can become increasingly difficult as performers must verbally state an action opposite their behavior, think of new actions, and remember actions they have already used.

Note: This activity exposes the performer and the group's inhabited spaces. Ideas, references, and actions serve as indicators. The questions and responses will vary widely, reflecting social and cultural contexts. The main goal of the exploration is to uncover the gestural and verbal spaces being conveyed. The dissonance between action and speech is used as a distraction to prompt instinctive responses and intentionally disrupt censorship, conscious thought, or premeditated expression. Discussion follows.

Forming a circle creates a temporary communal space, what Hakim Bey calls a Temporary Autonomous Zone. A circle is where the group gathers, collaborates, shares, and explores. It cultivates a group's identity and purpose, serving as an example, a way of thinking and communicating, a foundation for our work, and a constant reference point. It functions as a visual and conceptual guide, an evolving template of performance. The circle acts as a living mnemonic, providing a structure for discussion and embodiment. The circle has no hard edges, energy circulates, and there are no corners where things can hide. It is where everyone can see each other, eliminating hierarchical barriers and removing the distinction between performer and audience. The circle arrangement discourages ego-driven agendas, encourages consensus, and reduces the anxiety often linked to speaking in front of others. Ultimately, like ripples spreading across water, a single thought or feeling can radiate outward, reaching the wider community.

Through collaboration, clarity and voice naturally emerge.

Pay attention to what people and the work are saying.

Roles and responsibilities will evolve, and that's perfectly fine.

Issues and challenges will arise.

Remain open, honest, and willing to share.

Collective insight is powerful and revealing.

Be a caring guide, a keen listener, a curious questioner, and a thoughtful reflector.

Authentic leadership is rooted in patience, wisdom, guidance, and the grace to let go.

Control is just an illusion.

Empower and trust others, and you will discover deeper parts of yourself.

It is a collective journey.

The **Creative Circle**, initially discussed in the Body section of our work, is now developed and broadened.

FIGURE 2.1 The author participating in a *Creative Circle* at the Korean National University of Arts, Seoul

Language

Upcoming explorations will use the **Creative Circle** to foster focused efforts to examine the spaces. The first to be explored is the space of language.

TALKING CIRCLE

The group forms a circle. As they take turns, each performer states their name and shares what's on their mind in one sentence, using any form of expression they choose.

Exploration Variation: This exploration often encourages longer discussions. Over time, participants may adopt a format like the "Talking Circle," a Native American custom. In this arrangement, a feather or a similar object, like a stone, is passed among the group. Each person holding the object can share their thoughts or pass it along without speaking. The object holds the energy and spirit of what was said.

Note: This exploration effectively addresses and resolves both group and individual issues, fostering trust and a deeper understanding among performers. It creates a shared space through the use of language.

The following explorations were developed with Lul Theatre, a group in Addis Ababa, Ethiopia, as part of a series focused on language. The circle format facilitated interactions between sound and language, fostering dialogue and encouraging participants who are shy or hesitant. The activities can be done in sequence or separately, and they can be repeated.

CIRCLE SOUND GAME

The group forms a circle, and one performer begins by tossing an imaginary ball.

When passing to the left, each person makes the sound "burr."

When passing to the right, the sound is "baaa."

If they pass the imaginary ball across the circle, the sound changes to "sissss."

Any performer who passes the imaginary ball without making the correct sound is eliminated from the circle.

The exploration continues until only two performers remain.

Exploration Variation: Rather than simply making a sound, the performer must select a verb when passing the ball to their left, a noun when passing to the right, and an adjective when passing across the circle. If they forget or make a mistake, they are eliminated from the exploration.

ENDINGS AND BEGINNINGS

In small group circles, one performer begins by saying a word to any other member in the circle.

Example: The game starts with a player saying "Airplane." The next player must use the last letter of that word, "e," to say a new word that begins with "e," such as "entertainment." The following player then uses the last letter of "entertainment," "t," to form a word starting with "t," such as "telephone," and so on.

Initially, each player has five to ten seconds to respond with their word.

If a player fails to respond in time, they are eliminated from the circle. As players get better, response times can be shortened to four, three, or even one second per turn. The exploration continues until only one player remains.

Note: The exploration can happen in the local or native language, depending on the cultural context or the group's familiarity with that language.

CIRCLE STATEMENTS

In a group circle, a performer starts by saying a brief phrase or statement of four to six words to another participant. This phrase may, or perhaps should, be a cliché, banal, or commonly used expression relevant to the cultural context.

Examples include: "All for one, and one for all," "Kiss and make up," "With your tail between your legs," "And they all lived happily ever after," "Has the cat got your tongue?"

The recipient performer chooses one word (any word, except the same initiating word used by the previous performer) and creates a new statement. This statement is then directed at another performer, allowing the inciting cliché to be passed and transformed throughout the circle.

Example: "Kiss and makeup" might be responded to with "I love to kiss." Then, the next performer might respond, "Love is all that is important." Then to another, "Honesty is important," then "That's an honest face," then "Sometimes you must face facts," and so on.

If a performer cannot respond within a specified time, they leave the circle. The exploration proceeds until only one performer remains.

SMALL STORIES

Circle. Performers share stories about what they have learned so far from the workshop.

Then, share stories about first memories, first date, first time they flew on a plane, their first dog or cat, when they first learned about death, sex, heartbreak, and their first big disappointment, among other life experiences.

Exploration Variation: Share stories about fantastical things, blending imagination with reality. Describe what it was like to travel to the moon, ride a dragon, be a dragon, be a dog, an eagle, a fish, a rock, a cloud, and so on. Performers can adopt characterizations and voices.

Note: This exploration begins with the personal, real, and immediate, then extends into memory, and finally into the imaginative, helping develop the performers' storytelling abilities and comfort with language and presentation.

NEVER STOP

Two participants step into the center of a circle to simultaneously share their personal "monologues" about their immediate thoughts, feelings, desires, and aspirations on any topic or whatever is on their mind now.

Both performers should speak consistently, starting with their individual monologues. After establishing this, encourage them to shift their focus from listening to engaging with each other. Guide their monologues to evolve into dialogues, taking them on an imaginative journey.

They must continue the conversation while remaining focused on their monologues or dialogues. If any performer loses focus, the facilitator promptly replaces them, allowing the exploration to either continue from where it left off or switch to a new monologue or dialogue.

Exploration Alternative 1

The activity may also involve partners delivering their monologues with intense focus, trying to distract the other. When a partner loses focus, they must step out of the center of the circle and be replaced by another performer.

Exploration Alternative 2

The two main performers start at the center of the circle, as mentioned earlier. In this variation, the surrounding performers can help support them by providing lines, images, and ideas to maintain the flow of their performance. The facilitator can also step in to replace the center performers if needed. The primary aim is to encourage playful listening, imagination, trust, interaction, and a feeling of group identity and shared expression.

Exploration Alternative 3

A solo performer starts with a monologue on a self-selected topic or about their day. The other performers act as a chorus, echoing and emphasizing the ideas and images they find important.

For example, if the central performer says, "It was a beautiful day today," the circle performers might reply with "It was beautiful." The goal is to encourage listening, responding, and supporting each other. At first, the circle performers may respond too eagerly and lack coordination. The aim is to create a dialogic exchange where each leading performer and the surrounding performers engage in rhythmic interaction that develops throughout the monologue. An improved version could involve establishing a call-and-response rhythm, a pattern often found in many religious, traditional, and Indigenous cultures.

Highly Recommended.

The next exploration, inspired by Peter Brook, significantly improves listening skills, awareness, and collective mindfulness.

COUNTING TO 20

Part I

The group forms a circle and takes turns counting aloud in order, saying "one, two, three..." around the circle (either clockwise or counterclockwise) until reaching 20.

Part II

Then, ask the group to count from 1 to 20 again, but this time without considering their position in the circle. Anyone can begin, and anyone can call out the next number. The key rule is that no two people should speak simultaneously, as this could result in some numbers being called out more than once.

If two performers speak at the same time, the group must restart from the beginning, marked as "1." They can restart as many times as needed, even if they reach "19" and someone interjects at "20"; they will have to start over. It's about maintaining honesty and not giving in.

Complete freedom allows each person to declare a number at their discretion; however, two conditions impose discipline: Maintaining an ascending order and refraining from simultaneous speaking.

Note: This exploration highlights the connection between concentration, active listening, and personal freedom.

Performers succeed by paying attention to the space between calls and by attuning to the group's awareness and rhythm—the group space.

Intuitive sensing uncovers the emotions and thoughts we are taught and conditioned not to value.

Pride and Prejudice

During these next explorations, encourage performers to reflect on what they share with fellow group members. What cultural, political, economic, demographic, and spiritual similarities exist? What is obvious? What does this suggest, and what remains unspoken? Ask performers to consider how shared understandings and agreements between them and their fellow group members originated—whether they were taught or learned from family, teachers, peers, media, or other influences. What do their agreements, behaviors, and assumptions reveal and imply about our current moment? And how do these observations and affinities connect and network with different spaces and systems of understanding?

As we continue with the circle, our focus will be on how spaces carry inheritances—social and cultural genetics of a sort. The three explorations that follow move into potentially unsettling areas. It's essential to explain and prepare for **Nice or Nasty** and **Victim Circle**, or **The Gauntlet**, beforehand, outlining their objectives and how they are a step in the process of our work. Refer to the **Aggression** section in the **Body** section of our work.

These explorations acknowledge the prejudices, perceptions, biases, and behaviors embedded in inherited spaces. Addressing these inheritances is a crucial step in disrupting negative infectious cycles. It fosters clarity and healing, guiding the performers and group toward development and evolution.

These explorations are to be used only on a need basis.

Discuss with the group beforehand.

They are presented here as optional explorations and are not required.

If you choose to use any of them, do so mindfully.

They are grouped for convenience and not meant to be carried out sequentially.

As with all activities, explorations, or games in this book, performers may choose not to participate or withdraw at any time for any reason.

A discussion must follow each of these explorations.

NICE OR NASTY

1 Performers stand in a circle. Someone starts by giving a compliment to their neighbor on the right—saying something Nice. In turn, that performer passes the compliment along and builds upon it, then passes it to the person on their right. This activity proceeds clockwise around the circle.

Stress that the performers refrain from expressing their feelings and instead focus on their cultural or social evaluations of the other performers. This approach will enable performers to tap into deeper currents, views, prejudices, and stereotypes, as well as the legacies of inherited social, cultural, gender, racial, political, or economic groups.

2 After the compliments have traveled around the circle, the circle begins as before, saying something Nice to the person to their right.

This time, however, the recipient responds by saying something Nasty to the next performer. Who, in turn, responds to the next person with something nice. This alternating pattern—saying something **Nice or Nasty—**continues around the circle.

3 Reverse the circle's direction, so those who previously said something nasty now say something nice. Continue alternating between Nice and Nasty, moving around the circle.

Ask performers to mark how they feel when something nice or nasty is said to them.

4 For this time around the circle, performers say only Nasty things.

Negativity tends to escalate and gain momentum as it travels around the circle. It shows how hate and prejudice possess a dark emotional power—almost like they have a life of their own—something to explore later.

Let Nasty circulate around the circle several times, then switch directions.

5 Restrict comments to Nice, positive, if not loving, for the next round.

Ask performers to express what they genuinely like about the next person.

Note: Performers tend to be courteous when expressing harsh critiques. This behavior is anticipated, as it involves a delicate subject. As they should be, performers are more attuned to each other's emotions. Allow the initial one or two rounds to unfold naturally.

A discussion should follow, focusing on how the performers felt, the sources and histories of their comments, and their awareness of negativity and how it can take on a life of its own. Are the origins familial, social, or cultural? Where did they come from, and how did you learn them? How best do you respond to this in your everyday life?

These explorations act as a form of purification and cleansing, clearing the spaces between performers and reducing lingering social and cultural tensions. The ability to engage with and discuss sensitive topics is a shared experience that marks growth in group sensitivity, connection, and identity, building strength and courage. They provide a clearer understanding of the performer's role as a fearless advocate for their community. They built the group space. The worst has been said and done, and with that, performers are more sensitized. As our work progresses, awkward and delicate issues will arise; these explorations prepare and empower performers.

THE GAUNTLET

The circled group disperses into two parallel lines that face one another.

Each performer walks the gauntlet alone, facing verbal taunts and insults from the rest of the group. Participants may use any form of abuse, slur, gesture, or epithet they choose. All aspects of the individual, such as their physical appearance, personality, political beliefs, economic status, and tribal, ethnic, racial, sexual, religious, and cultural identity, are vulnerable targets. **There are no limitations.**

Physical contact is not allowed, but the performer walking the gauntlet can respond. When they reach the end, the walker joins the line and participates in mistreating others, who then proceed to walk the gauntlet.

After all performers finish their gauntlet walk, they repeat it. During this round, each performer receives a warm shower of praises, compliments, and accolades. Performers in line may offer a handshake, a pat on the back, hugs, a gentle embrace, apologize, ask forgiveness, applaud, bow, or blow kisses as they pass through the gauntlet.

At the end of the exploration, the line forms a circle and ritually signifies unity.

Exploration Alternative: The same sequence as before, but this time the performers assume the identity of an animal or other non-human being relevant to the project, context, or what is generally observed. They take on the challenge of embodying this being, following the same guidelines as before, but as a non-human entity. This approach explores how our perceptions and treatment of other beings are influenced and shaped by our relationships and projections. They may choose to adopt specific characteristics of the non-human being, or not.

Exploration Alternative: The group is divided into two lines facing each other, distinguished by external characteristics such as sexual, racial, or cultural traits. One group then makes comments—sexual, gender-based, racial, and cultural—toward the other, which remains silent and listens. The provoking group may use gestures, vocal tones, and physical closeness, with the facilitator moderating the intensity. Afterwards, roles switch so the other group takes their turn. This process can repeat with a similar setup, but now the focus shifts to positive praise and appreciation related to those external traits.

This alternative is inspired by the *Challenge Drumming* ceremony of the Greenland Inuit, which traditionally helps resolve personal conflicts. It also takes inspiration from Battle Rap.

This alternative involves performers facing each other in the center of a tightly formed circle. Each performer uses a hand drum or tambourine-style instrument to talk-sing (or rap) their differences directly to the other, who must listen without reacting. While there is no physical contact, the talk-singing performer can taunt, insult, mock, boast, and invade the other's personal space. After two or three minutes (you determine), the other performer has a chance to respond.

Throughout the challenge, the circled group can react however they see fit—cheering, jeering, applauding, or encouraging one or both performers. The back-and-forth between the performers continues in rotation until it is exhausted or one of them breaks, in anger, humiliation, or humor.

Note: The Challenge Drumming alternative addresses current group objectives or resolves conflicts among members. Any issues between members impact the group and must be addressed and acknowledged collectively.

In 1993, the Natal Performing Arts Council in Durban, South Africa, commissioned me to create a street performance to tour Natal Province and Zululand, educating blacks about democracy and voting. It was a tense and uncertain time. The first multiracial elections were scheduled for the spring of the following year, and widespread political, social, and cultural change was underway. Many black South Africans, especially those in poorly educated rural areas, lacked experience and knowledge of voting and democracy, despite having been granted the right.

During my time in South Africa, I observed how my white skin affected my interactions with black South Africans, carrying the legacy of apartheid. Blacks often showed deference, stepping off sidewalks or falling silent as I approached, their bodies shrinking and becoming submissive. This troubled me, as I realized my white presence held symbolic weight and negative associations.

When developing the performance—Makanda Mahlanu—my initial instinct was to play the clown. Under apartheid, whites were seen as distant authority figures representing power. I played the role of a humorous, friendly, approachable trickster, aware of stereotypes and humor to challenge fears, ultimately revealing the white "boogeyman" as a deception. The play aimed to dismantle the mystique surrounding whiteness, showing whites and blacks working and laughing together to emphasize unity. It also featured an example of voting—pausing the performance to simulate the act—before resuming. By demystifying my whiteness and revealing my vulnerabilities and humanity, I became just another voter among many, with the performance embodying and demonstrating democracy.

FIGURE 2.2 The author performed in *Makanda Mahlanu* outside Pietermaritzburg, South Africa, at a school for children orphaned by township violence

Source: Photo: Shelley Konjstad.

Our work views the body and its numerous socially and culturally shaped spaces it inhabits as a living text to be interpreted as a dynamic form of communication. We are all influenced by others, acting as carriers and agents of meaning. Uncovering bodily inheritances is a crucial step toward healing and remediation. **The Gauntlet,** in particular, serves as an initiation rite by ordeal. Afterward, performers and the group feel they have undergone a test. Many unspoken, personal, interpersonal, and cultural conflicts are revealed, often astonishing performers with how quickly and easily they emerge—what lies within must be understood, addressed, and transformed into a positive, actionable awareness. The performers are a community's avant-garde, venturing into dangerous terrain, returning, and then leading by example and knowledge.

From the Tuma Theatre Journal

In our version of **The Gauntlet,** performers stood in opposite lines—Alaska Natives on one side and non-Natives on the other.

After months of working together, the group grew more cohesive, demonstrating awareness, care, and respect for each other. However, the acceptance within the group did not mirror the outside reality.

A recent news report about an assault on a native teenager in Fairbanks by white teens sparked a serious discussion. While the participants showed openness and empathy, each had their own and divergent viewpoints.

Facing each other in a line, each side exchanged racial slurs and insults. At first, there was hesitation, but once provoked, a collective voice, like a demon let loose, emerged. Emotions intensified, and an unthinking, insensitive crowd mentality took over. Anger, hostility, and hatred manifested physically, lashing out at the other group, prompting similar responses.

Emotions ran high, the performers appeared fragile, and the atmosphere was tense and uncertain. Recognizing the need to redirect these negative feelings, I asked the group to say something "nice." Soon after the Alaska Native line began exchanging kind words, the two groups embraced, apologizing and hugging, with several in tears. Many were astonished and ashamed by what they had said and felt.

During our talk, performers gathered close, holding hands. Many expressed feelings of fragility and shame, regretting their hurtful thoughts.

The most shocking part was how quickly and powerfully their emotions surfaced, seeming to take on a life of their own. "It felt like a dark spirit was unleashed," said Paul Mountain, an Athabaskan Indian. These were educated, sensitive individuals opening to trusted people. Several recognized that their words were wrong even as they spoke them.

> "Hate is such a strong feeling. It can't be controlled," Erma Hooper, a Yup'ik performer, said.
>
> "It wasn't me talking," the usually shy Ralph Elook added.
>
> "A feeling of adrenaline; even though I felt guilty, I liked the feeling as I hated it and knew it was wrong," Melanie Brown offered.

Where do these perceptions originate? How do they hide, and how do we confront them? Hate and anger are demons within us. Through this process, the unspeakable and worst feelings were voiced. Now, what do we do with them, and how?

Each performer learned how the demon in others perceives them and, just as importantly, uncovered parts of themselves they had hidden, avoided, denied, or been unaware of. It was a humbling awakening.

After that experience, no topic or emotion seemed too sensitive or off-limits. The experience deepened trust, resolve, and confidence in their roles, thereby increasing the group's sensitivity to themselves and to each other.

FIGURE 2.3 *Imipashi* (the spirits) performing at a bus station outside of Lusaka. The performance was devised and directed by the author and toured throughout Zambia

Source: Photo by the author.

From the Zambia Journal

During a break, I noticed two workshop participants talking in the distance. Although they were speaking in Bemba, I could easily sense the energy of their conversation. Timothy Hugala spoke firmly, using animated, angular gestures to address Fraxton Phiri, who stood submissively, nodding in response.

When I asked our stage manager, Light Musonda, about the situation, he shook his head in disdain and said, "He is calling him a dog." I was surprised. "What? Why?" Light explained that Timothy was from a tribal group that had defeated Fraxton's tribe a century and a half ago, which allowed Timothy to reprimand Fraxton and refer to him as a "dog."

After the break, I inquired whether anyone felt rivalries or identified others as traditional enemies within our group. Surprisingly, some participants openly acknowledged existing tribal or inter-ethnic tensions.

I emphasized that our project's aim is to depict Zambia as a united nation and people through performance. We then talked about how inter-tribal conflicts obstruct Africa's progress, with Africans frequently divided, leading to significant suffering. Everyone concurred that ethnic and inter-tribal conflicts are damaging. I reiterated my view that artists are leaders who should demonstrate and embody change.

Next, I brought together traditional rivals and required them to work together by exchanging and teaching each other their traditional dances. Their traditional enemies would perform their rival's dances in the performance. This led to a flood of complaints. However, over time, a spirit of collaboration grew.

In this situation, I leveraged my outsider status to request actions that insiders might not have considered. By participating in a dance from a different tribe, barriers broke down, allowing performers to expand their performance vocabulary and deepen their understanding and appreciation of other cultural practices. Friendships evolved.

Sharing dances created a new perspective and approach that influenced the spirit of the workshop and performance. Traditional performance emerged as a universal language, open to everyone, rather than being confined to specific tribal identities or genetics. Tribal expression can serve to reinforce differences—I convinced the group to see it as an opportunity to change perceptions not only for the group but for their larger community. The fluid exchange between tribal cultures and performers was ultimately recognized as a source of strength.

Imipashi traveled across Zambia, visiting many different ethnic and tribal areas to perform outdoors for crowds of thousands. Audience reactions were overwhelmingly positive. Seeing traditional rivals dance together was symbolic and powerful, breaking down culturally and historically inherited barriers by setting a positive example.

> "I didn't know they could do that," shared one community worker in Chipata on the border of Malawi.
>
> "Now there is hope for Zambia," said a local official at our Victoria Falls performance.

For our young and impressionable audiences, seeing rival tribes come together through dance was deeply meaningful. A performance models and is part of an ongoing process of understanding, reconciliation, and healing.

The next exploration expands on **Nice or Nasty** and **The Gauntlet** by introducing a different level of physicality. The **Victim Circle** should not directly follow these activities, but be used separately and only if suitable. This exploration dives deeper into the psychophysical aspects of negativity, prejudice, violence, group thinking, and mob mentality. As with the previous two explorations, firsthand physical, psychological, and emotional experiences and reflections help build trust, expand emotional and intellectual range, and improve performance vocabulary—all crucial for our work.

VICTIM CIRCLE

A performer steps into the middle of a circle formed by the group, who hold hands to try to trap them. In the center, the performer faces teasing, ridicule, frustration, and verbal abuse. The aim is for the person to break through the linked hands, which may shift and block their escape. If the performer succeeds, the group member who let them go must then face the same treatment.

Before the **Victim Circle**, either in a previous session or right before it, I sometimes facilitate **Circle Pulls**, which are found in the **Body** phase of our work. This simple, physical activity encourages participants to engage together, revealing subtle negative undertones. It also works well as a warm-up, whether or not you plan to use the **Victim Circle.** Circle Pulls is a great way to boost group focus and physical involvement, helping to lower inhibitions and promote a playful sense of aggression and competition among participants. A similar exploration is **Pig in the Middle** found in the **Body** phase of our work.

Rhythm

Every space has a rhythm.

We are all born into a world of rhythm. Our surroundings are vibrant with rhythm, encompassing the meta-rhythms of day and night, changing seasons, migrations, the sun, moon, stars, clouds, and weather. Each day follows its rhythms, from dusk to dawn and night to day, along with the subtle details in our lives—wind, rain, the play of light and shadow by the sun, and the actions of many creatures like animals and insects. All are distinct spaces of feeling and information. These spaces respond and interact with other spaces, each with its own unique rhythm. They might include the aging of trees, the growth cycles of plants, the slow deterioration of stone, and the shifting of soils. All are spaces, cycling, being, becoming, and decaying, moving to their unique rhythm.

The dance of birds, bees, and flowers unfolds rhythmically. Our bodies resonate with this rhythmic world, helping us to create time, organize calendars, and structure workdays filled with the rhythmic mechanisms of cars and trucks, or the spinning of a computer hard drive. A town or city is an orchestration of multiple and varied rhythms; for some, it is a cacophony of traffic, activity, and people all moving through the rhythms of their day. We exist in these rhythms as we hold and express our own body rhythm. We inhabit a rhythm reality.

We express our rhythm reality through dance, gesture, voice, rattles, and various instruments—both percussive and non-percussive, including body slapping and foot stomping—spanning genres from ritual to folk, rock, symphonic, and hip-hop.

FIGURE 2.4 Jimmy Toolie and his wife, Gertrude, performing at the Festival of Native Arts, Fairbanks, Alaska

Source: Photo by the author.

For Indigenous people, rhythms are drawn directly from the natural world. Rituals, social dances, and songs reflect their world. When an Indigenous group maintains a continuous, intimate relationship with the Earth's rhythms for hundreds or even thousands of years, a sophisticated rhythmic dialogue develops. Techniques such as drumming, rattling, stomping, vocalizing, or strumming engage in a conversation with the surrounding environment. Body rhythms interact with spatial rhythms, creating harmony between the two. This relationship fosters identity, community, society, and culture.

FIGURE 2.5 A Theru Koothu performer portraying Krishna in Tamil Nadu, South India, accompanied by drummers and musicians

Source: Photo by the author.

Rhythm expression techniques vary globally, including throat singing from central Siberia, Slavic hand-clapping and thigh-slapping, Indian tabla drumming, Zulu ankle rattles, electronic drum machines, and the vocal percussive styles found in rap and hip-hop. The technology utilized for rhythm is closely tied to the resources available to each community. For instance, the Inupiat traditionally used walrus-gut drum covers. At the same time, the Zulu use unsheared cowhide, the Sakha prefer mare hide, and modern Europeans and Americans often rely on synthetic materials or electronic instruments. Each culture's drumming style, purpose, and instruments reveal significant insights into their relationship with the land.

The short, durable, and hairy milch mare is unique to central Siberia, and its milk was a traditional staple for the Sakha people before the Russians introduced cattle in the 17th century. The mare frequently appears in Sakha mythology as magical, prophetic, and even speaking; to this day, its fermented milk is still revered as a traditional intoxicant. It is no coincidence that the skin of the all-important mare is used for the Sakha shaman's drum. When the shaman drums, they "ride" the drum into a trance state and another reality. The drumsticks are shaped like a horse's hoof, and the beat resembles the galloping of a horse. The rhythm of this horse carries the shaman into the cosmos. Rhythm holds a space and provides the connective tissue linking spaces.

FIGURE 2.6 Archival photo of a Chukchi (eastern Siberia) shaman and his drum

RHYTHM SEQUENCE

Performers form a circle, each receiving a number.

The group then creates a four-beat rhythm using finger snaps, claps, stomps, vocal sounds, or other methods. Feel free to invent your own rhythmic pattern. First, get comfortable with the four beats, then reinforce and repeat them to establish a "baseline" for more experimentation. Practice and repeat until the group naturally internalizes the sequence and moves in unison.

Rhythm Sequence Example: Both open hands slap the thighs (beat 1), then hands clap together (beat 2). Next, the right hand moves to the side of the torso and snaps fingers (beat 3), followed by the left hand doing the same (beat 4).

To begin, you announce one of the assigned numbers for the performers. Once a performer's number is called, the group maintains the original four-beat rhythmic sequence—the "baseline"—before improvising their own sequence, still following the original rhythmic pattern. While the rhythmic sequence and pattern stay the same, the physical expression varies through improvisation.

The other performers carefully watch and master the improvised physical expressions of the rhythmic patterns. They work together to perform and refine the improvisational rhythm sequence as a group. At first, the improvisational variations are repeated several times to help performers improve their responsiveness and become more proficient at playing off the basic rhythm.

After each successful improvisational rhythm performance, the circle resets and re-establishes the four-beat "baseline." The performer who just improvised the four-beat then announces another number, and that performer, without breaking the beat, transforms the rhythm into their own improvisation. This process continues, with each performer improvising at least once, until everyone has participated. The next performer's number is called on beat 3.

Exploration Evolution: The "baseline" four-beat can be entirely omitted, allowing the four-beat to flow smoothly from one improvisation to the next.

The development progresses rapidly, fostering an instinctive sense of rhythm and responsiveness within the group. The four-beat pattern can be modified to fit various cultural contexts.

The objective of **Rhythm Sequence** is to develop rhythmic expression and skills, coordinate those expressions and abilities, and then play, create, and communicate collectively through rhythm. Rhythm serves as a group language, the ether that connects and synchronizes bodies immediately, intuitively, and psychophysically.

RHYTHM DIALOG

Partners: Assign one performer to be the first to move. Instruct that performer to execute a single, isolated, and precise body movement initiated by a handclap, vocal cue, drum, or stick.

Example: Pointing a finger, extending an elbow, thrusting the chest, or pelvis begins a movement, which is sustained. With each clap or sound, the partner responds, keeping actions precise and isolated. As the sequence unfolds, performers engage in a body-based dialogue, responding to one another's movements.

Once set, the initial sound establishes a rhythmic pattern or cadence, developing into a lively physical and rhythmic exchange. When well-practiced, performers traverse space while maintaining interaction, transforming into a dance-theater expression. Patterns may be traditional, cultural, pop-cultural, or hybridized originals, evolving through repetition and variation and forming the core of rhythmic expressions that drive the performance.

Exploration Variation: After establishing rhythmic patterns, the externalization of the rhythm may stop, with performers continuing internally. This exploration could serve as a reference/cadence for smooth, steady vocal progressions, structured text, movement of lights, media, objects, or scenes.

Cultures

In the Republic of Sakha, located in central Siberia, greetings differ from the typical "hello." People say "Capsay" (what is your story?), and the response is "En Capsay" (tell me your story). This exchange reflects a core Sakha worldview: Reality is a vast web of interconnected stories (spaces) that populate and flow through our lives, shaping it.

We are all entwined in multiple narrative spaces simultaneously—your family's history, your friends' experiences, the forests you have known since childhood, your home, your high school years, and your city. These overlapping stories define who you are. The most meaningful way to connect with the world is to understand and engage with others' stories.

Cultural narratives—mythologies, religions, cosmologies, and histories—are collectively shaped, evolving spaces. Consider how religious texts such as the Bible, the Koran, and the Ramayana have influenced human thought, emotion, laws, behavior, and daily actions over the centuries. Histories, poems, novels, films, songs, paintings, dances, and natural landscapes all carry stories that evolve, intertwine, and inspire across time and multiple spaces.

Stories are not only human. Trees, wind, rivers, animals, and spirits all have narratives. While they express them differently from humans, they participate in multiple narrative spaces just as we do.

The following explorations aim to improve rhythm awareness and expression using vocal and linguistic tools—essential resources for telling human stories. Performance involves drawing from and reinterpreting the sounds and language of our surroundings, blending them into a rhythmic experience. By linking emotions with physical gestures, performers create harmonious connections with their audiences and the communities they serve.

RHYTHM EXPRESSION

This exploration builds on **Mantra**, introduced in the **Body** phase.

Select and assign each performer a sentence from a novel, book, poetry, play, newspaper, or similar source. The line should be brief enough for quick memorization. Choose a line that aligns with the group's age, skill level, and the performance's goals in terms of length, content, and vocabulary.

In multicultural settings where English is not the dominant language, this activity enhances performers' confidence in their language skills.

Example: I applied a randomly selected line from an Italo Calvino novel in Zambia. In Poland, I used only one Polish phrase. In diverse groups and US educational environments, I have employed a wide variety of texts—from poetry to technical journals to legal documents. Each activity yielded successful results.

Part I: Encourage each performer to thoroughly examine their lines, focusing on all possible pronunciation nuances. Suggest they take their time by breaking the line into parts and emphasizing each syllable, vowel, and consonant separately. They should concentrate entirely on the sounds, not the meaning.

Part II: Have each performer create a personal emotional and psychological link to every sound in the line or sentence. Prompt them to explore variations in volume, intensity, repetition, rearrangement of words, and the individual sounds within words.

Part III: Encourage performers to connect with each sound physically, letting them explore how to express themselves emotionally and psychophysically. Suggest that performers form associations between body parts and specific sounds.

Suggest associating sounds with non-human entities in nature. Does the sound resemble the rhythm of a river, ocean wave, a tree, a cornfield, a rambling bear, a prancing deer, rainfall, or wind?

Part IV: Encourage the performers to develop a cohesive pattern that integrates their vocal and physical expressions, creating a dynamic, rhythmic performance. This should include language, body movements, and rhythmic elements.

Part V: Ask performers to develop and establish a steady, rhythm-driven performance.

Part VI: Performers present their work to the group. After all presentations, encourage the group to share their observations and feedback.

Highly Recommended.

FIGURE 2.7 *Rhythm Expression* exploration during a workshop. Kuskokwim College, Betha, Alaska

Source: Photo by the author.

RHYTHM CHANT

Identify a sound or a group of sounds; you can initiate it or ask the group to develop it. The sound might originate from culture, the environment, the animal, or the natural world. Avoid sounds created or inspired by technology or machinery.

It should be a sound, not words that carry meaning.

Example: With Tuma Theatre, the sounds were drawn from Alaska Native culture. *"Hey Hey Hey Hey Hey Hey Hey" and "Unga ya Unga ya Niy aye."*

Encourage performers to explore their sounds or sound groups and to experiment with cadence, rhythm, inflection, and inherent variations. Suggest that they expand, elongate, contract, and extend these sounds.

Side coach, suggesting a variety of emotions to guide the exploration.

Example: Happy, angry, devout, humble, crazy, etc.

Encourage performers to explore and find their unique sound expressions.

Instruct performers to repeat sounds until satisfied with the pattern and flow. Once established, invite them to explore the emotions behind the sounds, allowing them to arise naturally. Encourage focus on sounds and emotions, allowing vibrations to resonate throughout their bodies. Spend 10–15 minutes on this to create an emotionally engaging, repeatable vocal experience.

Return to the chant and develop it further if needed, useful as a warm-up or meditation tool.

A connected vocal expression becomes ritual-like, emotionally powerful, and a pathway to deeper self-awareness or group connection, facilitating sensory self-awareness.

Exploration Alternative: Start with a single idea or image from nature to inspire sound creation. Give performers one concept or image to focus on, guiding them to produce sounds that express this. The group then explores related sounds together, culminating in a collective chant.

Chanting effectively stimulates your body's memory on a psychophysical level, producing vibrations and internal resonance that activate your organs, energy centers (chakras), and pathways (meridians). Specific vibration frequencies trigger the glandular system, including the pineal, pituitary, and endocrine glands. This process releases endorphins in your brain and body, creating an ideal environment for holistic psychophysical performance. When performers chant, they should observe how it affects different parts of their bodies. The vibrations and sound communicate beyond words, connecting both performers and the audience. This heightened awareness can revive the potential for deeper communication, reestablishing vital links with our community that may have been overlooked or neglected.

CULTURAL RHYTHM

Establish a familiar rhythm for performers. In cultural settings, create a traditional rhythm (e.g., two beats for Yup'ik Eskimos, three for Russian Slavs, or a mix for Koreans, Nepalese, and the Miao) and convey it live with sticks, drums, or other methods. If no shared rhythm exists, begin with a heartbeat or group rhythm and explore.

Once the beat is set, performers are encouraged to improvise with their shared rhythm. Begin seated, asking them to isolate and explore the rhythm of various body parts—head, face, arms, shoulders, chest, feet, legs, hands, pelvis—and then the entire body while seated.

Afterward, stand and connect with the cultural rhythm through movement, incorporating different isolated body parts.

When ready, invite the performers to move through the space inspired by their cultural rhythm, using these expressive suggestions:

Walk	An animal	Dance with spirits
Journey	A bird	Trance dance
Praise	Grass	Bringing spirits in
Water	Clouds	Calling spirits
Sun	Moon	Flying

Define each expression clearly to build a vocabulary rich in emotional, psychological, and personal connections. Have performers develop a basic dance using only their feet.

Once set, ask them to apply explored elements to create a full-body rhythmic dance with any body parts. Encourage creating a dance reflecting their identity—concise, expressive, rhythmically connected, and easy to repeat—to improve creativity, self-editing, and awareness. Present for comments.

Exploration Alternatives: Using their personal rhythmic dances, invite performers to develop a dance scenario, whether working solo, with partners, or in larger groups. Dance scenarios can be suggested or connected through themes.

Example: A Transformation can involve an adventure, an animal, a spirit, or a mythological figure. It might also reference a well-known story or myth. Including elements from popular culture is acceptable, as pop culture often draws from stories with mythological origins.

Note: In this manner, the original rhythm acts as the foundation for both individual and group explorations, potentially guiding the creation of a rhythmically connected performance. Over time, the fixed beat transforms into an expressive motif or structural component.

RHYTHMIC MOVEMENT

Guide performers to create and explore a simple, steady beat using their bodies, inspired by cultural elements or tailored to the context. Begin with a straightforward rhythm such as a one-two or one, two, three pattern.

Next, encourage them to move individual body parts in harmony with this steady rhythm, highlighting how each part can express itself—including fingers, facial features, feet, toes, and more.

After exploring the body with the beat, allow for spontaneous reactions and expressive movements while maintaining the rhythm. Finally, prompt performers to contrast their expressions within the same rhythm—for example, the forehead showing anger, the right arm expressing love, and the hips conveying passion.

Note: Many cultures focus their music on simple rhythmic beats, which I call primary beats. These beats often form the basis for dance and performance movements. Recognizing and expressing these rhythms with different body parts, even if they do not follow traditional methods, is key to transforming traditional beats into a modern, broader context.

From the Siberian Journal

The performers represent a variety of cultures and engage in social rhythms. Traditional Sakha culture was disrupted by urbanization and suppressed through Stalin's "Russification" during the Soviet era. After the Soviet Union's collapse, modern influences, capitalism, and popular culture became integrated into their cultural identity, creating an overlapping of shared spaces.

The Sakha were shaped and suppressed for political and cultural reasons, with their individual thoughts, imaginations, actions, and spirit being influenced by the rhythmic patterns of other cultures. This can be likened to a racing heartbeat or a Frankenstein heart transplant—the rhythm is not their own, yet still links everything.

Most performers breathe shallowly, primarily using their upper lungs, which symbolizes a literal and metaphorical disconnection from their bodies. Instead of embracing the deep rhythms of the natural world, they are caught in the tempos of automobiles, buses, and social systems—the rhythmic spaces of the modern.

The Sakha seem like walking wounded, suffering from cultural trauma and wandering, not only because of external political, cultural, and social pressures but also because they have lost direct contact with their personal and cultural rhythms.

Restoring a rhythmic way of life is a brief and subtle effort—something each person perceives on a non-material level. This is essential to our work and reimagining their connection with the world.

PHYSICAL LANGUAGE

Performers receive a line from any source—such as a play, book, poem, or newspaper—and examine it closely. They analyze each syllable and sound, paying attention to where they resonate physically, emotionally, and intellectually, as well as the feelings or thoughts they evoke.

Performers develop a walk and body rhythm by exploring the line, coordinating language and movement to provoke each other and build a persona through the rhythm of language.

Allow time to explore and develop.

Performers select a single gesture from their exploration—one movement that best captures their evolving persona. Encourage them to explore this gesture with different emotions, ideas, and images, while varying movement speed.

Offer side coaching with suggestions. Reinstate the line and integrate it with the gesture, exploring how they coordinate. If not already, performers should move with coordinated movement, gesture, and language.

To conclude, ask the performers to fade their rhythmic, gesture, and line into stillness.

Exploration Alternative: The exploration may include a culturally specific phrase or line, such as a prayer, political slogan, sports cheer, greeting, obscenity, oath, or curse. Similarly, the exploration creates a rhythmic movement and gesture based on that phrase or line.

From the St. Petersburg Journal

Their interest in Slavic rituals conflicted with their training, which focused on theatrical techniques, character portrayal, and Western practices. They explained that they turned to theater because there were no practical avenues for exploring ritual expression.

Is theater suited to the objectives of ritual? There is some crossover.

Without a coherent Slavic ritual language or elders, they relied on theater as a substitute for elders and tradition bearers.

They sought to assemble fragments to recreate ritual. Western practices have similarly drawn upon traditional and historical shamanism, witchcraft, healing, and paganism in the effort to create identity and community through ritual.

The Metamorphosis members, like other Indigenous groups I've worked with, had lost part of themselves due to 70 years of Soviet rule that dislocated them from their culture and feelings.

Beginning with what remains, they explored their cultural fragments as a foundation to reimagine self, community, and their world. The self-aware performer acts as an agent of change.

Author's Note: For further details on work with Metamorphosis Theatre, see Thomas P. Riccio. "Finding a Way Back: Reimagining Ritual and Trance in Post-Soviet Russia." *Arts (Basel)* 15, no. 3 (2026): 62. https://doi.org/10.3390/arts15030062.

RHYTHM MEMORY

Phase I: The group forms a circle and establishes a four-beat rhythm, with each performer stomping their feet to feel the beat within their body.

Next, guide the performers to internalize the four beats by silently moving only their hands or heads to the rhythm, as if the sound has been muted.

Then, reintroduce the sound and keep the four-beat steady. Feel free to add variations such as foot stomping, hand clapping, finger snapping, or thigh, head, and stomach slapping. The group should mirror your expression of the four-beat.

Phase II: After establishing the sound off–sound on pattern over four beats, pass the leadership to another performer in the circle.

Each performer then takes their turn to lead the group in a unique expression of the four beats. Start with sound on, and once the pattern is set, switch to sound off.

As before, the entire group mimics the leader's expression of the four beats, both when the sound is on and off. Continue this process until every performer has had a turn to express and lead the group in a four-beat pattern.

Phase III: Next, introduce an evolution of the four-beat expressions by adding different physical or vocal accents to showcase the beat externally. The group follows your lead.

Once established, a performer embraces the four-beat pattern and infuses it with their unique expression. The beat can be conveyed through a deliberate gesture or spontaneously adopted by an inspired performer.

In either case, the other performers follow suit, allowing the four-beat pattern to flow around the circle in succession as each takes their turn.

Phase IV: Present the next phase of the four-beat expressions, utilizing any mix of physical and vocal accents. However, this time, one of the beats remains silent.

For example, the first beat might be a handclap, the second a "ma" sound, the third silent, and the fourth a snort.

The four-beat pattern is mirrored, and leadership moves around the circle.

As the exploration progresses, the performers articulate only two, then one of the four beats. The key point is that the group internalizes, shares, and maintains the four beats in all their expressive variations.

The exploration concludes with the group coming together to silently express the four beats.

Note: In certain cultural settings or when aiming for specific goals, a traditional or alternative beat sequence can be used, such as a variation of 5 to 10 beats. Going beyond this range may become cumbersome and challenging to manage and maintain.

The previous explorations highlight the body as a tool for integrating, transforming, and expressing rhythm. This enables the body to perceive and interact with the various spatial rhythms of the world, including inherited cultural rhythms and personal rhythms. Much like in ritual performances, rhythm reorients both human and non-human communities, as well as the material and non-material realms. Rhythm acts as a synapse, catalyst, and conduit, connecting different realities and communicating with what lies beyond the material.

SHAMAN RHYTHM

This exploration stemmed from collaborating with the Sakha people of Siberia and the Miao of SW China to connect with their traditional rhythms. It demonstrates engagement with and use of these rhythms, providing a foundation for further exploration and customization.

This process helps evolve traditional rhythms through improvisation, adapting them for modern use. Feel free to make this rhythm uniquely yours.

Central to the rhythmic heritage of Sakha and Miao are shamans' drumbeats for evoking the spirit realm.

Using sticks or percussion, establish three distinct rhythms for performers.

1 Dull strokes that mirror a sluggish, ominous movement.
2 Light taps or rumbling sounds.
3 A series of quick beats followed by a single abrupt drumbeat.

After establishing the vocabulary of the shaman's drum and dance, engage in improvisational dance. Experiment with different rhythms, intensities, and sequences. Investigate the emotional and mental responses elicited by variation and repetition.

Note: Bold strokes depict a slow, ominous dance, whereas soft taps or a deep rumble suggests a delicate, bird-like motion. Quick beats ending with a sharp strike indicate high jumps. The dancer and drummer may be the same person or different individuals. The purpose is to help the performer acquire a vocabulary of cultural rhythms drawn from a specific tradition or of their own making.

FIGURE 2.8 *Shaman Rhythm* exploration during a workshop at the Sakha National Theatre. Kuprion Milkhalov, whose grandfather was a shaman murdered by the Soviets, plays a traditional horse hide drum to "ride" it to the spirit world.

Source: Photo by the author.

Sound

CREATING SONG

Circle the group.

Start by choosing a performer to share a song fragment. The group then responds in a call-and-response way, echoing the fragment to grasp its basic rhythm.

At first, the song fragment might just be sounds, gradually adding words over time.

Next, another performer, inspired to join in, adds their own fragment to the original. This new part either complements or changes the initial sounds and rhythm.

Together, the group sings each fragment, blending them into a larger chant. Everyone contributes their part, helping the song grow and find the best combination, order, and rhythmic variation.

With each new fragment, the group repeats and builds on it.

As the activity continues, group discussions and adjustments shape the fragments into a cohesive, repeatable song or chant. Feel free to add or remove elements as needed.

Technological Space

Technology is an active and integrated part of our community, now expressing itself in a distinctive voice that both aligns with and differs from humans. It primarily aims to enhance and accelerate the human body rather than provide comfort.

From cell phones and computers to cars and medicines, technology is evolving to interact more independently with humans. It influences and shapes our environments, helping us become skilled at absorbing, improving, flattening, and controlling the rhythms within various spaces.

These technological rhythms surround us; the atmosphere resonates with hums, clicks, drones, and pulsing beats—both obvious and subtle. We engage in conversations with technology, which responds—it's alive with its own voice. As a result, our bodily rhythms are affected, shaping the rhythmic character of different spaces.

The world exists vibrantly beyond its inhabitants, each influencing the other. Some see this relationship as causal, others as spiritual; perhaps it is both or the same. The world is a space enveloped by an aura around the physical earth, with both elements interconnected and shaping each other. It embodies all life, a cycle seeking balance through rhythm.

VOCAL SEASONS

Circle. The performers lie on their backs, heads facing inward. For comfort, using mats or soft surfaces is recommended. With eyes closed in a dark setting, focus is enhanced.

This exploration relies solely on voices to convey the subtle details and complexities of the seasons. Perceptions and expressions of the seasons differ widely across regions. Interactions between various bioregions and their cultures shape unique seasonal responses, which become an essential part of cultural identity and performance.

Each season evokes specific emotions, memories, visual imagery, and physical sensations. Interpreting and representing the season is crucial to a culture's identity and self-expression, experienced personally but shared communally. A season serves as a collective space.

Begin with Spring, or depending on your location, the season that signals the start of the growing season.

Recommendation: This thoughtful, reflective journey can last anywhere from an hour to longer. Seasonal explorations can be spread across four distinct sessions, or these elements can be condensed into shorter sessions that cover all four seasons. The preferred sequence is Spring, Summer, Autumn, and Winter.

Phase 1: Examine how a particular season affects mental, physical, and emotional states, including when participants notice autumn, weather changes, rainfall, cloud formations, sky colors, daylight shifts, and sensory experiences. Consider activities, responses, clothing, eating habits, sleep patterns, overall mood, animal behaviors, home routines, communication, plant and insect changes, water bodies, societal shifts, relationships, festivals, seasonal foods, memories, spiritual essence, and whether it feels extroverted, introverted, reflective, nostalgic, or melancholic.

What additional features set this season apart and give it a unique character? What physical sensations do you notice during this time? Consider the smells, tastes, textures, sounds, and surfaces. Which rhythms, cycles, and harmonies are unique to this season?

Phase 2: After performers reflect on their mental and emotional experiences, guide them to produce non-verbal sounds that represent seasonal elements. Allow enough time for each exploration, encouraging sounds to surface and resonate within their bodies. Transition smoothly between elements, revealing a vocal depiction of each. Adjust the list as needed.

Geography
Sky
Clouds
Wind
Day
Night
Night into Day
Trees, Plants, Leaves
Animals
Insects
Movement and migrations
Humans and human life
Rhythm(s)
Passing of the Season
Feeling
(add, subtract, or alter order as you see fit)

Phase 3: After reviewing the list, have the group revisit the seasonal element they feel most connected to and express it, exploring its variety, tone, color, and rhythm. Then, prompt each performer to listen to others' vocalizations, creating a tapestry of sounds that reflects the group's seasonal understanding. Encourage performers to focus on one another and aim for a unified expression of the season's rhythms, movements, and flows. Repeat for each seasonal space.

SOUND CIRCLE

Performers form a tight circle—standing, sitting, or lying—with heads toward the center and each other. Eyes are closed, and the space is darkened to enhance focus on sound.

They explore vocalizations—any sounds with emotional meaning that are rhythmic and repetitive but not recognizable language.

While vocalizing, performers must listen to each other, establish their own sound, and interact through call-and-response, harmony, melody, and following the leader. There is no single correct method—focus on harmonizing, encouraging creativity, and letting the group's dynamics guide the natural flow of sound.

Exploration Alternative: It can be repeated with variations and with an alternative objective as a preface. For example: Using certain sounds, working to harmonize, following a specific external beat, ensuring the sound is shared, and making the sound continuously change.

From the Siberian Journal

Observational notes from a summer solstice festival, a demonstration of shamanic rites, and project work.

Vocal imitations are part of the traditional culture:

- Bird sounds
- Horse sounds—both vocal and hand-on-thigh galloping noises
- Cow sounds
- Chicken sounds
- Raven call
- Wolf howls
- Fire sounds created through popping vocals and finger-clicking
- Low, sighing chants
- Excited calls—one woman calling out enthusiastically, possibly in an ecstatic trance
- Bird chorus including geese and cuckoo birds

General observations and ideas:

- Standing with mouth open, as if yawning, to invite spirits
- Using hyperventilation techniques during some singing
- Predominance of "HA" sounds
- Exploring the basic sounds of Sakha singing, rooted in the complex Sakha shamanic trance tradition
- Developing vocal explorations using call-and-response patterns

Shamanic trances are facilitated by repetitive droning sounds. The temir komuz, a mouth harp, varies in size to attune to different frequencies for healing and trance, and is commonly used in Sakha shamanic rituals.

There is potential for performance by developing a temir komuz chorus and exploring its trance potential. Invite a player to educate performers and a shaman to talk, demonstrate, and advise. Women singing accompany men with simple sounds. Deep throat singing, shared among Sakha and Tuva, reflects their intertwined Turkic cultural roots, shamanic practices, and spiritual exchanges.

BURSTING WITH SONG

This practice originates from the traditional Yup'ik and Inupiat custom of collectively creating songs while sitting in a village steam bath. It is also known as "sitting," where individuals listen to their own feelings. The group discusses the purpose and placement of the song before gathering and sitting comfortably in a dark room.

When a performer feels a song from within and can no longer hold it back, they "burst" into singing a line, rhythm, sound, or phrase.

Other performers are free to interact with the song, offering contributions and improvising, helping bring the song to life. If no one else joins in, the song will eventually fade. Some performers might join for a while, then fall silent.

This musical burst can happen several times with different performers. The focus isn't on individual talent but on group exploration and collective creation. Performers practice active listening and maintain respect and openness.

A session lasts about an hour. The songs remembered are those that resonate most with the group, leading to surprising insights and seeds for future performances.

Note: Initially, only vocal sounds are used. Language usually shifts focus from feeling to intellect, which can hinder creativity. As the group develops, language may be introduced; if so, remind performers to stay emotionally connected.

SOUND CONDUCTING

An exploration involving partners or groups of three to five, led by one person acting as the "composer-conductor." This activity can also be done with a single performer leading a larger group of 25 or more.

A "conductor" is chosen, while the other performer(s) form the "orchestra." A collection of sounds is either supplied or selected by the group, often reflecting cultural, community-specific, or contextual significance.

The conductor and orchestra interact solely through vocal and sound vocabulary.

- Develop a sound vocabulary.
- Establish the gesture-sound connection.

- The conductor creates a song.
- The conductor's techniques affect the quality and intensity of the "sound composition."
- Using hand gestures, facial expressions, movements, and other non-verbal cues, the conductor "composes" with the orchestra.

Recommend that the composer-conductor design sections that intertwine and harmonize sounds.

Suggest that the conductor mimic the gestures of a symphony conductor to produce sound, silence, movement, tempo, and intensity. Initially, imitation is the most effective method, which can then develop into the conductor's own creation.

Encourage "orchestra" performers to listen to one another to coordinate and harmonize their sounds.

Highly Recommended.

Each culture, whether Indigenous, traditional, ethnic, modern, or urban, has its own unique sound vocabulary. This section focuses on identifying, exploring, and artistically expressing these sounds. Recognizing and vocalizing them captures the community's collective spirit. Building a shared vocabulary will serve as a foundation for the performance work.

SOUND GUIDANCE

Partner or small-group exploration involves choosing a performer to lead others through space using only vocally produced sounds—tones or pitches—that can be culturally relevant and either repetitive or varied.

No language is used; instead, the goal is to create an emotional soundscape or journey.

Responding performers interact through sounds, gestures, movements, and actions, remaining attentive to the leader to foster communication. The sounds are meant to guide an emotional journey, tell a story, and evoke feelings like discovery or amazement. Repetition, experimentation, singing, chanting, movement, and gestures enhance expression.

The focus shifts to nuances in sound modulation and expressiveness, with responders reacting subtly to auditory and gestural cues, thereby encouraging dialogical action and group connection.

Exploration Variation: In addition to using a culturally recognized sound vocabulary, include cultural movements, gestures, musical expressions, genres, and styles.

Checking In

A project is a specific event within a dynamic ecosystem that includes social, cultural, and physical spaces where performers and participants interact with non-human entities—such as environments, atmospheres, feelings, emotions, thoughts, and energies—that together create a holistic system. The interaction among these elements shapes the event you oversee. Like our ancestors and Indigenous peoples, who used performance to experience and affirm the wholeness of place, a project reaffirms community and connection to a place—an ordered yet adaptable system that needs balance and rhythm. Unfortunately, performance has often been marginalized, controlled, colonized, and commercialized, reflecting a broken anthropocentric worldview.

This awareness, developed through my work with Indigenous groups worldwide and driven by urgent global issues today, inspires both me and this book. It aims to encourage others to see, create, and act differently. What I have received from others is not meant for me to keep but to share freely. All you really have in life is what you can give.

Think of a project as a temporary, adaptable container—fluid and resilient, not tied to any specific ideology, aesthetic, or method. An accumulation of spaces in the process of becoming. Rigid processes or fixed expectations can hinder a project's vitality. This flexibility helps the project stay autonomous and responsive to its environment, circumstances, participants, and goals. This view, developed from my early experimental work, was enriched through collaborations with Indigenous groups and continues to evolve through site-specific performances with my Dallas-based performance group, The Dead White Zombies (DWZ). Indigeneity must take root and be replanted in many forms to face the urgent challenges we encounter as stewards of life on Earth. Humans, the most enabled and greatest beneficiaries of the Earth's bounty, bear the greatest responsibility for its preservation. The time to act is now.

The Dead White Zombies, founded in 2011, is an ongoing, post-disciplinary praxis and research project. The project is inspired by and draws on my nearly 40 years of experience in Indigenous performance, combined with my career-long work in experimental performance and my scholarly work in ritual, shamanism, and performance ethnography. DWZ, at this writing, is preparing for its Spring 2026 performance of *Big*

Bird, a site-specific event installed and performed in a 100-year-old former ice cream factory near downtown Dallas. As this book progresses, a more detailed explication of DWZ's work and methodology will show how the Cosmocentric can be applied in urban and diverse population settings. My ambition is to offer an alternative way of thinking and doing theater and performance, and a means by which to help transform the world.

The Dead White Zombies (DWZ) began as a research and development project exploring alternative expressions grounded in Cosmocentric views of being in the world.

From the DWZ manifesto, "We see ourselves as ironic and fun-loving agent provocateurs in an era marked by intellectual and cultural homogenization, conformity, franchising, and mass marketing. We are Dead because we are White, pale echoes of what once was. We are Zombies because we are still wandering, seeking life."

DWZ views performance as a key tool to confront global crises such as environmental collapse, species extinction, resource depletion, rapid technological change, and the commercialization of nature and reality. These interrelated issues form a complex challenge with no simple solutions.

DWZ uses its tools and resources to emphasize the social and cultural significance of performance as a means to foster change and transformation.

All constraints—personal, cultural, social, political, or economic—are not obstacles but essential information for growth. Every project is alive, reflective, and unique, appearing as a temporary site—an interweaving of bodies and spaces that establishes a sense of place. Our work is a process, a way of living, and a shared way of being.

The process requires you to be both participant and observer, knowing and unknowing. It demands your presence and reflection—remaining engaged with the site while maintaining a level of detachment. The process is temporary, shaped by the situation, and open to adjustments.

Throughout, systematic organization and orchestration are essential for success. You begin by mapping the project's landscape, then serve as a navigator and negotiator to balance the needs of performers and spaces involved.

We focus on engagement as a form of agency and use evolutionary methods, leveraging the power of interactive and inclusive processes. Our approach allows for endless variations of core structures—the entities and spaces a project interacts with—highlighting that the whole is greater than the sum of its parts.

Reminders

- **The method that shapes and guides a project recognizes and balances opposing forces.**
- **Always stay alert and critically re-evaluate established methods and aesthetics.**
- **Stay loose, relaxed, aware, and alive in the fullest sense.**
- **A project has no fixed landscape beyond itself—it's a journey of discovery and of the spirit.**
- **A project will reveal its methodology and clarify its process.**
- **A project and our work are organic beings. They are alive, speaking, feeling, and understanding each in their own way.**
- **Listen. Stay alert and available. Take care of yourself by being healthy, eating well, sleeping well, and working to maintain clarity and focus. Don't forget to have fun. You are doing what you love. Yours is a calling.**
- **It is not about solving practical problems but about fostering vibrant personal, social, cultural, and spiritual engagements and experiences.**
- **Performance achieves this by its ability to consider and engage with the tangible, intangible, and symbolic.**
- **Material and nonmaterial presences are coequal. Consider them flexible, without hierarchy, capable of adjusting as needed to serve the moment's needs.**

Balance

Balance operates fundamentally as a relationship between spaces. When we discussed space as units of information—atomized packets that give form to feelings, carry meaning, and propel life forward—we established that these spaces are in constant movement, continuously evolving, merging, coming into existence, and fading away. Balancing occurs not only within individual spaces but in the dynamic relationships between them, in how they synchronize, oppose, complement, and transform one another.

Consider the performing body: It simultaneously inhabits multiple spaces—the biological space, the cultural space of inherited gestures, the emotional space of present feeling, the social space of audience relationship, the architectural space of the performance venue, the historical space of tradition, and the imaginative space that begets affirmation or transformation of a community. Balance manifests when these spaces achieve the right relationship, when they communicate and coordinate rather than conflict and cancel.

This is why we started our work with the circle. The circle is not just a shape but a way of thinking, a form of aspiration, and a model for balance spaces. When performers arrange themselves in a circle—whether by height, by birthday, by lightest to darkest, by most calm to most anxious—they are practicing the fundamental skill of our work: Recognizing the multiple spaces each person carries and finding ways to organize these spaces into equilibrium. The circle has no hierarchy, no front or back. Energy circulates rather than accumulates. Everyone sees everyone. The circle physically manifests what balance requires: Relationship, visibility, and circulation.

We inherit spaces of balance and imbalance from those who came before. Cultural patterns, behavioral habits, social structures, and performance traditions all carry within them specific ways of organizing relationships between opposing forces. Some of these inherited patterns sustain life and promote flourishing. Others perpetuate harm, rigidity, and stagnation.

Cycles and Balance

The Taoist tradition, as documented by Kristofer Schipper in The Taoist Body, articulates this clearly. The core teaching concerns "hsiu-yang"—to arrange to smooth down any roughness or irregularities by repeating an action many times in harmony with the cosmic order until perfection is achieved" (42). This is not about imposing control but about recognizing and aligning with inherent and established natural patterns. Balance emerges with attunement to larger cycles rather than the imposition of external forces. Each space has its own way—multiple spaces interacting create their own unique way, which in turn aggregates into a greater totality. Each space is specific and balanced in itself but becomes something different and greater when interacting. It becomes a place. Each unique place is a thing, an idea, an energy, and a spiritual vector that influences and radiates outwardly. The balance of the whole depends on the balancing of its parts.

When a Miao xianniang spirit medium or a !Xuu Bushmen healer's journey into the spirit realm, their goal is to identify the imbalance causing illness in a community. The illness could involve a community member or relative, whether deceased or living, who is discontented, has broken a taboo, or has otherwise disrupted the community's order and harmony. It could also stem from an event such as mistreatment of others, harm to the forest or animals, a neglected observance or ritual, all of which can disturb community and family balance. When the Boers and English invaded and colonized what is now South Africa in the 19th century, the Zulu saw it as not just a physical event but also a widespread cultural sickness, *indiki*. An epidemic-like illness affecting their body, mind, and spirit, disturbing their cultural and social harmony, and causing symptoms such as depression, anxiety, paranoia, and panic. For the Zulu, it was spirit sickness, possession, and an imbalance caused by white colonizers.

Western cultural frameworks for understanding reality are rooted in the material, from which all else flows. Physical and material spaces take priority over the immaterial. In contrast, the Cosmocentric worldview, vividly exemplified by the Taoist concepts of yin-yang spatial relations and wisdom, understands reality as equally material and immaterial, in constant interaction and cyclical change. Yin-yang is a philosophical concept and a practical tool for understanding how opposing forces interact to maintain balance. "The shady side and the sunny side of the mountain" (35)—this simple image reveals that opposition is not conflict but complementarity. Cold and hot, moon and sun, soft and hard, feminine and masculine, death and life—these pairs represent different phases of a single cyclical process. "When yin reaches its apex, it changes into yang and vice versa" (35).

For performers, this means developing sensitivity to when to push forward and when to yield, when to concentrate energy and when to disperse it, when to speak and when to inhabit silence. The skilled performer does not force balance; they recognize where they stand in natural cycles and act accordingly. Balance requires participation in cycles rather than an attempt to transcend them.

This directly challenges the extractive logic dominating contemporary culture—the imperative to accumulate, to possess, to remove things from circulation for private benefit disrupts balance on a small and large scale. In performance terms, it means knowledge, techniques, and creative insights must circulate to remain vital. Hoarding creates stagnation. Balance requires a generous exchange.

Rhythm and Balance

Rhythm is the heartbeat of balance—the organizing pattern that coordinates multiple spaces into a coherent relationship. We explore rhythm extensively in our work because rhythm reveals and creates balance simultaneously. The two-beat pattern of Yup'ik drumming, the three-beat of Russian folk forms, and the four-beat we established in explorations like **Rhythm Sequence**—each rhythm creates a specific pattern of expectation and fulfillment, tension and release, that guides participants into harmonious relationships.

The body itself operates rhythmically. Breath cycles in and out. The heart beats—neural patterns fire and rest. Hormonal systems rise and fall. Sleep and waking alternate. These biological rhythms interface with environmental rhythms—day and night, seasonal changes, tidal movements, and lunar cycles. Cultural rhythms—work and rest, celebration and ordinary time, gathering and dispersing—overlay and interplay with these natural patterns.

Balance means recognizing these multiple rhythmic spaces and finding ways to accept and coordinate rather than contradict them. The Sakha performers I worked with in Siberia had been shaped by Soviet "Russification"—the imposition of foreign rhythms on their Indigenous patterns. They breathed shallowly, using only the upper lungs, literally and metaphorically disconnected from the deep rhythms of their land and bodies. Restoration of a balanced relationship required reconnecting to their own rhythmic inheritance while acknowledging the other rhythms now present in their lives.

James Bridle's *Ways of Being* offers a relevant perspective: "What matters resides in relationships rather than things, between us rather than within us" (13). Balance exists in the quality of connections between elements—between performers, between performers and space, between performance and audience, between the immediate event and its cultural context. These connections require rhythmic coordination to function.

The Body's Inner Landscape

The Taoist understanding of the body's "inner landscape" provides a practical methodology for achieving internal balance that can then radiate outward. "'The human body is the image of a country,' says the Taoist" (Schipper 100)—mountains, rivers, palaces, and inhabitants all mapped within. The head becomes a mountain with the Hall of Light as the central palace. The heart functions as the Scarlet Palace. The spleen operates as Yellow Court, "the body's ritual area and the meeting place of its inhabitants." The navel opens to "the Cinnabar Field, the source of all life" (100).

This elaborate geography serves as a navigational tool for balancing the body's energies. The practitioner learns to direct breath and attention to different internal regions, understanding how an imbalance in one area affects the entire system. For performers, this develops sophisticated proprioceptive awareness—the ability to sense and adjust movement and internal states in real-time during performance.

The principle extends beyond individual bodies. The performance space itself acts as a collective entity that needs balance. Just as the Taoist body has a ritual area where all inhabitants gather, performance creates a similar space where all elements come together—performers, audience, environment, history, imagination, and the unseen forces performance calls upon. The performer must first establish internal balance before this balance can radiate outward to serve the collective space.

Breath is the primary vehicle for circulating balance. "Exhale the impure chi and inhale the pure ones. This is called 'exhaling the old and inhaling the new'" (Schipper 1993, 138). The practitioner visualizes "the Original Ch'i of the Great Harmony" inhabiting and descending through the body "like the mountain that absorbs the clouds, like the earth that drinks up the rain" (138). Each inhalation draws in possibility, each exhalation releases completion. The pause between breaths—when inhalation has ended but exhalation has not begun—marks the balance point at which transformation occurs.

Process

A crucial teaching concerns the danger of premature intervention. "To intervene means in effect to violate nature by going against the spontaneous process with actions inspired and influenced by our partial view of things. Imbalance results from this intervention" (Schipper 1993, 103). True balance emerges not from control but from non-interference with natural processes.

This does not mean passivity. It means developing sensitivity to distinguish between actions that support natural unfolding and actions that disrupt it. "It is pointless to try to act on the world outside oneself so long as one's inner order is not established; otherwise, whatever is achieved will not endure" (103).

For performers, this suggests that technical mastery alone—precise execution, perfect timing—remains insufficient without inner cultivation that allows external actions to emerge from balanced internal states. A performance may appear technically accomplished yet feel hollow without this deeper work. Conversely, a performance with technical imperfections may resonate powerfully if it emerges from genuine internal balance.

"The only valid law is that of nature [...] To rediscover spontaneity, a human must first come to know the laws and secrets of the universe, beginning with those of their inner universe" (103). This points toward the Cosmocentric way: Using rigorous observation and contemporary understanding to rediscover, through practice, principles that were foundational to traditional cultures.

Reciprocity and Exchange

Yunkaporta identifies the imbalance's causation: "I am greater than you. You are less than me. This is the source of all human misery" (26). This narcissism disrupts reciprocal relationships sustaining communities and ecosystems, destroying "the basic social contract, that reciprocity, which allows people to build a reputation of generosity based on sharing to ensure ongoing connectedness and support" (28). Balance requires the right relationship with all beings. This is not moral philosophy but practical reality—we exist within interconnected systems where actions ripple outward and return.

Yunkaporta continues, "There is no way to be an outside observer in this system. You have to place yourself in it to see it in three dimensions, and you must move around and connect within it to see multiple other dimensions. From the outside is just a flat image" (40). We cannot comprehend performance balance from the outside. Balance emerges from participatory engagement, from placing ourselves within the system.

A stone provides a metaphor for balance across extended time. It "represents the Earth, tools, and spirit...simultaneously permanent and changing, resilient yet responsive to erosion" (32). Practices maintain integrity across generations while adapting to circumstances. "A stone teaches us that we should be strong no matter what tries to crack us or wear us down. Keeping an unbreakable core through your culture and your beliefs" (34). This suggests identifying and protecting core principles while remaining flexible about surface forms.

Place as Balanced System

Place emerges when multiple spaces achieve organization through intentional relationships coordinated by rhythm, narrative, and communal agreement. This process requires balance at every level.

First, the individual performer must achieve internal balance—coordinating the multiple spaces they carry (biological, emotional, cultural, historical, imaginative) into coherent presence. The explorations we have covered—breathing work, visualization, rhythm explorations, movement vocabularies—all serve this fundamental goal.

Second, the ensemble must achieve collective balance—finding the right relationships between different performers' spaces, performance traditions, and cultural inheritances. The **Circle Definitions** exploration, where we arranged ourselves by various criteria, revealed how multiple spaces coexist within any group. Balance means acknowledging these differences while finding ways to coordinate them.

Third, the performance must balance with its environment—the architecture, the audience, the community context, the historical moment, and the unseen presences. Performance doesn't occur in a vacuum but within nested systems, each of which requires respectful engagement.

Fourth, the work must balance inherited tradition with contemporary necessity. Neither pure preservation nor complete abandonment serves. The task is recognizing underlying patterns that make balance possible across changing contexts, distinguishing eternal principles from historically contingent forms.

Ritual provides models for achieving balance through performative action. The ritual area becomes "a mapping of 'Heaven,' a Mandala" (Schipper 1993, 76), where "all the time cycles and spatial categories of the space-time continuum are integrated" (76). Ritual creates a perfected microcosm in which cosmic forces align in proper relationship.

The process involves installation (preparation of the space), ritual actions (the performance itself), and dispersal. Energy must return to circulation. Performance must end, space released, concentration dispersed back into the larger flow. Attempts to hold onto the magic violate cyclical balance.

Contemporary Challenges

Contemporary performers face unprecedented challenges. Mediated lifestyles and urban environments disconnect us from seasonal rhythms. Global networks cause a constant clash and blending of place-based aesthetic traditions. Technologies mediate and alter performances in unpredictable ways, driven by consumer and capitalist demands. Audiences split along competing ideologies and value systems.

Yet, fundamental principles remain: Putting the world in the body is to follow the world and not to go against nature (Schipper 1993, 154). What matters are relationships, not things. What matters is what lives between us and not within us.

The task is recognizing underlying patterns, making balance possible across changing contexts. This requires what Schipper calls "stealing the secret from the Tao" (42) through patient observation; recognizing "knowledge is stored in every waterway, in every rock" (Yunkaporta 21); developing "an ecology of technology" (Bridle 2022, 11)—approaching tools as participants in larger webs of relationship.

Balance as Practice

The explorations in this book cultivate balanced awareness through direct experience. **Balancing With Others** develops physical sensitivity to how bodies distribute weight and energy in relationships. **Balanced Space** trains visual awareness of how presence affects equilibrium. **Center of the Circle** improves reflexive adjustment to changing configurations.

The **Column of Light** coordinates group movement without designated leaders.

These are not just hermetic explorations. They are technologies for developing the fundamental capacity our work requires: The ability to recognize and adjust relationships across multiple spaces in real time in service of one's larger community of place. This capacity operates at every scale—within individual bodies, between group members, and in relationship to the environment and audience—engaging tradition and innovation.

Balance is not a destination but a manner of traveling—the quality we bring to movement through space and time, the attitude toward forces that shape us, the way we participate in cycles larger than ourselves. "We don't have a word for nonlinear in our language because nobody would consider traveling, thinking, or talking in a straight line in the first place. The winding path is just how a path is" (Yunkaporta 2020, 18).

When we achieve balance, even temporarily, we become spaces through which forces flow, sites where transformation becomes possible, and sanctuaries where all beings who come into contact might find renewal. This is the performer's function—not to display individual virtuosity but to create and maintain conditions where multiple spaces can achieve temporary, living equilibrium. This equilibrium we call Place.

BALANCING WITH OTHERS

Performers move through the space, observing others and the environment—whether indoors or outdoors—its size and details. If outside, agree on space boundaries.

After familiarizing themselves, performers should gently touch others, such as on shoulders, backs, or arms, then move on.

Then, each performer pauses, approaches the nearest person, and leans on them, forming pairs. Maintain enough tension to stay upright.

Once balance is established, repeat and move to someone else.

Explore different points of body contact, ways of leaning, and striking a balance.

Next, performers form groups of three or four others to lean on, gradually involving more until the entire group can balance together.

Promote awareness of body dynamics, balancing naturally with people of different heights and weights, adjusting intuitively and effortlessly.

Exploration Variation: Adjust the movement speed in the space, providing options for slow motion, half speed, or fast motion.

BALANCED SPACE

The performers gather around the edge of an area. Identify the center with a mark or an object to designate it, and then visualize the space as a large disk. This marked center acts as the axis and balance point for the entire space.

Initially, the space is empty and "balanced."

Identify two groups—Group One and Group Two. One aims to destabilize the space, while the other works to restore its balance. Verbal cues are unnecessary.

The "unbalance" group disrupts the space and, after a five-count, freezes in their unbalanced stance.

The "balance" group then uses five counts to observe how the space was disturbed and works to restore its equilibrium.

Repeat the unbalancing and balancing actions. You can shorten the five counts. The goal is to develop an awareness of space and the visual weight of others, influenced by clothing color, weight, height, movement, and resting position.

Encourage performers to develop an instinctive and reflexive awareness of spatial balance while working with their group.

Once the exploration is established, have the groups switch roles: The unbalanced group focuses on balancing, and the balancing group on destabilizing.

Exploration Variation: Adjust individual group members to change group compositions. As the exploration continues, eliminate the five-count cue and enable both unbalanced and balanced groups to transition smoothly into ongoing actions and reactions.

Encourage the groups to stay in constant motion—shifting between balancing and unbalancing as they move. The unbalanced group shifts, while the balanced group adjusts to the space as the other group becomes unbalanced.

Slow down or pause movements if the exploration becomes too difficult to follow.

Encourage exploration to foster a dance-like, choreographic feel.

If suitable, produce a sound as they move and find a stance. Once exploration begins, introduce music or rhythmic cues to help synchronize their movements.

Exploration Variation: Include objects like rocks, grass, branches, or animals such as dogs, chickens, cats, mice, or lizards to balance and become aware of the rhythm and energy of non-human beings. This creates a unique experience with unexpected, fun, and humorous outcomes.

Highly Recommended.

CENTER OF THE CIRCLE

A performer moves to a new spot in the space and stands. The others quickly reposition themselves to form a circle around the performer, who acts as the focus and center.

At your prompting or once the new balanced circle is formed, another performer moves to a different position.

Repeat this sequence until each performer has had a turn to initiate a move with a circle forming around them.

Aim to improve speed, agility, and intuitive response.

Exploration Variation: Like above, but instead of a circle, use shapes like squares, triangles, curves, designs, or other patterns that reflect the cultural context or current goals.

Exploration Variation: The exploration can be used with character, gesture, or movement work.

A performer steps out of the shape (circle, square, triangle, etc.) and performs a specific movement or gesture (personal or cultural), and others follow with the same movement-gesture, forming a balanced tableau.

Once a group tableau is formed, another performer steps out, and the activity repeats with each performer enacting their unique movement-gesture.

Note: An effective way to promote an instinctive bodily response, strengthen community identity, boost cohesion, and improve the sharing of expression.

COLUMN OF LIGHT

Performers position themselves at the edge of a space, facing inward. They then identify a central point, imagining it as a column of light.

Moving as slowly as physically possible, performers approach the imaginary center.

Performers move to the center and form a circle around the imaginary column of light.

As performers arrive at the center, they organically vocalize a pulse—the "rhythm" of the column.

The pulse will evolve and change as others gather around the column of light. The pulse becomes an individual, then a collective, response to the light's pulse.

Advise performers to allow the column to respond to the group's presence.

Ask: Does the pillar reinforce, change colors, modify temperature, waver, or fade?

The performers, now circling the light, contact each other and slowly develop a movement—without a designated leader or follower—inspired by a collective understanding of the column of light.

The group can adjust its speed and movement in response.

Exploration Variation: Conduct the exploration entirely using sound or rhythm. Alternatively, once the group gathers at the column of light, introduce a rhythm.

Exploration Variation: If appropriate, ask the group to develop a repeatable movement pattern or ritual around the column. If appropriate, encourage vocalization to enhance movement and/or the "column" ritual.

Note: This exploration greatly boosts concentration, movement, teamwork, and creativity—essential elements for performance and ritual awareness.

From the Ethiopian Journal

Sample Work Session

Four-hour session, 1–5 PM, beginning with a group warmup.

Objective: Preface to establishing a performance vocabulary.

Circle Warm-Up to get the group tuned into the work. Start with what we know, then move into a quick **Movement in Space,** followed by **Balancing the Space.** Repeat this several times to develop peripheral awareness of others, body weight, and distance. Since the group has worked together long enough, it becomes easier to act and react intuitively, leading to **Mass Movement,** which builds on collective body movement.

Returned to **Movement in Space,** returning to neutral, regaining breath, one's own personal rhythm.

As they moved, I explained **Push Pull Spin**, initially reducing movement to half-speed to prevent injury. I then went into the exploration: Starting with Push, then Pull once familiar, and finally Spin after mastery. The process went smoothly, so we added Fall as an extra element. I coached them to stay at half-speed, since there is a tendency to speed up, giving them time for exploration to feel natural and intuitive, with a focus on action and reaction.

The work went very smoothly. It established the idea of a functional performance vocabulary that closely resembled dance. Scenario themes were included to guide the exploration, such as romance, war, ballet, village life, traditional life, city life, and downtown Addis Ababa.

Finished discussing how a basic performance vocabulary can be effective for performance expression, endlessly recombining to serve themes and storylines.

Took a Break.

Formed groups of three and four for **Being Watched**, with each performer taking their turn. After about 15 minutes, they used up the obvious tricks and their usual performance methods. In the first round, Daniel stayed focused, while Michael acted out scenes and familiar scenarios. Enday, however, ran out of ideas and started doing animal impressions. Nuredin, on the other hand, was very physical and stylish, ending his exploration simply and fully embracing the moment.

We returned to the idea of developing an Ethiopian Performance Vocabulary for our project, focusing on analyzing and exploring existing cultural languages of gesture, movement, and sound, as well as individual languages and their interpretations of inherited traditions.

Observation

In these upcoming explorations, the focus moves from physical and spatial studies to internal, personal, and emotional domains. Restoring harmony within the human-animal-spiritual-natural system involves the performer adopting and embodying the role of a participant-observer. The performer's awareness and presence are essential for finding and maintaining balance. Begin by evaluating your current state, the people around you, and everything that surrounds and makes up your environment. Everything you need to know is already present. Simply observe.

INTERVIEW

Performers are asked to find partners.

Within 5 to 10 minutes, they interview each other to gather as much information as possible. The focus is on key details such as birth date, hometown, education, siblings, favorite color, preferences, dislikes, and opinions about the workshop they are attending.

Each performer observes, records, and practices, as much as possible, their partner's walking style, speech, and unique habits, gestures, and quirks.

Then, each introduces and performs as their partner, embodying their partner's distinct personality and style, regardless of gender.

Based on the brief interview, it is likely that the partner's presentation may deviate from the facts. The speaker might forget details, misinterpret information, or confuse facts. If this occurs, they fill in gaps by extrapolating, inventing, and making assumptions. Encourage the partners to think creatively and to have fun.

Exploration Variation: Perform the exploration with one partner behind the other, slipping their hands and arms under the front person's armpits. In this setup, the partner behind acts as the expressive arms of the person in front. They can operate puppet-like or serve as the other's alter ego or coach if the front partner forgets or makes a mistake. Partners can also create a comic routine with back-and-forth dialogue.

Highly Recommended.

FIGURE 2.9 *Interview*, workshop with the Mandala Theatre, Kathmandu, Nepal. Performers Vikas Joshi interviews Nasir Hussain.

Source: Photo by the author.

OBSERVATION

Partners. The observing partner sits and observes every detail of their partner as they perform a series of activities for 25 minutes. Reading, walking, talking, exercising, their appearance, including clothing, hair, energy, and gestures such as movement, speech, posture, lying down, dancing, singing, and a range of emotional and facial expressions.

The performing partner offers and enacts as much information as possible.

The performing partner is prompted to act out a scene from their life while the observing partner takes observational notes and asks questions.

The observing performer then leaves, and the performing partner changes a set number of details—ten, for example.

The changes are presented in the scene from their life. The changes can be big or small. They might include how the belt is fastened, how the shirt is buttoned, their walk, expression, habitual gestures, hairstyle, the speed or rhythm of their speech, or the rhythm of their movement, as well as elements of the story that are added or left out.

Then, the observing partner returns and tries to identify the changes.

The exploration repeats, with partners swapping roles and following the same process.

Exploration Variation: This exploration is repeated with modifications to the room or space. A partner or group is asked to observe an indoor or outdoor environment carefully and in detail. They leave while the space or environment changes, then re-enter to spot the differences.

A combination of both variations can be used.

Training observational awareness is essential because it demands intentional effort within our busy and overstimulating environment. Observation is the foundation of understanding, and individuals must practice engaging their senses—sight, hearing, touch, and smell—to interpret the world, find their place within it, and make sense of the overlapping spaces that clutter modern life. Effective observation involves both active participation and passive openness, requiring patience, presence, and a non-judgmental attitude. It enriches our experience, encourages reflection, and improves quality of life. As Evelyn Alexander, an Athabaskan elder, dear friend, and collaborator from Minto, Alaska, often reminded me, "When you need to know, just go and listen. You'll know when you need to know. You'll see what you need to see. Just keep your eyes open." (Alexander 1997)

DRAWING FACES

Provide large sheets of paper and markers, crayons, or pens.

Performers face each other, with one partner serving as the model.

The model performer closes their eyes while their partner sketches their face, maintaining eye contact without looking at the paper.

Instruct the artist to focus on capturing the face's details, contours, and expression, starting with the most prominent features.

Set a time limit of about five minutes for the drawing.

The process is repeated with the other partner. At the end, gather all the drawings for a group show.

Ask the group to observe, interpret, and comment on the artworks, which will be overwhelmingly abstract and Picasso-like, revealing surprising and unexpected results.

This exploration emphasizes working with abstraction, impression, and imperfection. Observing the world through their bodies.

Exploration Continuation: Ask performers to select a drawing that speaks to them—it does not have to be theirs. Let the drawing serve as a guide and reference, and develop a character.

The performer draws from the drawing as inspiration for enactment—treating it as if it were a map and a text. This helps dispel the idea that a "script" is solely word-based, highlighting instead that anything imaginative, visual, auditory, or somatic can be a source of creative inspiration and departure.

Encourage performers to focus on the drawing's key features and embody its traits, emotions, and expressions. Assist them in recognizing the possibilities present in curves, lines, shading, and scribbles.

Show how a face and body can be simplified into a mask, forming the basis for creating a physical character—where movement can shape and inspire a psychophysical view. Drawings can serve as choreographic maps, helping to transform abstract ideas into tangible forms. Develop a choreography or character expression.

This exploration is an introductory step towards more in-depth work with mask characters, specifically **Archetype** and **Character Mask** explorations referenced later.

Exploration Variation: Performers, using the same drawing setup and not looking at their work, now depict a non-human subject such as an animal, landscape, flora, or another relevant element. Then, follow the same process to bring it to life and perform. This variation expands the performer's expressive range and subjects, blending realistic and abstract portrayals.

From the India Journal: Pondicherry University

After today's introduction, we started with breathing explorations, then moved into the **Mountain Heartbeat**, which led into **Leading Walks**, and finished with exaggerated, cartoon-like walks. Students often hesitate to express their individuality, mainly because of a strong conformist culture rooted in Hindu and Tamil traditions. This culture is shaped by family and community influences and is impacted by shame, social and economic status, and personal goals. Embracing individuality goes against these norms.

The students tend to be conservative and self-censoring. Unwilling to challenge norms or to creatively step out. We discussed my observations. They possess a rich, highly theatrical Tamil performance tradition, which they see as a duty rather than personal expression. Acknowledging their self-censorship highlights how they were raised in a highly structured, religious, educational, and socially restrictive environment.

We took a break.

Afterward, our focus shifted to **Whole-Body Communication.** They utilized their understanding of traditional performance, highlighting expressive elements such as big eyes, surprised looks, winks, eyebrows, and inhaling, which smoothly transitioned into **Body Part Drama.** Our conversation clarified the distinction between the codified, prescribed traditional expressions and their personal emotional expressions—drawing from tradition while actively evolving it. Many students combined traditional *Terrukuttu* performance with pop culture elements that appeal to them.

Following a short discussion, we moved into the **Mirror Sequence**. We began with performers limiting their movements to well-known traditional dance and performance gestures. Some performers have backgrounds in traditional dance, and I encouraged them to add personal expressions and influences from pop culture, pointing out both their differences and similarities. Traditional movements are essentially codified methods of perceiving and engaging with the world—a world that is conscious and interacts with entities beyond just humans.

FIGURE 2.10 Pritham Mangai and Mirnalini Krishnan are exploring Body Part Drama in a workshop at Pondicherry University, Pondicherry, India.

Source: Photo by the author.

HUGS

A group of equal numbers facing each other from 10 feet (3 to 4 meters). Partners are identified.

Talking is not necessary.

With your signal, partners walk across the space and hug each other. Their hugs are influenced by an emotional space they inhabit and share.

To start, they hug:

Awkwardly

After each hug, partners return to their line, facing each other, and close their eyes, holding onto the emotional state they experienced.

With your prompting, a new emotional state is suggested. The partners move across the space to hug again,

As if a long-lost friend has been found again.

Then, turn away, create some distance, and respond to each of the suggested prompts (adding or subtracting as needed):

- Like a relative
- As if you are comforting them
- As if you are congratulating them
- As if out of obligation
- As if you are in a dream
- As if you will never see them again
- As if they are preparing for death
- As if you love them deeply

Note: This exploration observes how intention influences, deepens, and transforms emotion. And how emotional spaces are carried within us and how they can come together, change with suggestion, and evoke a history of feeling.

Exploration Variation: One partner approaches and they hug, expressing a particular emotion, while the other shows the opposite or a different one emotion. The emotional spaces are contrasting, mismatched, and unbalanced. Have performers observe how their emotional states interact with their partner's, both when they are aligned and when they are mismatched. Reflect on emotional balance and imbalance, and the thoughts and feelings expressed.

FIGURE 2.11 *Hugs* during a workshop with the DirectorsLab Chicago

Source: Photo by the author.

EYE-TO-EYE CONTACT

This exploration is about sensing and observing the space between people.

Performers face each other from a distance of approximately 15 to 20 paces. They decide who will remain stationary and who will be active.

The stationary partner will remain observant and maintain emotional and physical neutrality throughout.

Throughout, they keep a steady, soft gaze eye contact, without speaking or making sounds.

The active partner gradually moves closer to the stationary partner, taking three slow steps forward, then pauses to observe how their spatial relationship shifts.

Take another three steps, stop, and again focus on the physical shift. Proceed until the active partner moves to the distance of social acceptance. Then to the distance of intimacy. Back away and observe how the distance between two bodies shifts.

All the while, the stationary partner observes silently, observing internally the shifts of distance and body proximity and how it changes them emotionally. When do they feel comfortable, uncomfortable, threatened, intimidated, intimate, etc.?

Ask the active partner to move freely and as they will, adjusting and exploring the distance between them and the stationary partner. Explore variation of speed—zooming out then quickly in. Move around the stationary partner and observe how it makes them feel. Move and try levels, moving on the floor, skipping as they approach.

Add or remove prompts as needed. Allow time for the exploration to develop.

Ask the active partner to observe,

- The changes in physical distance and how it shifts.
- The point of intimacy?
- The movement of threat?
- Identify major shifts in the physical relationship.
- Move back and forth between these moments.
- Create as much distance as possible between you and your partner.
- What kind of physical relationship has evolved?
- Approach your partner quickly.
- Observe how the physical space changes with different distances and speeds.
- Vary your speed by moving slowly, then quickly, back and forth.
- Get as close as possible to your partner.

- What does the physical relationship suggest?
- How has the space influenced the relationship?
- Notice how each change in distance affects emotional responses.
- Explore the connection between physical and emotional spaces.
- Observe how speed impacts internal and emotional changes.

Partners switch positions. The mobile partner becomes stationary, and vice versa.

Discussion follows.

Note: Encourage performers to become aware of and observe spatial relationships between themselves and others—humans, non-human entities, objects, architecture, and the environment. Integrate this observational awareness into your daily life.

Exploration Alternative: Explore spatial relationships with objects, such as a chair, car, table, group of people, piece of architecture, tree, cloud, or animal, either in a subsequent session or as a follow-up to the above.

Highly Recommended.

THE PERSON OPPOSITE

Performers stand at arm's length facing each other, maintaining eye contact and relaxing with a soft gaze.

Start by instructing them to carefully observe each other's faces, paying attention to the fine details.

They should not respond vocally to the questions posed.

Then, ask the following questions, giving performers time to reflect and answer inwardly:

- What do you value most about the person opposite you?
- What do you think is their favorite thing about you?
- Which part of the person next to you would you like to touch?
- Where do you think they would want to touch you?
- Where would you like to go with the other person?
- What activities do you imagine doing there?
- What do you think they might want to say?
- What advice do you think they might give?
- What kind of comfort would you like to offer?
- What would you want to protect them from?

Throughout the exploration, continuously question:

- Why did you smile?
- Why did you fidget?
- Why did you turn your eyes away?

GUIDANCE

An elaboration on the **Finger Follow**.

Partners face each other to decide who will be the "leader" and who will be the "follower." The follower's main task is to mimic the lead partner's right hand, moving exactly as it does.

For example, when the leader extends their hand wide, the follower responds by enlarging their body. A wiggling finger and clenched fist might be interpreted by the follower as a head wiggle followed by a clenched body position. Through trial, error, and exploration, they develop their hand-to-body communication. There is no right or wrong way. Discovered through experimentation.

The leader's goal is to guide steadily, maintaining the connection, without relying on verbal communication.

Part I: The leading partner uses their right hand as a joystick to steer the follower through space. They help the follower by maneuvering over and around obstacles, guiding them to sit, controlling dance steps, and adding actions such as hopping, bouncing, wiggling, leaning, and falling. Once the main hand guidance is set, the leader should also use finger movements for more precise and intricate actions.

Part II: At this stage, the follower has more freedom of movement, responding more abstractly to the hand instead of strictly following its lead.

Part III: If suitable, encourage performers to include culturally specific hand gestures. Alternatively, they can be suggested to incorporate these gestures rhythmically, regardless of cultural origin. Invite the partner to actively respond with their whole body while interpreting the gestures and/or rhythm.

Remind the guiding partner that if they misuse their power, their partner may seek revenge when roles are switched.

Switch Partners: The leader becomes the follower, the follower becomes the leader, and repeat the process of exploration.

Exploration Variation: The leader employs both hands—one to steer a part of the body and the other for a different part. Alternatively, one hand can act as the emotional guide, with the other serving as the physical guide.

Exploration Variation: A leader guides two performers, one with each hand. As this variation progresses, instruct the leader to create a scenario by guiding the following performers.

Exploration Alternative: Encourage the primary partner to use different body parts besides the hand, such as the left eyebrow, mouth, or right shoulder, to guide the partner by combining movements involving the forehead, neck, and chest. The follower's goal is to interpret, imitate, and foster focus, trust, and playful communication, developing shared communication spaces through body-to-body dialogues and interactions that establish reciprocity and balance.

Mimesis

The mirror explorations that follow will be familiar but will soon develop into a series that goes beyond simple observation and imitation into sympathetic magic, a term coined by James George Frazer in *The Golden Bough* (1889). Our series of explorations will focus on how like produces like, or how an effect resembles its cause, and on how a person or thing that has previously been in contact with another can continue to influence it from a distance after the contact has ended. In this way, when the performer connects with the real to carry and transform its essence into the world, it constitutes a form of sympathetic magic. To perform is to enter and participate in a space of magic, conflict, and contradiction. The performer draws on and imitates reality—an inauthentic act—in pursuit of authenticity.

Like a magician or shaman, the performer can produce an effect they desire by imitating it. The interaction with another person or object will affect the person with whom they were once in contact equally, illustrating that spaces are mutable and porous entities—transfer portals, in a way—through which life and we move.

French philosopher René Girard argues that human behavior is rooted in mimesis, and that imitation can generate conflict—a productive competition between the real and the mimetically imagined (Girard 1987, 7, 26). A space in which things are worked out and, in turn, advance the human project.

The performer and the space they inhabit, where the clash between the real and the mimetic occurs, functions as a means by which the world grows and learns. In earlier times, this would have been seen as a sacred and magical act, with the performer and the performance space becoming what Grotowski and many Indigenous cultures regarded as a sacrificial act occurring in a sacred site. A further granular explication of this phenomenon deserves more, but for now, be conscious and aware of the seemingly banal and benign nature of our work. Enjoy and observe banalities; they reveal volumes.

The series begins with observing an individual and their interactions, then gradually proceeds into the abstract, emotional, and metaphysical.

MIRROR SEQUENCE

Phase I: Group members partner up, preferably with someone they know least or haven't worked with much. Partners introduce themselves and look into each other's eyes, stating their intent to observe. They should feel comfortable making eye contact and establishing physical connection. They decide who will lead first and who will imitate and reflect.

Phase II Begin: The lead partner moves slowly to maintain visual contact with the partner. Highlight that the goal is not to lose their partners but to achieve synchronization and flow.

Request the partners to move gradually, and as they do, observe the contrast between their anatomies and body movements.

After several minutes of exploration, they switch roles and repeat.

Phase III Rhythm Awareness: Switch partners and return to the initial leader-follower roles.

The lead partner is asked to restart the movement with greater ambition. The movement should not be overly complicated or too fast, as this could risk losing their mirroring partner. If the mirroring partner is not performing their movement accurately, the lead should adjust by simplifying or slowing down.

Again, the goal is not to lose your partner or to show off how tricky or elaborate a movement can be. Instead, the aim is to explore and gradually improve partners' ability to adapt to and follow each other's movements and rhythms.

The leader alters the movement's rhythm, tempo, and cadence. Understanding that rhythm and movement are variables to be managed is essential.

Continuing seamlessly, partners exchange roles, with the initiator becoming the mirror and vice versa.

Repeat this role-switching several times while maintaining a natural, rhythmic flow between the partners.

Phase IV Movement in Space: Once it is clear that the partners are coordinating their movements, ask them to move in the space. They should start slowly to stay in sync and maintain a rhythm together.

Partners switch and continue their interaction smoothly from where they left off.

Repeat if needed to develop mirrored rhythmic movement between partners moving in the space.

Encourage the lead to "Take their partner on a trip," or "Show their partner something," or "Look for something in the space/room," or perhaps even "Dance with your partner."

Remember, partners should always keep the main goal in mind: Not to lose their partner by going too fast or making their actions or rhythms too complicated.

Phase V Sound Mirrors: Encourage the partners to incorporate vocalizations into their movements. These vocalizations don't need to be elaborate, but they should naturally develop and remain consistent with their movement and rhythm.

Avoid using words, as they can shift the performer from an instinctive to a more intellectual interaction.

The lead partner should not force the sound or vocalization but allow it to happen naturally.

The mirroring partner is now challenged to adjust their voice and sound production to match their partner's; just like with movement and rhythmic mirroring, they should be patient and do their best to adapt.

Hand clapping, foot stomping, body slapping, and other non-verbal expressions are also encouraged.

Phase VI Free Exchange: At this stage, partners should smoothly adapt to each other's actions and rhythmic changes. Ideally, they will also build confidence, curiosity, and creativity.

Partners switch roles quickly and repeatedly. Repeat this process several times, encouraging performers to pick up where their partners left off to ensure a seamless flow.

Maintain vocal and sound expressions throughout. Once this is established, allow partners to interact freely on their own.

At this point, the concept of who is the leader and who is the mirror becomes irrelevant. The partners engage in a free, dynamic exchange of movement—both leading and following—creating an expressive duet with its own creative energy.

Offer guidance when needed but avoid instructing the partners on how or why to move. Their goal is to become two parts of the same whole. If the partners meet another group, they interact naturally.

At this stage, the partners should move through the space smoothly, rhythmically, and vocally synchronized.

Phase VII Creative Dialogue: The partners are to stop mirroring each other or trying to synchronize in any way.

Encourage the partners to converse, make sounds or vocalizations, and move or find a rhythm together without imitating one another.

Urge partners to move and vocalize together within the space. Instead of one person leading and the other following, promote a continuous, natural exchange where both lead and follow, making it impossible to tell who is leading.

Phase VIII Sympathetic Magic: By this stage, partners should be more intuitively aware of each other's movements, rhythms, sounds, thoughts, and feelings that they now share. Encourage them to interact openly and naturally.

Motivate partners to maintain their interconnection as they move across the room or space. Encourage performers to move through a series of personal thoughts and emotions while retaining their shared connection.

Exploration Alternative, Sound and Language: Introduce sound interaction, associating sound with the emotions expressed. One partner may use sound while the other does not, then switch roles. Afterward, both partners can use sound without language.

Another option is for one partner to use only emotional sounds while the other responds with movement. Partners can respond freely, choosing to react with sound or movement.

An additional variation is for partners to open and interact freely with others in the group. This can lead to seeding performance development or scene partner work.

Exploration Alternative, Violent Mirrors: This variation can be used in conjunction with the previous mirror series sequence or in a separate session.

Two performers mirror each other as they duel with imaginary swords, judo chops, boxing moves, wrestling techniques, guns, and more. Of course, the partners must keep a safe distance and avoid touching each other.

Exploration Alternative, Mirror Distortion: Performers mirror one another, then add a funhouse distortion segment to the exploration. The goal is to explore the potential of elastic, cartoon, and Commedia body and facial expressions.

Exploration Alternative, Group Mirror: Wrap up the mirror exploration series by having partners pair up to mirror each other. Slowly expand the number of pairs until everyone is involved in mirrored movements. Follow some of the earlier steps but modify them to suit the group's needs.

FIGURE 2.12 *Mirror Sequence* in a workshop with Metamorphosis Theatre, St. Petersburg, Russia. The author is at the center

Source: Photo by Olga Chernyarskaya.

Dualism, contrast, contradiction, the binary, and conflict are inherent in every aspect of our thoughts, feelings, and imagination. Every person, being, concept, feeling, and idea implies its opposite. The yin and yang, good and evil, democrat and republican, conservative and liberal, past and present, male and female, gay and straight, war and peace, love and hate, justice and injustice, right and wrong, rich and poor, black and white, day and night, heaven and hell, awake and asleep, off and on, top and bottom, inside and outside, and life and death. It is the all-encompassing drama of being, the constant conflict that draws us in, holds us, troubles and pleases us, what we seek to resolve and celebrate.

All beings and things strive for balance, seeking ways to achieve it—such is the process of living on Earth. A human, like an animal, a tree, a forest, the oceans, or the weather, is in a constant state of flux and tension between balance and imbalance. Adjusting, accommodating, and adapting to the ever-changing flow of life, this is the essence of being. Duality shapes everything and is everywhere. Performers and performance are avatars and exemplars, mediators of balance. Whether it is a Siberian Shaman, an Ancient Greek, a Shakespearean or Arthur Miller Tragedian, a Marvel Superhero, or Marina Abramović, all are defined and enacted by contrast and conflict, seeking a momentary resolve—a moment of balance, a glimpse of perfection.

The explorations in this section work well at different levels to improve: (1) observational, mimetic, and responsive abilities; (2) physical and technical control and expression; (3) the range of emotional, physical, and psychological expression; (4) focus and precise concentration; and (5) the pursuit of balance and harmony.

From the Siberian Journal

We moved to the ballet room, which became available. After yoga, we did another mantra exploration. I asked them for the Sakha word for "happiness." Nothing is simple; they provided three different words. It also seems the words are often presented in context and hyphenated. We chose three words, with participants selecting their preferred one (spelled phonetically in English):

Joel Songoo
Verroo Katoo
Jechai Chochai

We had explored this before and jumped right into it. Previously, the activity was seated. This time, I encouraged them to expand their movement throughout the room and to synchronize their vocalizations. I dimmed the lights, and the exploration lasted over 35 minutes.

Some observations about the exploration:

- Sound and movement flowed like waves—beginning loudly as everyone struggled to be heard, then softening into harmony, and eventually fading into quiet.
- This exploration takes time to deepen—progressing from obvious to exploratory stages, where new combinations and insights can emerge. It also needs time to explore emotional depths and build connections with others. Moreover, the activity must allow space for creativity and imagination to thrive.
- This practice offers a way to connect with their singing tradition—encouraging them to step outside it to create something personal, rooted in tradition yet spontaneous.
- Zoya mentioned that pure sound acts as an outlet for emotions, describing it as a form of cleansing. She experienced deep sadness and wept while using these sounds. Interestingly, the sounds were derived from the Sakha word for "happiness." Once free from their original meaning, they transformed into pure sound expression.
- The exploration provided unstructured free play, opening to abstraction, allowing their creativity to expand, and encouraging experimentation.
- The group is becoming more comfortable with each other. There is a noticeable level of independent and creative interaction among members. Anna and Sophia felt the exploration went on too long; I explained that they were hindering their own exploration of the words' deeper meanings and suppressing their creativity.

Rita is beginning to connect–I had my doubts, but over the past few days, with yoga, meditation, and explorations, she is discovering new connections. I see her as the villain in the story—she has a good, traditional singing voice, a strong focus, and excels at imaginative and technical work.

- It is fascinating to see how easily some sounds turn into songs, reminiscent of traditional Sakha singing.

Later, there was broad agreement that the exploration was successful—many felt it was just the right length and genuinely enjoyed it.

After a brief discussion, we started with **Eye-to-Eye Contact** and then moved to the **Person Opposite**—these two explorations served as preparations for the **Mirror Sequence**.

We completed the full sequence except for the emotional mirroring due to time constraints. The group demonstrated significant concentration, technical skill, and focus. However, they primarily stuck to concrete, precise mirroring, and I had to constantly ask them to move beyond it. They appeared hesitant, possibly even afraid, to venture beyond established, self-enforced boundaries. The sequence lasted 45 minutes, but the same participants—Anna, the two Mishas, and Maria II—got tired early again. I advised them that they could rest while still mirroring each other. Kuprion, Marie I, Nina, and Kiril performed very well.

Broke for lunch. Resumed at 6 PM.

Many people voluntarily started their stretches, which was a positive sign. We introduced the **Disembodied Head** exploration late Friday after a long week. This exploration built on the alignment work from the previous day, and participants found it enjoyable. However, many hesitated, holding their heads and reluctant to trust anyone else. We formed a line, with each person taking a turn to feel the weight of my relaxed head, guiding it as I surrendered to the holder's touch.

TOUCH

Partners. One partner guides the other with a gentle touch, using only the fingertip of their index finger. The leading partner touches any part of the other's body, and that touch guides the follower.

For example, when the leader touches their partner's head, elbow, stomach, nose, knee, or back, the partner responds by following, being guided by the touch, and moving around the room. Start slowly, then gradually increase the speed and communicate only through touches.

As partners become more comfortable with the exploration, the leader can use the index fingers of both hands to guide, move, shape, and articulate their partner's actions.

When two partner groups meet, leaders guide and direct followers in a puppet-like manner, with followers interacting.

The leaders assume the role of a controlling spirit or ancestor.

Ask: How does it feel to be manipulated, with your body obedient to someone else's commands? What spaces, images, ideas, and feelings does this exploration evoke? When does guiding someone cross the line, and how does it affect you?

Exploration Alternatives: Encourage leading partners to rotate, switch, and lead different partners. Alternatively, the exploration may involve several leaders and one or two followers, with the followers being pulled in conflicting directions. Remember to rotate leaders and followers regularly. Promote movement within the space and remind those leading to vary the rhythm, intensity, and their approach to lead, to diversify the exploration.

CONSCIOUSNESS TOUCH

Part 1

Participants sit cross-legged facing each other with their knees touching. They should close their eyes. The exploration should progress slowly and without rushing.

Encourage the performers to focus on their urge to touch their partner. Guide this impulse upward through focus and breathing, allowing it to manifest in the act of touching. The person who feels the impulse first should act.

Once the initial touch occurs, the first partner keeps their hand in place to transmit the sensation of their impulse to the other. The second partner should internalize their reaction to this initial touch, allowing it to inform their response. When they feel ready, they can respond with their own touch. The first partner then internalizes this response and reciprocates with their touch. This process continues, with each partner touching, reflecting, and then responding.

Touches must be gentle, non-sexual, and benign. If sexual touching occurs, redirect these individuals to touch more neutrally.

Note: Sexual touching can be culturally acceptable, as I observed during my work with the Zulus. In this context, Zulu men often freely touched their female partners' breasts without resistance, viewing it as a man's right, and women considered it part of their social norms. Conversely, some male Zulu partners avoided touching each other, fearing it would threaten their masculinity. This behavior conflicts with the purpose of the exploration. In such cases, the activity should be stopped, its purpose clarified, and the social and cultural context discussed collectively.

Encourage partners to gently touch arms, legs, shoulders, heads, and sides of the torso. Sensitivity to gendered relations and interactions must be considered. Remind them to move slowly and thoroughly absorb reactions, feelings, and sensory sensations before responding. Performers can use one or both hands. Hugging is also allowed.

This phase of the exploration continues until you feel it has reached its conclusion.

Part 2

Continuing, performers are instructed to respond with a touch that contrasts with their initial sensation.

For example, if they felt a gentle touch, they might react with a firm touch. If they experienced a weak touch, their response could be a gripping touch. A loving touch may be met with a perfunctory touch, an envious touch countered with a generous one, and so on.

Encourage a slow, thoughtful pace for the touches.

If performers seem to move too quickly, slow down the rhythm. Foster the meditative flow of their touch dialogue as it unfolds.

Exploration Variations

- Encourage performers to adjust their communication style, emotions, or attitude as needed.
- Invite them to tell a story through touch without using words. Incorporate natural sounds with each touch, possibly developing into musical interactions if the exploration leads in that direction.
- Prompt performers to move around the space, using their touch to guide, clarify, or express complex ideas, emotions, and movements.
- Expand the activity to involve more partner groups and, if appropriate, include all ensemble members.

Note: Discuss touch and what is socially and culturally coded and appropriate. How is touch influenced by social, cultural, and family upbringing? Were there any inhibitions, concerns, thoughts, and feelings evoked? How and when does a touch become sexual? How is the intent behind a touch shaped by the physicality of the touch itself? Can feelings be conveyed through touch? And if so, how?

FIGURE 2.13 *Consciousness Touch* during a workshop with the DirectorsLab Chicago

Source: Photo by the author.

SHADOWS AND HUMANS

Partners. This exploration starts with movement and develops to include sound and language. This exploration builds on and refers to the **Mirror Sequence.**

An outline of the sequence progression:

1 To begin, one partner is the *Shadow* and follows exactly the actions of the *Human*.

The Human is unaware, does not recognize, nor interacts with the Shadow.

2 Then the Shadow decides selectively and chooses what to shadow or mimic.

Think of the Shadow as a spirit, ghost, or ancestor.

3 The Human moves and talks about whatever interests them; it could be a description of their day, relationships, money, family, health issues, or any other daily concerns.

The Shadow determines how, when, and in what manner to engage with the Human.

4 The Shadow is free to engage in dialogue, make suggestions, point out things, and influence the actions of the Human. Again, the Shadow is invisible to the Human but influenced by its presence and interactions.

   The Shadow questions the Human, correcting and guiding their movements, feelings, and thoughts.

5 The Shadow guides and influences the Human, eventually taking control of them, so they now repeat, think, and act as the Shadow dictates.

Note: This exploration looks at how different cultures worldwide perceive and connect with spiritual spaces and beings. It illustrates a key aspect of cosmocentrism, demonstrating how ghosts, ancestors, and spirits are understood and how they influence the living.

The **Shadows and Humans** exploration was inspired by issues encountered during a project in Addis Ababa, Ethiopia. Through this example, I want to show how our work is reflective, converting ephemeral ideas, emotions, and concepts into concrete, actionable realities that are designed by and for performance. Below is an excerpt from my journal outlining the process of (1) becoming aware of the Space(s) inhabited, (2) Historical and contextual overview of the present, (3) identifying issues of concern and focus, and (4) the recognition and empowerment of the performer as creator.

Shadows and Humans provoked wide-ranging discussions, which, in turn, spurred further explorations and improvisations—each step building and referring to the previous steps, culminating in the performance of *Andegna*. The process, methods, and objectives were consistently refined and discussed, often in Amharic, with my role primarily advisory. Our work emphasizes transparency, collective ownership, and empowering performers to work collaboratively and independently as the need arises.

From the Ethiopian Journal

Discussion following **Shadows and Humans.**

After lunch, the afternoon session is from 2 to 6 PM. Held outdoors beneath a fruit tree (per traditional storytelling) on a beautiful day. The aim was to identify concerns within the group, explore themes and ideas that emerged from the exploration, and determine what they would like to see included in our performance.

A particular emphasis was placed on being shadowed by history, which haunts the imbalances in their daily lives.

A general, free-floating discussion. Fragmentary and highlighted notes:

- We were taught Western styles that were foreign, with success defined by criteria we did not understand.
- Recognizing the ways of the outside world is a source of power.
- The shift from feudalism to socialism highlighted the importance of global influence, politics, and economics.
- Tenant farmers believed in a hierarchy based on family and royal rights granted by divine decree—or at least that is what we were taught.
- The existing system failed, leading to the need for better alternatives.
- Feudalism lingered because there was no strong opposition or other option.
- The Sheba myth is connected to Judah. Emperor Tewodros, coming from a lower social class, took his own life when British forces attacked. He organized the military, incorporating a range of technologies and libraries. Instead of full adoption, he opted for a gradual approach to integrating technology.
- Western missionaries were dispatched.
- Although he sought technology, his request was declined. A blend of political and religious authority and control from the West emerged.
- Today, the church wields considerable influence.
- While a few individuals hold formal power, many others are involved in its work.
- Bishops face imprisonment, and orthodoxy is seen as the "correct way." This reflects the deeply ingrained belief that God provides and offers guidance.
- Religion has historically shaped and continues to shape both people and culture, helping maintain a low crime rate.
- Many find it hard to imagine a society without religion.
- Governance is rooted in ethnic and cultural norms that go beyond religion, culture, and politics.
- There is a principle of coexistence that promotes respect for different religions and individuals.
- Media and film have a profound influence on modern Ethiopia.

- It is common for families to have mixed faiths, including Islam, Orthodoxy, and Protestantism. Traditional Ethiopian culture emphasizes these diverse religious practices.
- Painting is portrayed in a storyboard style because many are poorly educated.
- The storybook format may serve as a potential motif for performance.
- Globalization and consumerism create a reality that values money over people, eroding traditional values.
- There is a migration from rural areas to cities.
- Modernization began during the Haile Selassie era, when students were educated in Europe and the West, marking the start of Ethiopia's opening.
- Compared to other nations, Ethiopia appears less developed.
- Their government focuses heavily on control, embodying an attitude of "eliminate what you cannot govern."
- Both internal and external controls are prevalent. Concepts of socialism, Christianity, and Western values, products, and lifestyles were imported.
- There exists a psychological fear that manifests itself in others.
- Today, after the terror of the Derg (the socialist regime from 1975 to 1991), the people have retreated from political discussion.
- We know we want to change, but instead, we accept and adapt. The government recognizes this and takes advantage of it.
- The trauma of the drought, Westernization, and modernization has challenged their long-held values and way of life.
- The present government presents an old way in a new bottle.
- People are self-censoring, as are the newspapers and media, in a silent conspiracy.
- The trauma of the Derg. The shadow of the Derg remains large and haunts their world.
- The Derg killed if you disagreed.
- The family had to pay for the bullet, and the bodies were left outside the family gate, unable to be mourned or moved. Otherwise, you would die as well.
- The federal system is based on ethnicity—preference based on ethnic identity rather than merit.
- Ethnicity remains a divide-and-rule strategy and policy to this day. Derg did the same. The present government pretends we are all part of an Ethiopian whole, but we are not. It is what they say for political reasons.
- Many are employed without qualifications; the sons and daughters of the politically connected get good jobs.

- Ethiopia's musical culture is losing its unique identity.
- We are losing our Ethiopian self—we accept the modern.
- English is the language of money and power.
- Ethiopia adopted the English and American educational policies wholesale and without qualification. In many international and foreign schools, children are not allowed to speak Amharic.
- We are constantly reshaping ourselves, struggling, and mixing to create change and survive.

FIGURE 2.14 *Andegna* (The First), in performance at Addis Ababa University. Kadist Girma as Lucy is at the center.

Source: Photo by the author.

Gifting and Exchange

Gifting and skills exchanges embody fundamental principles of reciprocity and generosity, reflecting humanity's harmonious relationship with the earth. Many Indigenous cultures hold gifting in high regard, recognizing that the Earth provides for humanity and that the appropriate response is reciprocity—giving of ourselves and our resources to maintain the world's balance.

Among the Inupiat, wealth is traditionally defined by one's ability to give, a tradition exemplified during the *Kiviq*, the Eagle-Wolf Messenger Feast, still celebrated in Barrow, Alaska. The entire celebration centers on gifting and exchange, rooted in creation stories where the Eagle Mother provided everything humans needed to thrive. At the same time, wolves offered the gift of dance, teaching humans to connect and communicate with the land (Riccio 1993).

These gifts of song and dance became humanity's means to honor the spirits and participate in the cycle of exchange. Gift exchange embodies a material, communal, and spiritual act designed to please the Eagle Mother, show gratitude to the wolves, and share. It transcends physical items—what matters is the spirit of giving, reciprocation, and belonging to something greater than oneself.

Lewis Hyde illuminates how giving and exchange merge spirit, material objects, and our way of being. These wider spirits are part of us but not "ours"—they are endowments bestowed upon us. To feed them by giving away the increase they bring is to accept that our participation carries an obligation to preserve their vitality:

> For where we maintain no institutions of positive reciprocity, we find ourselves unable to participate in those 'wider spirits' [...] unable to enter gracefully into nature, unable to draw community out of the mass, and, finally, unable to receive, contribute toward, and pass along the collective treasures we refer to as culture and tradition.
>
> *(39–40)*

These principles found practical application during my work in Zambia, where I led workshops uniting performers from different tribal and ethnic backgrounds. The participants included urban professionals trained in Western methods who performed community theater for international NGOs, addressing themes like HIV/AIDS prevention, political corruption, and social issues for rural audiences. They worked alongside traditional performers, including mask dancers, drummers, and storytellers from various ethnic groups from throughout the country.

My two-week workshop in Lusaka was followed by a six-week performance development phase that engaged nearly 40 participants from 32 ethnic groups across the nation. Our goal was to showcase the ethnic diversity that forms modern Zambia while addressing the ethnic and political divisions that challenge much of sub-Saharan Africa.

Urban actors trained in Western methods had limited experience with traditional performance, while traditional performers were familiar only with their own cultural practices. For this project, skills exchange became essential—the group needed to learn to give to one another before they could give to their community.

The transition from personal to cultural rhythms occurred naturally as members with rich ethnic traditions contributed dances, songs, and rhythms, enriching the collective vocabulary. This practical exchange not only fostered artistic expression but also deepened community connection and reignited the traditional values of gifting. Skills sharing served as a strong equalizer, building social bonds akin to those found in traditional communities, whether in a Miao village in Hunan or the Italian-American community in Cleveland, which influenced me. Generosity, care, and respect underpin our work.

All exchanges occurred within the group circle, which evolved into a sacred space of sharing. Given the participants' histories of inter-tribal conflict, initiating gift and skill exchanges was crucial. Group-wide sharing of dances, stories, and personal histories provided a foundation for deeper discussions about historical conflicts, helping dispel preconceived notions and prejudices.

Most powerfully, participants who had historical "enemies" within the group were encouraged to offer gifts to their former adversaries. The circle saw traditional opponents exchanging songs, dances, and apologies—a powerful display of reconciliation through mutual respect and reciprocity.

Inspired by this spirit of openness, the group recognized that ethnic conflict had hindered Zambia's progress. As artists, they now saw themselves as leaders with the opportunity to repair ethnic divisions by showing their unity as Zambians. This idea grew over time as they connected it to a widely shared creation myth that told of how all Africans were formed into one people but parted ways because of greed, fear, and pettiness.

The performance titled *Imipashi* incorporated elements from several tribes—including dances, ceremonies, masks, and songs from the Bemba, Luvale, Chewa, Ngoni, Tonga, Tumbuka, and Soli peoples. Former enemies collaborated by sharing their dances and representing their rivals' tribes in a nationwide tour, which drew thousands of spectators to fields and marketplaces.

Imipashi reinterpreted Zambia's rich mythology, music, dance, rituals, and ceremonies as a contemporary performance. Drawing on traditional performance practices, the piece explored the fading of traditional values in modern Zambia while highlighting a cultural revival that inspired new hope for a dynamic, multicultural nation.

Throughout this process, we maintained transparency through open discussion and reflection, clarifying rationale and assessing explorations. My aim was not just to facilitate a process—I wanted participants to learn how to share the process and explore new possibilities. A key goal was to empower, understand, and to expand on the work in their own way.

Our efforts were tailored to reflect the group's expressed interests and my understanding of their needs, with all discussions ensuring full transparency. My experience and insights either refined previous explorations or sparked new ones aligned with the project's goals. We demonstrated how gifting and reciprocity—an act of caring—can promote understanding and build new community connections. Sharing our skills, stories, and vulnerabilities helps create a spiritual bond that unites everyone with everything.

GIFTING

Performers are invited to present a gift of song and dance to the group. The song and dance may originate from their culture, tradition, or personal experience, or be an original creation.

A tangible object is also given to the group. It does not have to be large or lavish. It could be found or made, food, drinks, or a poem.

The main point is that the gift is made by the giver especially for the occasion.

It is given in a performative or ritualistic way (whatever that means to them), sharing its history, their relationship, and its significance, as well as how the offering contributes to our project and work.

SKILLS EXCHANGE

Ask performers to identify at least three skills they can contribute to the group, whether cultural, social, or personal. Each skill is acknowledged through a group-established ritual. After presenting and practicing each skill, a discussion follows.

Exploring

This section's explorations enhance and complement our previous work, laying the foundation for our performance creation efforts, which will be more fully addressed in the **Place** phase of our work. Many explorations intertwine, embodying two or more of our key motifs: **Circle, Rhythm,** and **Balance.** Consider this section integrative, a catch-all for various ideas and concepts.

BLIND MOVEMENT

All performers are asked to close their eyes, wear a blindfold, or work in a dark space. They should avoid any form of vocal communication.

Ask performers to carefully observe their environment and mentally map it. The aim is for them to cultivate a bodily awareness of how to navigate space without using sight.

Side Coach: *Move slowly through the area, using only touch to identify each other. Once you have done this, explore and mentally map their environment, paying attention to the textures, tastes, and sounds around them.*

Next, instruct the performers to respond to your voice by following a sequence of directions. For example, avoid your voice and congregate away from it.

Then, move toward your voice, or follow your voice as you move through the space.

Add obstacles, sounds, distractions of any sort, diverse textures, and emotional levels of your voice to enhance their sensorial exploration.

Consider leading the group through a story using a culturally significant myth or vivid imagery.

Embracing darkness encourages the imagination and activates the sensing body.

For example, the group might explore a dark cave, then sense something lurking in the distance. Then ask performers to crawl, leap, freeze in fear, run, hide, and call out in fear, anger, or contentment as they navigate an imagined narrative space.

Highly Recommended.

PSYCHIC SPACE

Partners face each other across an empty space. One partner wears a blindfold. An obstacle course made of chairs, tables, tree branches, hanging objects, and other safe debris is arranged in the area to challenge blindfolded participants as they navigate it.

When signaled, the sighted partner guides the blindfolded partner through the course without physical contact or speaking, solely by observing obstacles and conveying thoughts through nonverbal cues.

The sighted partner's primary task is to assist the blindfolded partner in navigating the course by careful observation and non-verbal projection. It takes a while to establish, so start slow. Remarkably, it then begins to work.

The partners develop awareness of thought or telepathic communication: The space of thought.

Both partners must genuinely believe they can communicate in ways that go beyond just verbal or physical interaction.

After the blindfolded partner completes crossing, switch roles. Rearrange the obstacles before the next iteration of the exploration.

OUT OF BODY

This is an advanced visualization technique and should only be attempted after completing several other visualization explorations.

Begin by helping performers achieve a calm and alert state of mind. A yoga or meditation session can serve as an ideal prelude.

Building on **Body Part Isolation** relaxation techniques to loosen all body areas might also be a helpful preparation.

Ask performers to lie on their backs in a quiet, preferably dark space. Encourage them to concentrate on establishing a gentle, relaxed breathing rhythm, focusing solely on their breath.

After establishing a rhythmic and relaxed breathing, prompt the performers to visualize their body shape. With each successive breath, this shape gradually fills with color.

With every breath, the colored body form becomes more distinct and prominent. Once you sense the group is prepared, invite them to envision their body form rising and distancing from their physical presence. Encourage performers to maintain relaxed and rhythmic breathing. There is no need to hurry.

Encourage the performer to let their colored body explore freely in any direction. Ask performers to mentally note their observations.

Give the performers time to travel, then invite them to reconnect their colored bodies with their physical forms while maintaining deep breathing.

To help them return, pinpoint specific details about their location to ground them in the present.

For Example, *you are in the rehearsal studio in the Jonsson Building at the University of Texas at Dallas on November 19. It is 4:38 pm, and you are lying on the floor listening to my voice.*

Once everyone has returned, invite the performers to sit and open their eyes when ready.

A group discussion follows.

Note: This exploration serves as a tool and resource to enhance imagination and creativity, while broadening sensory abilities beyond the physical realm. It helps build familiarity and confidence in understanding oneself and unlocking untapped potential. Along with other visualization techniques, it emphasizes that awareness is in the imaginal, as it is in the physical world.

SPACES WITHIN

Offer as an overnight or weekend exploration.

Instruct the performers to choose a space to observe and share with the group.

Encourage the performers to pick a location to sit and simply observe and be present in that space for 5 to 20 minutes. The space can be indoors or outdoors, private or public, and either familiar or unfamiliar.

For example, they might sit in places such as a classroom, a bar, a restaurant, a hotel lobby, a bedroom, a hospital emergency room, a living room, a public library, a bus stop, the woods, a shopping mall, a garden, among others.

Side Coach: *take a moment to observe and fully engage with the environment: Inhale the atmosphere, note the surfaces around you (whether they are soft, hard, comfortable, or uncomfortable), consider the temperature (is it hot, warm, cool, or cold?), observe the lighting (natural or artificial, its intensity, warmth, and direction), take stock of any people nearby (whether close or at a distance), and reflect on your overall impression of the space and its ambiance.*

Encourage performers to bring the space, its atmosphere, and details into them.

Performers then present their space to the group by simply sitting or standing, embodying the space they have observed without illustrating it. Simply be present.

Allow several minutes for each presentation and performer to settle into the space. Some performers will quickly relax and settle in, while others may take longer.

Observers are encouraged to notice and feel the transformation occurring within the performer. How the performer inhabits their space simply by being will become clear to the group.

Invite the presenters to remain quietly before the group and in their space as the following questions are asked. Aim for a consensus.

- Were they inside or outside?
- What kind of surfaces were present—hard or soft?
- Was the area public or private?
- Was there activity? If so, what kind—busy or distant?
- Were they in a space they enjoyed? How was the lighting? Was it bright or dim, natural or artificial?
- Were they alone, or were there others nearby?
- If there were others, how close were they? Were they interacting with the presenter?
- Where do you think the presenter was positioned?

Presentations will vary greatly; some will clearly reveal the inhabited space, while others might be less obvious. Spaces can sometimes blend indoor and outdoor elements, such as when traveling in a car, watching a film or TV show set grammar corrected outdoors, or flying on an airplane. There may also be times when performers are in public but lost in private thoughts.

This exploration aims to show how we carry spaces with us and how they physically inhabit and psychophysically influence us.

More generally, it will emphasize the different spaces individuals occupy, as well as the shared cultural and social settings they experience.

Note: While engaging with diverse groups, including the Zulu and Sakha, in places like Poland and Finland, this investigation initiated a broader conversation about shared spaces influenced by factors such as darkness and climate (Finland, Estonia, and Denmark), public anxiety and activity levels (Zulu and Burkina Faso), and the types of surfaces and cleanliness (Nepal, Chicago, University of Arizona, UC Berkeley, Sakha, St. Petersburg).

Highly Recommended.

CREATING SPACES

Form groups of three to six performers and ask them to select a location they are all familiar with. Once they decide, they discuss and concur on the sensory and emotional details of the space.

Space is essential to both the performer and the performance. This exploration builds on **Spaces Within,** offering another tool for anticipating performance making.

Simply, the group occupies a space. Observers outside the group watch and interpret its location, atmosphere, surroundings, energy, and the emotions it evokes.

Exploration Alternative: Have the group select a space relevant to the project, then fill it with objects, lighting, furniture, or other elements to create the desired environment. Recall our earlier discussion on sympathetic magic and endow the elements with relevant significance.

The use of substitute or placeholder objects is encouraged and can offer new perspectives and conceptualizations of the space and/or project. Unusual or chance juxtapositions may lead to unexpected and creative opportunities. Remind yourself to be open. Let chance occur. Do not judge, instead consider.

Once the space is set up, the group can begin exploring it via improvisation.

MOVIE POSTERS

Choose one performer to pose in a specific "movie poster" style from the options below. Once the stylized pose is set, others are encouraged to help complete the poster, which often serves as a cultural symbol.

This activity views the poster as a framed hieroglyph that hints at, reveals, and encodes wider social and cultural beliefs.

Performers do not need to physically interact for their poster but should connect through the idea, theme, or emotion it conveys.

Ideally, each poster should feature three to six performers.

Another movie poster idea is proposed, and a different performer is asked to develop the concept for it—then the process repeats. Use or create movie poster ideas based on need, taste, and project goals.

Ideas for movie poster theme prompts might include:

Horror	Sci-Fi	Zombie
French New Wave	Comedy	Japanese Anime
Slasher	Documentary	Bromance
Western	Nature	Adult
A night in India	Kung Fu	Disney
Romantic comedy	Rom-Com	Silent
Cartoon	War	
Thriller	End of the World	

After establishing the poster pose, ask the performers to transition to the next moment—one distinct movement. Then, proceed to the following moment and repeat this process to bring the poster—frame by frame—to life. Finally, reverse the process by moving back through each moment to return to the original poster pose.

Alternatively, invite the performers to animate the poster (hats off to Augusto Boal) by moving, drawing inspiration from the original pose, and improvising with one another.

Exploration Alternative: While performers are in their starting pose, encourage them to generate a word or brief sentence inspired by the pose and their relationship with others. Use this as a seed to develop a character or a scene. Listen to the pose and follow wherever it may lead.

Exploration Alternative: The same exploratory approach can be used for an internet meme, gif, an Instagram or TikTok post, a TV show or commercial, a music video, trailer, or a movie scene as a reference. Follow the same steps outlined earlier.

Note: This exploration offers an opportunity to investigate issues, concerns, and ideas that live within a historical moment, while also allowing for an examination of the group's cultural dynamics.

PHOTOS

Provide each performer with a photograph that is both character-driven and emotionally powerful. These images should depict individuals in unique settings, conveying a sense of emotional depth or a dramatic scene.

Ask the performers to interpret the depicted situation:

What is happening?
What has been said?
What needs to be said?
How is it being expressed?
Who is saying it and why?
What is the space inhabited?
Who are they, and why are they together and in that moment?
What are the inner thoughts, and what is the main person in the image saying?

- Consider what they want.
- Define the emotional atmosphere.
- Create a brief history, outline the context, and describe the primary individual in the image.
- Determine what the person in the photo is leading with.
- What are they holding?

- What is their walking rhythm?
- Explore their walk.
- How do they talk?
- What is their language?
- What is their educational level?

Identify gestures and actions, then analyze them and incorporate them into your body.

Give the performers five to six minutes to embody the photo.

Ask them to experience the depicted moment in time, along with its emotional, social, and cultural context.

Present the photo scenes to the group.

You and the group interview the performers. Ask:

What do you see?
Where are you?
What is your name?
What are you feeling?
Where are you going?

DEFINING SPACE

Create a clearly defined rectangular or square area with rope, tape, sticks, chalk, or other materials. For 10 to 15 people, approximately 10 feet by 8 feet is suitable. Adjust the size as needed based on group size to maintain a compact space.

The group gathers at the edge of this area, looking inward. When ready, each performer steps into the designated space.

Inside, they move freely, staying alert to the movements and presence of others, noticing how each new addition alters the overall rhythm and how the presence of bodies, and their sizes, shapes, and positions, impacts the space.

Encourage performers to enter and exit the space as they wish.

Notice how the rhythm, balance, and mood change as different performers enter and exit the space.

Ask performers to increase their walking speed until reaching the maximum comfortable pace for the space.

Then, reduce the speed and observe how the atmosphere changes.

Ask one performer to walk faster and another to walk slower. Observe how this affects the atmosphere, energy, and performative possibilities.

Remind performers to remain participant-observers.

Add or subtract performers within the space, then establish the tempo by tapping out a rhythm. Gradually increase the tempo, then slow it down. Ask performers to reflect on how bodily interactions, thoughts, and emotions adjust and align with the rhythm.

As performers leave one by one, observe how the space changes.

The walking pace inside the space decelerates with each departure. The exploration ends once the final performer exits.

Then observe the empty space and recall the actions that took place.

Note: This exploration draws inspiration from a workshop led by Ann Bogart and John B. Calhoun's "Behavioral Sink" theory. His experiments with rats and mice show how overpopulation leads to social shifts and eventual collapse. This framework helps analyze space, presence, rhythm, and the performers' emotional and psychophysical dynamics. The phenomenon was explored in DP92's The Dead White Zombies performance.

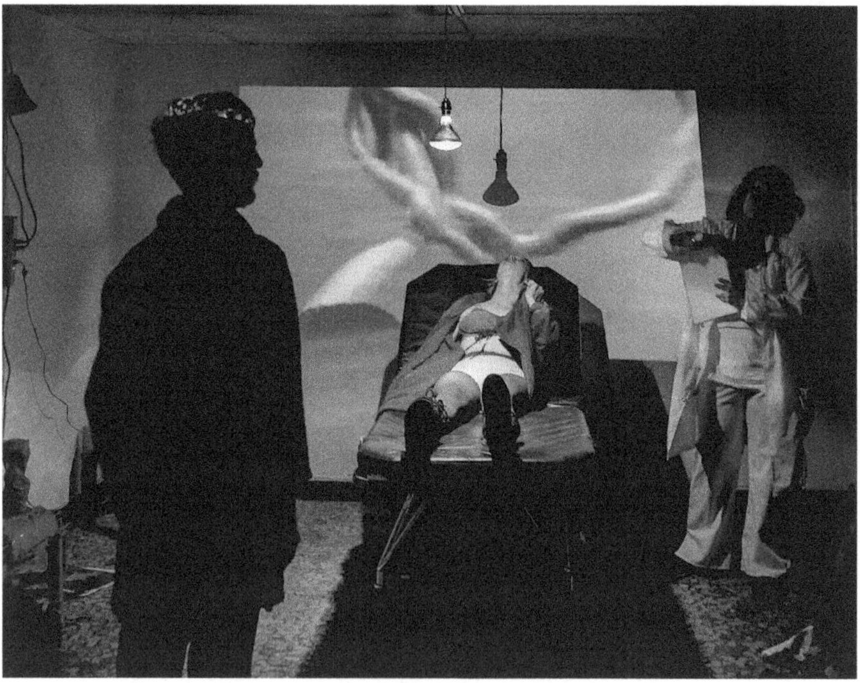

FIGURE 2.15 *DP92,* written and directed by the author, produced by the Dead White Zombies. From L to R: Bailee Rayle, Hilly Holsonback, and Ilknur Ozgur.

Source: Photo by the author.

Objects

The next few explorations view objects as carriers and conveyors of the space that created them. Each is endowed with the life that formed them, reflecting their history, context, and purpose. Objects surround us, yet they are mostly silent and subordinate to human needs. We interact with them constantly in our daily lives. How can we unlock their expressive potential, establish a dialogue, and deepen our awareness of the world through them?

IMAGINARY OBJECT

The group forms a circle. One performer begins by holding an imaginary object and exploring its qualities—texture, feel, taste, size, and weight.

After handling the imaginary object, they pass it to the next performer, who interacts with it, identifies it, and then transforms it into a different object.

Participants should let their tactile interactions guide the transformation into a different, imaginary object.

Once a performer is happy with the transformation and their imaginary object is finished, they pass it to the next performer in the circle, who, in turn, accepts, transforms the object, and passes it on. And so, it goes around the circle.

When discussing, think about how objects shape and maintain a space. Ask which space or spaces they are part of. Consider what the depicted objects represent. What social and cultural significance do they have? How do these objects carry history, emotion, and ideas? Do they hold sympathetic magic?

OBJECT IMAGINATION

Start with a single stick—an ordinary, roughly one-meter-long branch, preferably cut from a tree and full of aliveness. Give it to one performer, who uses it as a reference point and creatively transforms it into something else inspired by the stick. The performer then demonstrates what the stick has become.

For example, the stick might start as a gun, then the performer aims and fires, marches with it over their shoulder, and stands at attention with it. The next performer takes it as a gun, then transforms it into a telescope and looks through it, before passing it on to the next performer, who changes it into a golf club used for putting. The following performers shape it into different objects, such as a ballet barre, a cane, a barbell, a snake, a spear, and more.

Allow the stick and its handling to inspire its transformation, as a shaman conjures something into being.

Give time to feel and interact with the stick, letting the physical interaction and body memory guide its transformation.

Introduce another object into the circle. The more unusual the shape of the object, the better. For instance, a piece of plastic plumbing, a seashell, a rock, or a uniquely shaped piece of driftwood. Then, following the process outlined above, transform it imaginatively into something else. This exploration can continue with a variety of objects.

Exploration Alternative: Transform a stick or an object into something of ritual or spiritual importance. Examples include a magic wand, a musical instrument, a mythical spear, a divining rod, a totem, or an offering.

Note: This exploration seeks to assist performers in recognizing the expressive qualities and potential of any object. Cultivating this awareness fosters a psychophysical dialogue with non-human entities, enriching and expanding the performer's imagination.

Every object, along with everything else, possesses physical, imaginative, meaningful, and historical aspects. Deliberately engaging with objects creates a space that moves beyond the simple binary of magical versus rational perspectives. By identifying and exploring the expressive possibilities of objects, performers can enhance their creativity and access the transformative energy inherent in all things.

OBJECTS

Gather everyday objects that have moving parts or can be moved in different ways. Create a large, varied collection and place them throughout the space. Encourage performers to pick one object to work with, or assign specific objects to each performer.

Objects assigned should match the performers' needs. For example, performers with broad movements might get a small, intricate object, while those who are stiff or awkward could be given a flowing cloth or similar item.

Past examples include a working Zippo lighter, a door hinge, a large piece of cloth, a rattle, a large set of keys, pliers, a Swiss Army knife, miniature car toys (like Hot Wheels), a rotary telephone, a squeaking stuffed animal, an air pump, a stapler, scissors, and more.

Once their objects are chosen, ask performers to simply sit and examine them.

Part I: Ask the performers to find a space away from others. Instruct them to place their objects in front of them for close observation and to make mental notes about what they see. The following questions can help performers with their observations. Allow the performers time to respond to your suggestions and questions.

- What color is the object?
- What is its weight?
- What is its height?
- What is its texture?
- Where are its curves?
- Are there straight lines?
- Are there angles?
- Where is it sharp?
- Is it soft?
- How old is the object?
- How do you think it was made?
- What materials compose the object?
- What designs are present on it?
- Is there any writing on the object?
- Which parts move?
- Which parts are stationary?

Part II: A second set of questions and suggestions encourages performers to engage with their object. Give the performers time to respond to your prompts and inquiries.

- How does the object move?
- Start moving the object.
- What sound does the object produce?
- Create that sound.
- What other sounds can the object make?
- Experiment with different sound levels and variations.
- How else can you move the object?
- Can you roll it?
- Throw it?

- Drop it?
- Flip it?
- Experiment with different methods to move the object.

Part III: If the performers are not already standing, ask them to stand. Allow sufficient time for each question and suggestion to be thoroughly discussed.

- Move the object through space.
- How does it communicate?
- What sound does it produce? Voice?
- What kind of character or personality does your object have?
- If it encounters another object, let it communicate—each using their own object language.
- You may also choose to move together with the other object.
- How do your object and the other objects interact—do they get along, are they related, or share similarities?
- How does the character or personality of your object change through interaction?
- Meet, communicate, and interact with all other objects.

Part IV: Ask the performers to lie down, set their object aside, and close their eyes. Instruct them to breathe with the object, then guide the group through a visualization of their object.

Allocate enough time for the exploration of the following questions and suggestions.

Side Coach

See the object in your mind and slowly embody its characteristics in your body.

- *What does it symbolize?*
- *What images does it evoke?*
- *What is the object's most obvious or prominent feature?*
- *How can you express that with your body?*
- *How does your body move like the object?*
- *What is your object's rhythm?*
- *What does it repeat?*
- *What is its preoccupation?*
- *How does it seek balance?*
- *What are the equivalents of legs, arms, heart, stomach, genitals, mouth, nose, eyes, and head?*

- *Slowly start to move like the object.*
- *What other movements of the object can you show with your body?*
- *How do you express the object's personality or character? How is the object strong?*
- *Weak?*
- *Angry?*
- *Friendly?*
- *Sad?*
- *Confused?*
- *Comforted?*
- *Happy?*
- *Tired?*
- *Excited?*
- *How does the sound of the object come through your body?*
- *Use your body to express the sound or sounds of the object.*
- *Begin to move (if you haven't already) in the space to explore your object body.*
- *Walk as your object.*
- *Run.*
- *Your object is in a hurry.*
- *Your object is slow and lazy.*
- *Encounter other objects.*
- *Use only your object's sound and body to communicate with others.*
- *What is repeated?*
- *What are its rhythms?*
- *How do they vary, and what does the variation express?*
- *Have a conversation with other objects.*
- *Take other objects on a little journey through the space. Which objects does your object enjoy?*
- *Dislike?*
- *Friends?*
- *Try to make contact and communicate with each object.*
- *Depending on the workshop goals, the exploration may develop into a performance development improvisation, introducing a situation, objective, or action.*

Allow time for performers to interact, explore, make discoveries, and engage in play.

End the exploration by instructing the performers to gradually sink to the ground and close their eyes to relax.

Guide them to gently let go of the objects from their bodies and return to their original shapes. Prompt them to consider how their bodies have changed and how they relate to the objects, taking and being affected by their attributes.

Exploration Alternative: Use objects that are specific to the cultural, social, political, or spiritual contexts of your project. Apply this to tribal, ethnic, or pop culture objects. The possibilities for imagination and creativity are endless.

Note: A similar process, with some modifications, can be applied to embody and interact with animals.

Highly Recommended.

Character

The human body's shape and movement are essentially universal, with everyone sharing similar biology and anatomy, albeit with some differences. This universality enables us to empathize and connect psychophysically with others. Exploring these aspects helps performers see how daily gestures, when repeated, develop into a standardized vocabulary for performance. It provides insight into how the human body communicates through motion as it interacts with and navigates spaces.

GESTURE EXPLORATION

Invite each performer to select a simple movement gesture. This can be anything randomly drawn from everyday life, or, if working within a play or performance development context, a gesture that is specific to a particular situation, culture, or environment.

The performer then repeats the gesture without any emotional attachment or rhythm variation.

Allow it to become automatic, abstract, and disconnected from its original meaning, transforming into a neutral expression.

Side coach and encourage the performers to repeat the movement while exploring and responding to the following qualities in their own way, for example:

Confidently	Confusedly	Strongly
Cautiously	Tenderly	Floatingly
Angrily	Sharply	Heavily
Shyly	Confidently	Fragilely
Suspiciously	Deceptively	

Add or subtract qualities based on need.

Repeat and explore the gesture, encouraging it to inspire movement in other parts of the body.

The performer then allows the gesture to inspire an emotional response.

For example, the repeated gesture of clapping one's hands together, when accompanied by an attitude of strength, will likely evoke movement in the chest and shoulders. At the same time, an emotional or mental attitude of power, arrogance, or revenge may arise.

Move through each quality, each time letting it inspire movement and emotion.

Repeat the exploration with another movement gesture following the process noted above.

Exploration Variation: The same exploration can be used to analyze and expand the qualities of culturally specific gestures. Ask the performers to demonstrate different dance, ritual, and performance gestures, then have them pick one to explore for its expressive potential.

Pop culture and social media offer many examples, including Hip-Hop, TikTok, Instagram memes, music videos, video games, sports, and politics. Examining these cultures directly highlights the expressive power of certain gestures, which can then be incorporated into a larger creative framework. This process helps develop a performance vocabulary based on these movements, as they symbolize and communicate different spaces.

Every day, we move through layers of overlapping and intersecting spaces. Our task is to identify and explore their expressive possibilities.

Explore a variety of social and cultural spaces, a suggestive offering:

Religious	Sexual	Spiritual
Consumer	Family	Working Class
Athletic	Corporate	Homeless
Meditative	Leisure	Wealthy
Ethnic	Gender	

Highly Recommended.

CHARACTERS FROM LIFE

This exploration builds upon and expands the concepts presented in **Leading in Space,** mentioned above. It is highly recommended to revisit that exploration in preparation.

This exploration offers a methodology of creating a character externally from the body rather than internally, following various psychological and/or memory-based methods.

Performers observe three individuals in various everyday situations, such as at a store, on the street, with friends, or within their family.

Or you may confine the observational reference to one location. A favorite of mine is having students "capture" three individuals at Walmart, known for attracting a wide variety of extreme body shapes, movements, and rhythms.

Performers observe and record with their bodies. First, identify where and how the person leads, for example, from their forehead, pelvis, stomach, neck, or similar area. Refer to **Leading in Space** exploration.

A counterpoint balances each lead—another body area bearing tension, forming an opposition or harmony. Locate where the body is held.

For example, if someone leads with their forehead while walking, they may hold their shoulders still. Pay attention to these oppositions and observe how different body parts interact and balance each other.

Performers objectively observe how each person moves and how they coordinate their walk. They begin by walking near, behind, or alongside the individual (be respectful if it is a stranger in a public space), identifying and walking with what they lead with.

Side Coach: *Next, identify where the person holds their body, often marked by stillness. Observing, focus on the person's walk and what they choose to hold.*

For example, someone leading with their forehead might also support their lower back or shoulders. A weightlifter leaving a gym might lead with their chest, while holding their lower back or glutes.

Walk their walk in your body; doing so will reveal the rhythm of their walk.

The three key elements for capturing a character from real life are:

Lead, Hold, and Rhythm.

Performers are encouraged to rehearse and refine each of the three individual body leads, holds, and rhythms. In doing so, a psychophysical understanding begins to develop, offering a foundation for further explorations.

Side Coach: *Some questions for further in-depth explorations:*

- *What gestures originate from and are natural to the individual?*
- *How does that person see the world differently?*

- *How does this change your emotional and psychological outlook? What organic activities, gestures, rhythms arise from your exploration?*
- *How does the individual develop into a character, taken from the real world, interpreted, transported, and transformed to enter the space of performance?*
- *What is the voice of the captured character?*
- *What emotions, thoughts, and images are evoked?*

Invite performers to present their three characters to the group. Ask the group if they can identify the lead and hold.

Ask the performers to describe their characters, including how they move, where they lead, how they hold themselves, and the rhythm they embody. Explain how they observed and "captured" the character and what insights they gained.

Exploration Evolution: Invite performers to choose one of their captured characters to work with and interact with another performer's captured character. They improvise by greeting each other and starting a conversation, letting the scene develop naturally.

Exploration Alternative: When performers observe people in real life, they capture one or two overheard phrases or sentences as part of their study.

Incorporate these phrases into the character movement exploration.

By incorporating character language, performers will gain deeper insights into the person being observed, enhancing understanding and providing a pathway to develop their character. This is somatic fieldwork.

This exploration can develop into a physical character study and is a crucial investigative tool for revealing community spaces through character. It has been successfully applied in a wide variety of specific, mixed-cultural settings and university courses and productions.

This observation-to-performance development process is inspired by and common to Indigenous, place-based performances.

Highly Recommended.

Sharing

This next series builds on **Body of Sound** and other visualization explorations, delving deeper into the spaces that comprise a character by exploring emotional and psychological range and complexity. Central is the question: How are these spaces socially and culturally constructed?

MONSTER WITHIN

This exploration is most effectively presented after a visualization session or a meditative moment in our work.

Performers start by either lying down or sitting.

With eyes shut, performers are asked to consider what they conceal from others and themselves. What inner monster(s) do they hide from the world, and how and when do they believe the world perceives them as monsters?

Monsters symbolize and embody exaggerated forms of greed, hatred, loneliness, anger, violence, cruelty, prejudice, vengeance, envy, spite, fear, and darkness. What is evil in you?

Side Coach

- *What actions make you feel ashamed or guilty?*
- *What are your most hidden and guilty habits and fears?*
- *What transgressions have you committed that you regret, feel ashamed of, and hide from others and the world?*
- *What distortions and compensations do you use to conceal these traits?*
- *What masks do you wear to gain social acceptance?*
- *If you were to express them physically, what would they look like?*
- *Where and how do these monstrous traits show up in your body, voice, and movements?*
- *Can you identify the origin of each monstrous trait?*
- *Was there an event, a moment, a series of events that created your monstrous traits?*

Encourage performers to imagine their feet transforming into their monster form. Let the leg change in shape, affect their movement, and emit a sound.

Progress through each major body part, allowing each to manifest and express calves, thighs, pelvis, stomach, chest, hands, arms, neck, and, finally, the head, envisioning its transformation and its monstrous sound.

Limit performers to nonverbal sounds.

Side Coach

- *How does each body part reflect the inner monster?*
- *What are its colors, textures, shapes, and details?*
- *How does your monster use these features?*

Encourage performers to assemble their monsters and explore a vocabulary of movements.

Ask the performer to move, sit, crawl, and, when ready, slowly stand and move in a space conditioned by their monster perception.

Side Coach

- *What does your monster want?*
- *What is it looking for? Why must you find it?*
- *What are you thinking?*
- *How do you feel?*
- *How do you see and relate to the world and others?*
- *What attracts or repels you?*
- *What do you seek?*
- *Where do you find comfort?*

Slowly transform your sounds into word fragments, then, if you feel the need, express in words and actions your monster feely to one another.

Consider it a confessional, a vomiting of the monster—an exorcism that frees your body, mind, and spirit of the monster within.

Once the monsters have been established, suggest interaction with other monsters.

Exploration Variation: Refer to a monster's description based on a culturally specific story, legend, or myth. This approach enables the entire group to develop their own interpretations of the same monster within a culture, exploring its significance.

Side Coach: *How is your monster like Frankenstein, Dracula, Satan, space aliens, and zombies in their various forms resonate and endure across cultures worldwide?*

What universals about our species do they reveal? How does their psychophysical expression act as a social and cultural reminder and warning? Why are they so prevalent in popular culture today?

Notes: The **Monster Within** explores the darker aspects of life—those hidden corners within ourselves and daily existence. Recognizing and managing these aspects are essential skills for the performer and act as a guiding reference during the development and performance of the work—Place. Our work demands that the performer be completely empowered, courageous, and capable of performing whatever is needed.

This exploration also aids in designing culturally specific, anthropomorphic, or otherworldly characters, especially when creating mask characters. Please refer to the mask section below.

When developing culturally specific characters, performers are encouraged to research written, oral, historical, and visual sources. For example, the Tokoloshe of Zulu culture or the Predator of contemporary film.

THE SITUATION

A performer is seated, with the group surrounding them. Two chairs and a table are recommended, but performers are free to move and do not have to remain seated.

The performer is neutral; then a second performer enters the space, bringing with them an idea of a character and situation to which the other performer responds.

The character and situation can be anything. An improvisation develops naturally between them, going wherever it leads.

Guide performers to practice active listening by paying close attention to one another's full expression, including gestures, rhythm, language, and emotions.

Performers should remain relaxed, alert, and flexible as a scene evolves.

If the first performer cannot understand the character or situation introduced by the second, they should honestly remain confused until more details surface.

The second performer entering must also adapt to their partner's responses. Often, the situation will develop in unexpected ways—it cannot progress if either performer is inflexible.

Remember, just as in life, no one completely understands a situation, their character, or their goals. Life is a continual process. We begin with assumptions based on what we know and keep learning along the way, requiring flexibility, openness, and engagement. The world and our lives are shaped through relationships.

Some Key Points

1. Do not block suggestions or cues from the other performer or the situation.
2. Instead, focus on building upon what is introduced, or organically transform it if needed, to support ongoing development. Like life, interaction is a flexible and flowing process.
3. When unsure, repeat what you know and where you are, and dive deeper into the moment until the next moment reveals itself. This is the way of meditation and ritual. Repetition creates depth, and depth unveils the moment.
4. If you observe that the situation or character is losing momentum, make a change.
5. If you feel lost or overwhelmed, pause to solidify your current stance before moving forward. Just as in life, it's best to stay put until you have a clear understanding of your circumstances.

Exploration Progression: When motivated, a different performer from the group steps in (or is prompted by you) and softly taps one of the two performers on the shoulder, signaling them to exit. The new performer then introduces a different scenario and character for the remaining performer to respond to, allowing the improvisation to progress. The new character/scenario should be unrelated to the previous one.

Note: I have used this exploration to accomplish different objectives across various cultures and contexts. You could restrict improvisations to themes like myths, current events, local issues, or particular subjects such as abuse or political corruption, specific to project needs. This allows improvisation to be customized according to the project's goals and interests, serving as a tool for investigation.

Discuss how the characters and situations captured interest, pinpoint moments when the flow slowed, and evaluate when performers ceased listening, along with the impact on audience engagement. Experiment by introducing additional characters that develop existing situations and by adding props when fitting.

Highly Recommended.

GROUP SITUATION

Two groups of equal size. This exploration is inspired by and expands upon the work of Augusto Boal.

One group is identified as *Initiators*, and the other, *Reactors*.

The Initiators discuss a situation and their characters.

They inhabit and shape the space with attitude, gestures, posture, and spatial relationships, reaching an intuitive consensus, then solidify the moment with a frozen pose.

The Reactors take a moment to observe and consider before entering the space to respond to the Initiator's pose and position themselves in relation to the established pose.

How they respond further defines and enhances who, what, and where both groups are in their relationships and situation.

Ask the Initiators to make one simple gesture, the next animating step extending from their pose.

The Reactor group responds to that gesture.

And so forth, building the gestural action-reaction between the two groups and creating a gestural and collective dialogue.

Allow the dialogic interaction to develop and, if appropriate, evolve its pace of action and reaction. If the dialogue loses its form and coherence, slow down the pace or step back into the previous pose, reclaiming the initial impulse.

The interactions between the groups may vary widely and evolve unexpectedly from the original pose and relationships.

Individual interactions may likewise diverge from person to person. This is all part of the improvisational, exploratory process.

Repeat the exploration with the groups switching: The Initiators becoming the Reactors.

Note: Encourage performers to attentively observe spatial dynamics and grasp the symbolic, comedic, or iconic significance of the poses they see. Guide performers to remain conscious of the changing action and to respond by adjusting their situations and characters accordingly.

QUESTIONS

Performers form circles with 4 to 6.

One performer begins by turning and asking the person to their right a question. The recipient listens, reflects, and internalizes the question without reaction.

That performer then turns to their right to ask the next person their own question. This cycle of asking a question without reaction continues around the circle.

Once around the circle, the process reverses direction. Asking a question that goes unanswered but is felt and considered by the performer on their left.

As questions evolve in response to previous ones, they foster a shared emotion and thought process.

An internalized reaction prompts the next question, creating a flow of feelings and thoughts that moves through the group.

Note: This activity deepens listening and responding skills and develops associative and collective thinking and feeling.

QUESTIONS AND ANSWERS

Two performers step into the circle, providing their spontaneous and unfiltered responses to one another to the questions you ask and elaborate on.

Side Coach

- *What are you thinking?*
- *What are you feeling today, right at this moment?*
- *What answers are you seeking?*
- *What do you desire?*
- *Where are you headed?*
- What concerns you?

This is not a hurried exchange. The goal is to establish a natural, instinctive flow of questions and answers.

Your questions are like subconscious prompts.

Questions may be repeated, and the performers' exchanges and responses can inspire you to ask additional questions.

Encourage performers to openly share their feelings, observations, thoughts, and their perceptions of the world and their lives, especially focusing on the spaces they inhabit.

Responses may sometimes seem nonsensical or fragmented, this is fine. The aim is to connect with their immediate experiences, emotions, thoughts, and each other.

Remind performers to stay present, authentic, and unfiltered to capture the moment's immediacy. Allow enough time for the exploration to develop and gain momentum. When it appears to have peaked, replace the current performers with two others and start the process again.

Exploration Alternative: Invite more performers to collaboratively develop questions and responses gradually, leading to a group-inclusive session.

Note: This process can be repeated as a way to "check in" and assess the group's pulse.

Discuss how the exploration uncovered the spaces performers occupy and share.

ONE WORD

Part I

Partners sit or stand facing each other. To start, one performer moves, and the other responds—not as a mirror, but with an intuitive gesture.

There are no words or sounds.

This is an emotional and physical exchange between the partners. Let it develop naturally. Avoid forcing or rushing the interaction. It is not a performance; it is a conversation.

Partners begin stationary, then move into the space as inspired. Moving and responding, acting and reacting are the main focus.

Part II

Change partners.

Performers communicate with only one word at a time. Let the one-word exchange flow naturally. Encourage the performer to listen and respond without rushing or aiming for a perfect performance. Focus on listening and responding with sincerity and immediacy.

Side Coach: *Consider the feeling before replying and be aware of generalized or default responses.*

Allow the underlying meaning and emotion to emerge naturally. Use the movement and gestures as you see fit.

This exploration is about the space between performers.

An Example recorded during a workshop at the University of Alaska Fairbanks:

A: Hi
B: Hi
A: Go?
B: No.
A: Why?
B: Feelings.
A: From?
B: Life.
A: Life?
B: Wow.
A: Sorry.
B: Yes.
A: Depressed?
B: Maybe.
A: Feeling?
B: Empty.
A: What?
B: Enthusiasm.
A: Hugs.
B: Thanks.
A: Walk?
B: Okay.
A: Com'on.
B: Now?
A: Now.
B: Thanks.
A: Come?
B: Okay.
A: Thanks.
B: Friend.

Part III

Advance to the next level by asking performers to use a brief phrase. Focus on maintaining authenticity and immediacy.

An Example from the same workshop:

A: Hiya doing?
B: Fine, and you?
A: Could be better.
B: What's up?
A: My aunt died.
B: Really?
A: Yeah.
B: Sorry.
A: All the sudden.
B: What happened?
A: Cancer, fast.
B: Her family?
A: Three kids.
B: Everyday a gift.
A: We have to help the kids.
B: Must hurt.
A: Nothing I can do.
B: How about therapy?
A: Done that.
B: No use huh?
A: Not at all.
B: How about sports casting?
A: Thought about it.
B: You'd be great.

Exploration Alternative: Create a dialogue with an animal or other non-human entity. Start the conversation, wait for a response—interpreting it as needed—and develop a unique way to communicate.

Note: This exploration aims to cultivate awareness of everyday conversations and personal spaces in exchanges. It encourages performers to be present for movements, gestures, words, emotions between people, beyond interactions, and with the environment.

In our fast-paced, stimulus-filled world, it is easy to default to superficial reactions. Performers' core responsibility is to remind our community of the vibrant life among humans and non-human beings. They are specialists and communicators, making the invisible visible, leading by example.

THREE LINES

Have performers select only three lines from a script, suggested, or any random text. Instruct them to move around the space and, when they meet other performers, to use any of these lines—either partially or fully—to interact.

Begin without any physical contact.

Performers may repeat entire lines, use fragments, or even just single words from their chosen text. They have the flexibility to modify or extend these lines or words as needed.

As the exploration progresses, encourage the performer to incorporate a touch or some form of interactive contact when engaging with others. This touch complements language and is organic to their meeting.

Encourage performers to explore their language and contact in various ways, such as softly, strongly, teasingly, or foolishly.

Performance

The next group of explorations is focused on the spaces of narrative, story, and performance.

NARRATIVE SPACES

When planning or developing a performance project, it is essential to first identify the story, its main ideas and goals, and outline the process.

Begin by sharing a story you or a performer wants to explore. Then, discuss and pinpoint the key moments—such as the highs, memorable images, or especially moving scenes.

Take notes or create an idea board to capture your insights.

Work with performers to identify each crucial moment in the story. Review the original context, modern reception, interpretations, and symbolism. Discuss personal connections, concepts, imagery, actions, emotions, and ideas that are evoked.

After the discussion, assign each narrative moment a simple, provisional name.

Then, position each narrative moment around the edges of the space. The narrative is physically represented as a large circle. You can mark the space using chalk or another method.

Then, ask performers to move to the narrative moment that resonates most with them.

Once there, they explore the issues associated with that moment: Actions, images, sounds, atmosphere, emotions, colors, movements, and related elements.

Some moments may have only one or no participants, which is normal and highlights the group's interests at that time. This will change as the exploration progresses.

Allocate 30 to 60 minutes for each group to discuss and develop their chosen narrative elements.

Afterward, groups share their insights and discuss their narrative moments and what they discovered.

Examine how and why these moments connect and resonate, including their symbolism and metaphorical meaning. And how best to develop the moment further.

Then, either in the same session or another session, return to the narrative exploration, encouraging performers to move to and explore another moment.

Repeat the process until all moments are explored.

You now have the seeds and outline of the narrative. You can choose to develop it further or use it as a performance creation methodology.

Note: This exploration moves us toward and anticipates performance development, which will be elaborated upon in the Place section of this book.

From the Ethiopian Workshop Journal

Lealem Berhanu, a 29-year-old workshop participant, playwright, and director of a children's puppet company, remarked, "Ethiopians are brought up not to question. Asking questions feels threatening." Nibret Baychenkin shared that he was taught to ponder "what needs to be done" instead of asking questions. Otherwise, he faced criticism or even punishment from his father, family, teachers, and the church. This is why we are the way we are.

Discussions and improvisations examining body codifications highlighted how religious conformity led to social, political, and imaginative uniformity. Unlike democratic societies such as the United States, Great Britain, and Western Europe, which have a rich history of individual rights and freedoms dating from the Magna Carta to the Universal Declaration of Human Rights, Ethiopia lacks such a legacy. Instead, Ethiopia's history reflects a series of rulers driven by religious motivations and political tyranny, stretching back to the 4th century. Haile Selassie's 44-year rule, ending in 1974, was akin to a medieval feudal system. It was followed by a civil war and 17 years of brutal communist rule, leading to the current ethnically centered, one-party "democracy." The norm of "not asking questions" remains widespread. Many ensemble members had relatives who experienced torture, imprisonment, or death during the terror ending in 1991.

Considering the frequent droughts, famines, civil wars, economic turmoil, and uprisings, it's clear that deep-rooted insecurities and traumas have been handed down through generations. Nevertheless, after 1,600 years of conformity—reflected in religious, educational, and political spheres—this rigid system is now being challenged by the egalitarian ideals emerging from a growing global culture, with its accompanying values and ambitions.

A recurring theme in our conversations, work, and improvisations was conformity. "We hesitate to ask questions and feel ashamed to accept others' opinions," said Nuredin Nesro, a tall Muslim performer. "That's why we stay inflexible; we're stuck in a single way of thinking and afraid of anything new," added Meaza Takele while breastfeeding her infant son.

Endale Berhanu, an Ethiopian Orthodox church deacon, shared his passion, saying, "People tend to focus on trivialities when they talk. They avoid discussing important issues, fearing to say anything meaningful. That's one reason Ethiopia struggles to progress; we're just fooling ourselves." Daniel Gumatain, a recent college graduate, expressed his frustration: "This is why mistrust is widespread among us; we doubt each other's intentions and often ask, 'What did they really mean?' It's a habit we've developed. We tend to stay silent because it feels safer—only speaking out on insignificant topics." The group described their culture and education as one that values rote learning, calling it a "mimic culture."

Creating our performance involved examining established modes and traditional forms, which serve both complementary and opposing roles. Traditional performance conveys identities, values, histories, and behaviors, offering a rich source of insight. However, they can also impose limitations. Cultural perceptions of self and body shape influence individuals' emotional, psychological, and imaginative experiences in life.

One could argue that traditional cultures persist through the ways bodies and their environments are expressed and represented. To inscribe upon the body is to exert control over individuals and society. Therefore, traditional legacies have a dual impact: They form a core part of identity while also imposing constraints, often upholding outdated norms that resist broader social and cultural change.

For more information about this project and production, see: Riccio, T. (2012). Shadows in the Sun: Context, Process, and Performance in Ethiopia. *New Theatre Quarterly*, *28*(3), 272–295. https://doi.org/10.1017/S0266464X12000450

GROUP STORY

Share a story orally.

The story can come from recent news, personal experiences, a legend, a folktale, or a myth. It should connect to the themes, interests, and context of your project. Choose a story with at least as many characters as there are performers in your group.

Let the performers know that they will create and perform the story. Assign all human and non-human roles.

Begin the story with you as the narrator. When it is a character's turn, they act and narrate their role in the present tense, with action. They can elaborate and add details as they wish. Keep the story flowing smoothly without interruptions. Continue this way, with each performer narrating in their own style and pace until the story ends. Then, the discussion begins.

Exploration Alternative: Once the story is told, ask the performers to retell it (from memory) using only illustrated actions. Performers enact the story until another performer continues.

A discussion then follows, focusing on identifying the story's plot, characters, actions, ideas, and performance styles.

Next, repeat the story in the same manner, with performers bringing an enriched and deeper understanding to it. The story can then grow and change, incorporating this greater insight, knowledge, and confidence in telling it.

Note: A great way to enhance storytelling skills, build confidence, and discover potential performance material.

All performance begins with a question.

From the Kenya Journal

CHAPS, Community Health Awareness Puppets, Nairobi

Notes from the opening session of a five-day workshop, lasting 8 hours with a 45-minute lunch break. The group included nurses, teachers, a mechanic, health workers, and puppeteers, with experience levels from one to seven years.

Participants came from different regions of Kenya, and most do not know each other.

Introduction

I started with a brief introduction and interviewed the performers to assess their experience and interests by asking questions, building rapport, and understanding their personalities.

I explained the process, guided the group through it, and clarified the rationale.

The focus was on finding common ground, which is vital for breaking down barriers among people from different regions, religions, political views, economic backgrounds, and tribal affiliations.

Many participants mentioned they come from rural areas and feel uncomfortable and anxious in Nairobi. I encouraged them to share their feelings.

Emphasize outlining the process since they will use and share our work with their groups. Our efforts are long-lasting and sustainable. We are all part of one community.

Key points introduced

- Foster your observation skills. Reflect on your process. Set clear goals and priorities.
- Your work should stem from your own experiences, understanding that it is often undervalued due to generational inequalities and stories that shape self-perceptions, educational, gender, and sexual insecurities.
- You have a unique contribution to make and are playing a valuable role in Kenya's growth and healing.
- If you believe what you are doing is worth it, then give it your best effort. Do it for others, your community, and yourself. Take pride in who you are and what you do.
- Explore and embrace your cultural heritage—it is a vast, rich resource.
- Our work is spiritual and cyclical, like life itself, a revisiting of ideas, feelings, and themes.

Workshop Exploration Sequence

- Circle
- **Mountain**
- Heartbeat and breathing awareness
- **Walks**—It is a physical and straightforward activity—simple, communal, non-threatening, and accessible; something anyone can do.
- Then, progress to more complex levels. Walks—emulate the walks of others—gain insight into their world.

Break

- **Mirrors Series**
- **Alternative Moves**
- **Three Moves**

Lunch Break

- **Babble** (as a prelude to accessing difficult subjects)
- **"I Am"**

Closing Discussion

Reiterated and reviewed the takeaways.

- Be mindful and trusting of yourself—recognize and value the opportunities that your environment and situation present.
- Listen to your own feelings to confront and counteract the "colonial doubt."
- Recognize and utilize your unique voice and your cultural voice.
- Become the change you envision.
- Working in the areas of violence against women, corruption, FGM (female genital mutilation), and AIDS awareness requires you to be mindful of your own well-being.
- Document your activities and thoughts, reflect on them, correct, and adjust as necessary.
- Talk to colleagues. Keep exploring and expanding your theater and acting techniques.
- Needs and situations vary and differ.
- Don't become fixed, lazy, or stuck in one approach or perception. Remain open and listen to yourself, others, and the world. Keep records and review them.

Expansion

All beings, human or not, emit a radiance that extends from their body, mind, and spirit into the surrounding world—beyond the physical, formless yet perceptible. This is the aura that envelops a person, an animal, an object, a house, or an idea—the memory of a moment or dream image. Each being and thing occupies a physical space shaped by the fleeting, chance interactions of a passing moment. These vectors radiate energy outward, pulsating with life and forming an aura. The more layers and spaces are vectored, the more complex the effect and the stronger the attraction to that entity. As the world becomes increasingly technologically fixed, regulated, and defined, knowing that the unresolved and inordinate still exist is a necessity and reassurance.

Celebrities, movie stars, fashion models, villains, criminals, artists, religious figures, politicians, and controversial personalities embody radiance—an elusive quality hard to define. Similarly, infants, the elderly, trees, mountains, the Amazon rainforest, dogs, and cats also radiate this presence. This sympathetic magic phenomenon arises from a blend of innate qualities, context, needs, and expectations that generate projections, speculations, and interactions. Openness, accessibility, and relatability are essential—remaining vivid and distinctive while allowing multiple interpretations creates a contrast that attracts and invites thought, feeling, and aliveness. A forest, in its own way, is like a movie star or religious figure; it feels authentic, present, and tangible, difficult to grasp easily, accessible through the complex interplay of familiarity and interaction. All are positioned archetypically and imbued with expectations tied to social, cultural, and historical encoding. They serve a need; they are indicators, symbols, and reassurances that life and the world are unresolved, mysterious, and ambiguous—and, in that, life and the world are still in the process of forming, and there is hope. It is this interplay of contrasts and contradictions that intimates possibility, the unexpected, and change.

Up to this point, our work has identified and explored various spaces. These include spaces of culture, society, imagination, body, inheritance, physical, personal, psychological, emotional, and group dynamics, among the many we inhabit. Now, we turn to and introduce expanded and radiant spaces. A topic worthy of a book is brought to your attention here.

RADIANCE

Starts as a visual exploration.

Performers lie on their backs with eyes closed. Once they have achieved deep breathing and relaxation, ask them to visualize ripples of light emanating from their core, resembling small concentric waves like those created when a pebble is tossed into a still pool.

Invite the performers to radiate color from the center. They may choose any specific color that inspires them, or a combination of colors.

Encourage visualizing with synesthesia, seeing the colors of your body's radiance pulsing into the world.

Allow sufficient time for the visualization to reveal itself.

Once the color radiating from the center has been established, ask the performers to visualize it spreading throughout their bodies.

Side Coach

- *Wave patterns shaped like your body radiate outward. The intensity of these wave patterns fluctuates with each inhalation and exhalation.*
- *Imagine the color flowing in waves around you and spreading into the surrounding space.*
- *The illuminance of your body radiates into the space around you.*
- *Your color and energy extend far into the universe. Feel what your energy touches as it extends deep into space.*
- *While your eyes remain closed, sit up and continue breathing the color from your body into the space around you.*
- *When ready, stand up and continue emanating your color(s) and visualizing the waves of radiance flowing from your body form.*
- *Observe how the radiance of your body form moves, then how it changes when it encounters and interacts with the radiance of others, objects, architecture, and your movements.*

Encourage performers to move slowly, expanding their colors and energy as they go. Remind them that their source of color always comes from the body's radiance.

Exploration Alternative: Use the same setup and progression, but instead of or alongside visualizing color, have performers vocalize a sound. This can be a personal sound or relate to the **Vowel Colors** exploration from the **Body** phase of our work.

Note: This exploration is most effective once performers are familiar and comfortable with guided relaxation and visualization techniques. It can then be used independently or combined with other visualization explorations, depending on the group's needs and goals.

Highly Recommended.

COLOR SPACES

Begin by having the performers lie comfortably on the ground. With eyes closed, guide them to establish a calm and steady breathing pattern, inhaling through the nose and exhaling through the mouth. You might also consider including a tension-release exploration as a prelude.

Performers are then encouraged to inhale a color—primary colors work best, but they may also include black, violet, orange, pink, and green.

As they breathe in, performers visualize the color filling their entire body down to their toes. When they exhale, the same color shoots out like a fountain from their mouth toward the ceiling.

As their breathing deepens, encourage them to exhale with a natural, unforced sound. When the color and sound become steady, invite the performers to stand. Then, have them open their eyes and begin moving, inspired by the color. If they lose focus—the connection between sound and color—they can close their eyes and resume movement once they are refocused.

Side Coach

- *Transform your color sound and flow with it. What emotion does it express?*
- *Start forming an abstract image in your surroundings, using the colors and sounds that originate within you.*
- *Your body embodies the color, and everything you touch radiates the color you inhale and exhale. Your feet and hands act as paintbrushes, allowing you to paint the room with color and sound.*

  Give the color painting time to develop.

  When the color exploration has peaked, ask the performers to **Meltdown** and settle on the ground.
- *With each breath, the color you were focusing on fades from your body, and you return to your normal, neutral breathing.*

Next, use the same method to add another color, similarly developing the color and movement throughout the space with sound to enhance the environment.

Exploration Evolution: Introduce a story or myth to the group.

Using the colors that resonate most with the group, guide them to visualize and illustrate the landscape, then the characters, actions, and story in storyboard form.

Note: The sounds and movements associated with specific colors can vary greatly across different groups. Each sound creates its own atmosphere and emotional tone. By paying attention to this, you can gain a deeper understanding of how a particular group or culture connects and interacts with the world. These insights can be surprising and are a valuable resource.

Using colors can offer valuable insights for future growth. Focus on experiencing the colors that resonate most deeply for an extended period. It's recommended to limit the initial session to no more than five colors.

Highly Recommended.

GROUP COLOR SPACES

Performers line up at one end of the space and are instructed to move according to a color: Red, Yellow, Green, White, or any other. The color choice can relate to the current workshop, rehearsal, or performance goals. For example, if the group is focusing on a mourning scene, they might explore black or white, depending on the cultural context.

On the other hand, for a spiritual scene, white can be used. This approach emphasizes the use of radiance consciously.

Ask performers in both lines to close their eyes.

Then ask the performers, in one line, to radiate a single color, developing it until it achieves a unified expression.

The performers in the opposite line remain accessible and receptive to the radiant color expression from the other line.

When prepared, the radiating line opens its eyes and gradually approaches the receiving line, ending up just a few feet away. I also recommend encircling the receiving line.

Without comment, the line of performers switches their roles and repeats.

Discuss afterwards.

Exploration Variation: Once performers are comfortable with the exploration, introduce a scenario featuring radiating colors, transitioning from one hue or shade to another with varying intensity.

For example, start with a color representing death and transition through feelings of grief, sadness, release, calm, gratitude, and memory. etc., Side coach to introduce and mark the shifts. And, if desired, includes additional images, actions, or ideas.

Exploration Variation: Guide the group in exploring and defining the movement(s) and gesture(s) associated with the radiant color.

Note: The exploration can develop into performers collaborating physically and vocally to create, for instance, a dance, sound, or song inspired by color or evolved from color.

AURA SPACE

Partners stand facing each other at just over twice arm's length, keeping their eyes closed during the exploration.

They gently extend their arms in front of them. Without physical contact, they are encouraged to sense each other's energy and align their open palms with their partner's.

After the initial attempt to match palms, instruct the partners to lower their hands to their sides, then turn around in place three times.

Ask performers to revisit the exploration, concentrating on sensing their partner's energy through their palms. They should feel the energy, then move their palms sideways away from the source, and then explore the dimensions of the energy space between them. Partners then open their eyes to observe the energy connection.

FIGURE 2.16 Aura exploration in a workshop with the DirectorsLab Chicago

Source: Photo by the author.

A TREE

Ask performers to find a tree where they can sit.

Guide performers to seek the tree's permission, either verbally or through gestures, before sitting beside it. If the tree shows approval—something they can sense intuitively—the performer may rest their back against the trunk. Encourage them to tune in, feeling the tree's support and vitality with their bodies. Allow enough time for reflection and mindfulness.

The performer imagines the tree, symbolically entering and exploring its interior, embodying it. Starting with the root system, visualize drawing water up through the roots and transporting earth minerals and nutrients throughout the entire tree and its structure.

Encourage the performer to feel the flow of energy as each cell of their "tree body" is nourished. Observe how energy moves through their human form, much like it flows within a tree.

The body's trunk channels energy throughout the tree, reaching the branches and extending to the performer's extremities. This energy sustains the leaves and fruit, symbolizing life and supporting your existence. Encourage the performer to feel their link to all living things and see themselves as part of the totality of life.

Encourage performers to develop an awareness, intuition, and continuing relationship with their environment, aiming to perceive their daily experiences as flowing through them. All phenomena—such as trees, clouds, birds, and others—exist both externally and internally within the performer.

Note: This exploration was inspired by my work with the Athabaskan people of Alaska's vast boreal forest. Before cutting down a tree, they place their hands on it, explain their reasons for doing so, what it will provide, and offer thanks. After the tree is felled, it remains whole and uncut to allow the spirit of the tree to depart—spruce is given a day, while birch, which has lived longer and has an older spirit, is given three days.

Archetypes

Jung and Freud's interest in myth and ritual stemmed from a deep understanding that human behavior follows universal patterns beyond individual experiences. Like shamans and healers worldwide, Jung and Freud believed that psychological healing involved connecting individuals with these fundamental archetypal structures. Deep structural patterns woven into both personal psychology and the collective fabric of society, including myths, laws, social hierarchies, political systems, religious practices, and economic relationships (Jung 1998, 40).

Today, critics often dismiss archetypes as overly simplistic, believing they reduce the complexity of human identity. However, this view overlooks their true purpose. Archetypes are not strict categories but dynamic templates—what Eliade describes as patterns that exist "in the Time of the Species— even of organic Life itself" (1959, 54). They function similarly to Platonic Ideas, creating a cultural legacy through which each generation interprets and reinterprets human experience. In my work with urban communities of Indigenous and mixed backgrounds, I've seen that exploring archetypes is crucial for understanding how people navigate, resist, and reshape social and cultural roles. Archetypes give depth to history, enabling individuals to evaluate, adapt, and distinguish their actions—they are the foundation of cultural identity.

We are currently experiencing an unprecedented transformation of archetypes. Our highly mediated global society is blending traditionally distinct, culturally specific archetypes into more uniform, universal versions. Hollywood franchises such as Marvel, *Star Wars*, and *The Lord of the Rings* play a key role in this, spreading Western narrative patterns and character types globally. Social media further speeds up this process by promoting archetypal forms that are immediately recognizable and easy to share.

A fascinating paradox emerges despite globalization bringing a wealth of diverse cultural expressions, which often tend to simplify into archetypal forms. Why does increased diversity lead to more sameness? Two reasons stand out. First, as cultures interact and influence one another, simplified forms may develop naturally to ease communication and build shared identities—a kind of cultural common language. Second, this tendency to essentialize might be a survival tactic against information overload, helping people navigate the overwhelming flow of images and stories by focusing on broad, easily understandable symbols. This could be a return to core values, a collective grounding, or an unconscious effort to create a new global mythology during uncertain times.

Jung noted that "archetypes are impressions of ever-repeated typical experiences... when an archetype appears in a dream, fantasy, or life, it always carries a certain influence or power, either evoking a mystical or captivating effect, or pushing to action" (Jung 1998, 78). During periods of change and anxiety, revisiting and reiterating archetypes provides reassurance—they symbolize simplicity, fundamental truths, and essential ideas.

This reveals an essential truth: Archetypes form the fundamental structures underlying all humans and cultures. We are essentially agent-avatars—vessels for inherited behavioral patterns that influence our world, whether we realize it or not. Recognizing this inheritance is crucial. My work with archetypes, involving both non-Indigenous and Indigenous application—including projects with the Chicago Director's Lab in 2008, Lul Theatre in Ethiopia in 2009, initiatives in Pondicherry, India, Nepal in 2013, the Miao (2001–2019), my work with Hanson Robotics, and with the Dead White Zombies from 2011 to the present—centers on this deep exploration. Using psychophysical responses and continuous inquiry, performers explore the patterns they carry and reassess key aspects of their biological, psychological, social, and cultural identities. These practices do more than analyze archetypes from a distance; they activate them, allowing participants to experience how these deep structures function within their bodies and communities. This process helps them regain control over the archetypal inheritance that might otherwise persist through them, especially as global media increasingly influences which archetypes dominate our collective imagination.

ARCHETYPAL SPACES

Invite the performers to select an archetype from their cultural heritage, popular culture, mythology, or classical literature.

Here are some suggestions to help you get started with your exploration.

Odin	The Dumb Blonde	The Hunter
Raven	Jesus	Ulysses
Coyote	Commando	Zeus
The Hero	Princess	Iron Man
The Damsel in Distress	Krishna	Neo (Matrix)
	The Zulu Warrior	Terminator
The Evil Villain	The Poor Mother	Robot
Superman	Werewolf	Zombie
Mickey Mouse	Vampire	

The performer is asked to develop the archetype's movement and walk from what they know and associate with it.

Side Coach

- *Their way of moving: Sharp? Curved?*
- *How is the body aligned? Bent? Straight? Where does the archetype's energy come from?*
- *Where is their center?*
- *What part of the body do they lead from?*
- *What is their attitude toward the world? Are they happy? Sad? Angry?*
- *Explore a variety of emotions.*

- *What attracts you to the archetype?*
- *How are you similar? Different?*
- *What comforts you about the archetype? What frightens you?*
- *What attracts you to the archetype?*
- *How are you similar? How are you different?*
- *What comforts you about the archetype?*
- *What frightens you?*
- *Is this your archetype?*
- *What space or spaces does your archetype inhabit?*
- *What kind of scenarios are suggested?*

Customize and modify questions to fit specific needs and relevant archetypes. No extensive prep work is required for the study; participants should depend on their current knowledge. If their understanding of the archetypes is unclear, encourage them to expand and clarify what they know.

This activity aims to identify which archetypes resonate with the performers. Multiple archetype explorations may occur within a single session.

Note any archetypes that particularly connect with individual performers.

This exploration can serve as a foundation for developing performances and characters or as a reference for future work.

Exploration Variation: Instead of relying on cultural stereotypes, you can portray characters from everyday life. Examples include postal workers, doctors, people who are overweight or slender, pilots, drinkers, seniors, infants, football players, fashion models, bar owners, street vendors, sex workers, farmers, police officers, politicians, and bureaucrats.

ELEMENTAL ARCHETYPES

This exploration is inspired by and builds upon The Hero and the Outlaw: Building Extraordinary Brands Through the Power of Archetypes by Margaret Mark and Carol Pearson (2001).

A solid starting point for exploring and understanding the archetypes that are fundamental to human societies, cultures, and personalities.

Innocent
Explorer
Sage
Hero
Outlaw
Magician
Average Guy or Gal
Lover
Jester
Caregiver
Ruler

Review the listed archetypes together, moving through each one. A detailed analysis will identify the archetype that best fits each performer.

Performers can begin sitting or lying down with their eyes closed. Guide them to visualize how the archetype transforms different parts of their body while keeping their eyes closed.

The progression starts from the feet and moves upward through the body to the head, including the jaw, ears, mouth, nose, forehead, cheeks, and eyes. Participants should stay aware of the transformation, breathing in sync with the archetype's rhythm.

Encourage performers to gradually explore their environment, paying attention to how their renewed body and awareness perceive and reshape the world, recognizing their space.

- Archetype explorations can be gradual, taking place over multiple sessions and revisited as needed, or condensed based on time constraints and requirements.
- As always, listen, observe, and adapt. There isn't a single correct approach to exploration.
- Use this initial exploration as a template for the subsequent archetype explorations.
- Apply the same template to guide the upcoming archetypal explorations.

Innocent

Side Coaching Throughout

The Innocent may also be known as Pollyanna, utopian, traditionalist, naïve, mystic, saint, romantic, dreamer.

Your Motif:	Free to be you and me
Your Desire:	To experience paradise
Your Goal:	To be happy
Your Fear:	Doing something wrong or bad that will provoke punishment
Your Strategy:	To do things right
Your Trap:	Naïve and can be duped or used by others
Your Gift:	Faith and optimism

Explorations

Continue side coaching, asking performers to respond, move, express, and explore. There is no right or wrong way to do so.

- Express a desire for purity, goodness, and simplicity.
- Convey a childlike simplicity: Naïve, dependent, obedient, trusting, and idyllic.
- Suggest renewal, cleansing, and re-entering the Promised Land.
- Evoke an almost mystical sense of oneness.
- Innocence stems from values and integrity, not external experience—being, not doing.
- Acknowledge denial, repression, and avoidance.

Reiterate the first section, including the motif, desire, goal, etc.

The Innocent Archetype

- Seeks a clear answer to a simple problem.
- Is connected to goodness, morality, simplicity, nostalgia, or childhood.
- Has functions related to cleanliness, health, or virtue. Is moderate.
- Demonstrates straight-arrow core values. Aims to stand apart from what is tarnished.
- Give ample time and promote the investigation of each archetype. Offer side coaching and reiterate as needed to aid the improvisation.

As the exploration unfolds, performers can be encouraged to 1) express vocally, 2) engage and collaborate with others, 3) utilize straightforward phrases or statements that can be easily repeated, and 4) construct a scenario.

Explorer

Known as the seeker, adventurer, iconoclast, wanderer, individualist, pilgrim, quester, antihero, and rebel.

Motif:	Don't fence me in.
Desire:	The freedom to find out who you are through exploring the world.
Goal:	To experience a better, more authentic, more fulfilling life.
Fear:	Getting trapped, conforming, inner emptiness, nonbeing.
Strategy:	Journey, seek out, and experience new things, escape from entrapment and boredom.
Trap:	Aimless wandering, becoming a misfit.
Gift:	Autonomy, ambition, and the ability to be true to one's soul.

Explorations

- Alienation, dissatisfaction, restlessness, yearning, and boredom.
- Hitting the open road, going out into nature, exploring the world.
- Seeking your own individuality, to individuate, to become fulfilled.
- Expressing individuality and uniqueness.
- Being so alienated, you cannot find any way to fit in.

The Explorer Archetype

- Helps people feel free, is nonconformist, or is pioneering in some way
- Is rugged and sturdy, comfortable in nature, on the road, or in dangerous settings or occupations.
- Lives an alternative lifestyle.
- Expresses and helps others to express their individuality.
- Is an on-the-go person.
- Is different from a non-conformist.
- Organizes their life to be an explorer.

Sage

Known as an expert, scholar, detective, oracle, evaluator, advisor, philosopher, researcher, thinker, planner, professional, mentor, teacher, and contemplative.

Motif:	The truth will set you free.
Desire:	To discover the truth.
Goal:	To use intelligence and analysis to understand the world.
Fear:	Being duped, misled, and ignorant.
Strategy:	To seek out information and knowledge, become self-reflective, and understand thinking processes.
Trap:	Can study forever and never act.
Gift:	Wisdom and intelligence.

Explorations

- Confusion, doubt, and deep desire to find the truth.
- Searches for absolute truth, desire for objectivity, looking for experts and knowledge.
- Skeptical, critical, and innovative thinking, becoming an expert
- Wisdom, confidence, and mastery.
- Dogmatic, removed, and disconnected from reality.

The Sage Archetype

- Provides expert information.
- Encourages others to think.
- Knows scientific breakthroughs or esoteric knowledge.
- Is supported by hard data and logic.
- Focuses on quality and is unquestionable.

Hero

Known as a warrior, a crusader, a rescuer, a superhero, a soldier, a winning athlete, a dragon slayer, a competitor, and a team player.

Motif:	Where there is a will, there is a way.
Core:	Prove one's worth through courageous and difficult action.
Goal:	Exert mastery in a way that improves the world.
Fear:	Weakness, vulnerability, and wimping out.
Strategy:	Become as strong, competent, and powerful.
Trap:	Arrogance, developing a need for there always to be an enemy.
Gift:	Competence and courage.

Explorations

- The bully kicks sand in your face, or someone tries to intimidate or abuse you; a challenge beckons; someone needs you to help defend him or her
- The development of boundaries, competence, and mastery expressed through achievement, motivated or tested through competition
- As with a soldier, doing your duty for your country, organization, community, or family.
- They use strength, competition, and courage for something that makes a difference to you and to the world.
- Ruthlessness and obsessive need to win.

The Hero Archetype

- Has the potential for invention or innovation and will have a major impact on the world.
- Helps people perform at their upper limit.
- Addresses major social problems and asks people to help address them.
- They have a clear opponent or competitor they want to beat.
- Is an underdog and wants to rival the competition.

- Has the ability to do a tough job efficiently.
- Has follow-through.
- Is a good, moral citizen.

Outlaw

Known as the rebel, the revolutionary, the villain, the wild man or woman, the misfit, the enemy, or the iconoclast.

Motif:	Rules are made to be broken.
Desire:	Revenge or revolution
Goal:	To destroy what is not working for the outlaw or for society
Fear:	Being powerless, trivialized, inconsequential
Strategy:	Disrupt, destroy, or shock
Trap:	To go over to the dark side, criminality
Gift:	Outrageousness, radical freedom

Explorations

- Feeling powerless, angry, mistreated, under siege.
- Identifying as an outsider, dissociating from the values of the group or society, and its conventional behaviors and morality.
- Behaving in shocking or disruptive ways.
- Becoming a rebel or a revolutionary.
- Criminal or evil behavior.

The Outlaw Archetype

- Is dissatisfied and disaffiliated from society and identifies with values at odds with those of society at large.
- Functions to destroy something, like a bulldozer or virtually like in many video games, and is genuinely revolutionary.
- They are not very good for people, thumbing their nose at society's ideas and what constitutes the norm.
- Helps retain values that are threatened by prevailing ones.
- Pioneers new and revolutionary attitudes.

Magician

Known as the visionary, catalyst, innovator, charismatic leader, mediator, shaman, healer, or medicine man or woman.

Motif:	It can happen.
Desire:	Knowledge of the fundamental laws of how the universe works.
Goal:	To make dreams come true.
Fear:	Unanticipated negative consequences
Strategy:	Develop vision and live it
Trap:	Becoming manipulative
Gift:	Finding win-win outcomes

Explorations

- Hunches, extrasensory or synchronistic experience.
- Magical moments and experiences of transformation.
- The experience of flow.
- Miracles, moving from vision to manifestation.
- Manipulation, sorcery.

The Magician Archetype

- It's about a transformative experience.
- Carries the implicit promise of transformation.
- Shapes the future and is a cultural creative.
- Expands and extends somatic mindfulness.
- Is user-friendly.
- It is spiritual and psychological.
- It is new and very contemporary
- Seeks quality and depth.

Average Guy or Gal

Known as the good ole boy or girl, the regular Jane, Everyman, the common man, the guy or gal next door, the realist, the working stiff, the solid citizen, and the good neighbor.

Motif:	All men and women are created equal.
Desire:	Connection with others.
Goal:	To belong, fit in.
Fear:	Standing out, putting on airs, and being exiled or rejected.
Strategy:	Develop ordinary solid virtues, the common way, blending in.
Trap:	Give up self to blend in and have only superficial connections.
Gift:	Realism, empathy, and lack of pretense.

Explorations

- Loneliness, alienation.
- The orphan, feeling abandoned and alone, seeks a sense of affiliation.
- The joiner, learning to connect, fit in, accept help, and friendship
- The humanitarian believes in the inherent dignity of every person, regardless of their abilities or circumstances.
- The victim who is willing to be abused rather than be alone.
- The lynch-mob members who are willing to go along with abuse to be one of the gang.

The Average Guy/Gal Archetype

- Is someone who helps people feel they belong.
- Function is common and everyday.
- Moderate and open.
- Down home.
- Positive and non-elitist.

The Lover

Known as partners, friends, intimates, matchmakers, enthusiasts, connoisseurs, sensualists, spouses, team builders, and harmonizers.

Motif:	I only have eyes for you.
Desire:	To attain intimacy and experience sensual pleasure.
Goal:	Be in a relationship with the people and experiences they love.
Fear:	Being alone, a wallflower, unwanted, and unloved.
Strategy:	Become more and more attractive-physically and emotionally.
Trap:	Doing anything to attract and please others, losing identity.
Gift:	Passion, gratitude, appreciation, and commitment.

Explorations

- Infatuation, seduction, falling in love (with a person, an idea, a cause, work, a product).
- Seeking great sex or great romance.
- Following your bliss and committing to whom and/or what you love.
- Spiritual love, self-acceptance, and the experience of ecstasy.
- Promiscuity, obsession, jealousy, and envy, Puritanism.

The Lover Archetype

- Helps people find love or friendship.
- Fosters beauty, communication, or closeness between people or is associated with sexuality or romance.
- Is intimate, elegant, organized collectively rather than as a hierarchy.
- Needs to differentiate from others that do not feel as deeply or are as committed as they are.

The Jester

Known as the Fool, trickster, joker, punster, entertainer, clown, prankster, practical joker, or comedian.

Motif:	If I can't dance, I do not want to be part of your world.
Desire:	To live in the moment with full enjoyment.
Goal:	To have a great time and lighten up the world.
Fear:	Boredom or being boring.
Strategy:	Play, make jokes, and be funny.
Trap:	Wasting away one's life with idle pursuits.
Gift:	Joy

Explorations

- Ennui, boredom.
- Life as a game, fun.
- Cleverness used to trick others, get out of trouble, and find ways around obstacles, and transformation.
- Life experienced in the moment, one day at a time.
- Self-indulgence, irresponsibility, and mean-spirited pranks.

The Jester Archetype

- Helps people feel they belong.
- Helps people have a good time.
- Is fun-loving, freewheeling.
- Is differentiated from self-important, overconfident, and established others.

The Caregiver

Referred to as the caretaker, altruist, saint, parent, helper, or supporter.

Motif:	Love your neighbor as yourself.
Desire:	Protect people from harm.
Goal:	To help others.
Fear:	Selfishness, ingratitude.
Strategy:	To do things for others.
Trap:	Martyrdom, entrapment of others.
Gift:	Compassion, generosity.

Explorations

- Seeing someone in need.
- Caring for and nurturing one's dependents.
- Balancing self-care with the care of others.
- Altruism, concern for the larger world.
- Martyrdom, enabling, guilt-tripping.

The Caregiver Archetype

- Serves the community and others
- Supports family, friends, and is associated with nurturance.
- Works in health care, education, other caregiving fields, and politics.
- Helps people stay connected with one another.
- Helps people care for one another.
- Supports causes and charitable activities.

The Creator

Known as an artist, innovator, inventor, musician, writer, or dreamer.

Motif:	If it can be imagined, it can be created.
Desire:	Create something of enduring value.
Goal:	Give form to a vision.
Fear:	Having a mediocre vision or execution.
Strategy:	Develop artistic control and skill.
Trap:	Perfectionism, mis-creation.
Gift:	Creativity and imagination.

Explorations

- Daydreams, fantasies, and flashes of inspiration.
- Being creative or innovative in imaginative ways.
- Giving form to your own vision.
- Creating structures that influence culture and society.
- Overly dramatizing your life, living a soap opera.

The Creator Archetype

- Encourages self-expression, choices, and options.
- Fosters innovation or is artistic.
- Works in fields like arts, public relations, and technology.
- Is a do-it-yourself type of person.
- Has time for creativity.
- Creates and organizes their life around creativity.

The Ruler

Known as the boss, leader, aristocrat, parent, politician, responsible citizen, role model, manager, or administrator.

Motif:	Power isn't everything. It's the only thing.
Desire:	Control.
Goal:	Create a prosperous, successful family, company, or community.
Fear:	Chaos, being overthrown.
Strategy:	Exert leadership.
Trap:	Being bossy, authoritarian.
Gift:	Responsibility, leadership.

Explorations

- Lack of resources, order, or harmony.
- Taking responsibility for the state of your own life.
- Exerting leadership in your family, group, organization, or workplace.
- Becoming a leader in your community, field, or society.
- Tyrannical or manipulative behaviors.

The Ruler Archetype

- Has a high status and is powerful.
- Wants to enhance their power.

- Helps others get more organized.
- Helps maintain or enhance power.
- Is distinct and a clear leader.
- It is reliable, stable, promising, and provides safety and predictability in a chaotic world.

Exploration Alternatives and Evolutions: After each archetype has been explored and discussed, several options for continuation arise. The performers may choose one archetype to explore in greater detail, defining its psychological, emotional, and physical characteristics. Once determined, the archetypes can work improvisationally to create characters and scenes for performance.

Improvisations can involve two or more archetypes or serve as a collective exploration where performers interact and shift among various archetypes.

When combined with a rehearsal process, archetype exploration can enhance character development.

Performers might explore several different archetypes or combine attributes from several archetypal explorations to create a character unique to their imagination.

Performers should reflect on their connections to specific archetypes, rather than others. They should consider what each archetype reveals about their own approach to life.

Questions to explore:

Which archetypal models influence your life and existence in the world?

When do you take on these archetypal roles?

Archetypal Ensembles

The 2005 performance, "There is Never a Reference Point," transformed a Dallas warehouse into a fragmented psyche. Based on Jamie Dakis's experience with Dissociative Identity Disorder (DID), commonly known as multiple personality disorder. The production, which I devised and directed, was inspired by years-long correspondence and interactions with Jamie, a former student of mine at the University of Alaska Fairbanks. The performance served as a template for my research and development work and for the founding of the Dead White Zombies, a Dallas-based, post-disciplinary performance group.

Producing a performance in a traditional theatrical space or dramaturgical form would frame, confine, and normalize the content, passion, and essence of Jamie's world. The form of the performance was just as important as its content. Placing the performance in a traditional theater—a space that has become institutionalized and corporatized—echoes a legacy from the Ancient Greeks and the European Enlightenment. This legacy involves a seated, detached body observing an illuminated mind, reinforcing the outdated split between mind and body. Such a setup is a remnant of a bygone era and is incompatible, if not obsolete, in our age of interactive engagement and empowerment.

After months of struggling, trying to find a form that would honor and best represent the content, I hit upon the concept of an immersive narrative. I was in the Natural History Museum in Prague, walking through a floor of themed displays by myself, when it struck me. The arc or frame of the event was shaped by my associations and my walking through it, influenced by my pace, personality, and what caught my interest, forming a narrative unique to me. My takeaway was that the audience was immersed in Jamie's psyche, where eight of her primary alters—out of nearly 40—were always present in varying states of engagement. Each audience member was free to move, shaping their own perception and interpretation. They saw and engaged with what they needed and wanted—this principle would go on to influence the Dead White Zombie aesthetic. The performance would be authentic and organic, true to Jamie's experience. Form and content were equally important, if not the same.

In a large, brick-walled, cavernous basement of a 120-year-old, former Sears and Roebuck warehouse, near downtown Dallas, we created eight distinct environments for eight personas—Jamie (the hub character), Marta, James, Rita, Patty, Magick, Misty, and Cleopatra—each with their own movement vocabulary, emotional territory, and spatial domain. Audiences walked freely among these personas, experiencing Jamie's persona not as unified but as an ensemble of voices in continuous, sometimes overlapping, dialogue.

The process of creating the work relied heavily on archetypal explorations. Just as performers investigating the Hero archetype discover a distinct center of gravity and worldview, each of Jamie's personas operated from an archetypal core. With research and consultations with psychologists and psychiatrists specializing in DID, culturally encoded archetypal affinities were how a person suffering with the condition ordered their world. James was an archetypal variation of the protective masculine authority, a complexity of the Hero and Outlaw. Marta was a distorted version of the Caregiver and Ruler, transformed into a judgmental mother. Cleopatra is a transhistorical Sage offering timeless wisdom, female empowerment, and acceptance. Magick expressed a transformative Magician mixed with Lover sensuality. Each served a specific archetypal function within Jamie's internal community, much like archetypes serve essential functions in human societies.

The crucial distinction: In DID, fragmentation emerges as a survival mechanism from severe trauma. When Jamie's uncle sexually abused her at two and a half years old, her persona splintered to survive what could not be integrated. This is clinical psychology, not metaphor. Archetypal performance work, by contrast, is conscious cultural technology. Performers deliberately access different aspects of human experience while maintaining meta-awareness and making conscious choices.

Yet, the parallel proves instructive. Both reveal the mind as operating as an internal ensemble rather than a singular self. Anthropological evidence supports this: Paleolithic survival likely depended on individuals embodying distinct archetypal roles. Communities needed the Explorer, who ranged beyond known territory, the Caregiver, who tended to the vulnerable, the Hero, who faced danger, the Sage, who preserved knowledge, and the Jester, who maintained social cohesion. These weren't jobs, but essential archetypal functions that ensured communal survival.

There is Never a Reference Point demonstrated how these functions can internalize as distinct personas. The performance's genius lay in spatial architecture. Rather than a linear narrative on a proscenium stage, each persona occupied their own environment: Marta's oppressive kitchen, James's bar with a poker table, Rita's desk piled with books, Magick's ritual altar, and Patty's staircase leading to the bedroom of abuse. Each space manifested an archetypal territory within the self.

Audiences navigating this space experienced what archetypal exploration reveals: Different personas perceive reality through fundamentally different lenses. The Hero sees challenges to overcome; the Caregiver sees needs to meet; the Outlaw sees systems to disrupt; the Innocent sees purity to preserve. Each archetype reorganizes perception, movement, emotion, and relationship.

Archetypal exploration develops performance vocabulary through this principle. When performers investigate the 12 elemental archetypes—Innocent, Explorer, Sage, Hero, Outlaw, Magician, Average Guy/Gal, Lover, Jester, Caregiver, Creator, Ruler—they build a repertoire of distinct ways of being. The Hero leads from the solar plexus with forward momentum; the Innocent moves with open, trusting gestures; the Outlaw's energy comes from rebellious defiance. These aren't abstract concepts, but realities gained through haptic, movement, and vocal interaction, which is central to our psychophysical approach to performance.

Jamie's eight personas were not random; each carried specific survival capacities. When Marta's judgment became unbearable, Magick's rebellion provided escape. When trauma memories overwhelmed Patricia, Rita's analysis created distance. When James threatened violence, Cleopatra's wisdom restored power. The internal community functioned as a survival mechanism.

A healthy persona operates similarly, though with more integration. We shift between archetypal modes daily: The Ruler organizes our schedule, the Caregiver responds to distress, the Jester defuses tension, and the Hero tackles challenges. The difference between healthy flexibility and pathological dissociation lies in conscious choice, memory continuity, and ability to integrate experiences rather than experiencing them as separate, amnestic personas.

This understanding transforms our work. Traditional Western acting assumes unified character psychology. However, archetypal exploration suggests that even within a single character, multiple archetypal energies can operate. A mother shifts between Caregiver and Ruler, between Innocent and Outlaw, depending on circumstances. Rather than seeking psychological consistency, performers explore how different archetypal modes emerge contextually, creating more complex portraits of selfhood, mind, and being.

The spatial architecture offers a template. Just as each persona inhabited a distinct environment, performers can ask: What spaces does this archetype inhabit? The Hero occupies battlefields and competitive arenas—the Innocent dwells in gardens and safe havens. The Sage inhabits libraries and contemplative spaces. Spaces shape the self, which, in turn, shapes the place of selfhood.

The performance also illuminated shadow aspects. When an archetype dominates without integration, it becomes tyrannical. Marta's Ruler energy became abusive authoritarianism. Magick's Lover energy led to self-destruction. James's Hero energy twisted into aggression. Health depended not on eliminating any persona but facilitating dialogue between them—what Jamie attempted when mediating the chaos of competing voices.

This mirrors the goal of our archetypal exploration: Not identifying with any single archetype, but developing fluency across the spectrum. A performer accessing only the Hero becomes one-dimensional. One who fluidly shifts between Hero, Caregiver, Jester, and Outlaw demonstrates the versatility that both artistry and life demand. The work recognizes the archetypal patterns that reside within you, which you may have suppressed, and how to bring them into balance.

Beginning with culturally familiar archetypes—such as Odin, Wonder Woman, Harry Potter, Cinderella, and Superman—performers access archetypal energies through recognizable forms. As we move deeper into the 12 elemental archetypes, we explore fundamental patterns that structure human experience. The questions lead to discovery: Feeling archetypes in the body, discovering movement signatures, emotional landscapes, gifts, and traps.

This creates a performance vocabulary—a set of bundled codes and mnemonics that form a larger, transcendental narrative. When performers deeply explore the Sage, they carry it as available vocabulary. They can access the Sage's center of gravity, vocal quality, and worldview without conscious thought. The archetype becomes not something performed but something temporarily and intentionally conjured into being out of necessity.

This addresses the contemporary paradox: While global media flattens archetypes into corporate-controlled forms—such as Marvel's Hero and Disney's Innocent—conscious archetypal exploration reclaims these patterns from within. The entry point is Iron Man's hero. Performers then find their own way into heroic energy, initially drawing on it, then transcending the forms of popular cultural semiotics, the hieroglyphics of the postmodern age. They seek genuine, if not universal, connections to archetypal forces by first reproducing franchised versions, then exploring and awakening deep structural patterns that have traveled and evolved through time. As with ritual performance, the past, present, and future align with the material and eternal, enabling a glimpse of the universal.

There is Never a Reference Point demonstrated that consciousness operates as an ensemble of archetypal performances, with different personas emerging depending on circumstances. Archetypal exploration enables performers to achieve greater fluency and choice in navigating these energies, transitioning from unconscious possession to conscious collaboration.

The goal is integration: Not eliminating the archetypal ensemble but bringing it into harmonious dialogue. Jamie struggled to mediate among personas, helping them communicate rather than war, moving toward what DID treatment calls integration—not fusion but collaborative coexistence. Similarly, archetypal work aims to develop the capacity for fluidly embodying multiple energies as life requires, while maintaining awareness of the larger self and containing and coordinating them all.

This echoes ancient wisdom that Indigenous performance traditions have long recognized: The performer connects different realities; multiple, not just one, spirits or archetypal forces flow through us; and health relies on having sustainable relationships and maintaining harmonious balance. The performer who can embody the Innocent in one scene and the Outlaw in the next, shifting from Caregiver to Hero to Jester (or a blending of archetypal complexity) as the performance demands, demonstrates the psychological and spiritual flexibility that defines human wholeness.

In an era when corporate franchises and algorithms dictate which archetypal narratives are dumbed down to dominate the collective imagination, archetypal exploration becomes a form of cultural sovereignty. By grounding work in somatic practice, local knowledge, and community-generated meaning, performers resist mass media's flattening forces. They discover that Disney or Marvel does not copyright archetypes, but rather living patterns from the deepest structures of human consciousness and culture—inherited patterns as old as Paleolithic hunting bands, patterns that will persist as long as humans seek meaning, connection, and transcendence.

FIGURE 2.17 Marta, performed by LeeAnn Herrington as the mother alter in There is Never a Reference Point. Written and directed by the author.

Source: Photo by the author.

Masks

The exploration of masks offers a unique perspective on our current focus on archetypes. I have worked with masks—adapting, designing, and performing them—in several cultural contexts. For a fuller understanding, I will highlight my work at the Korean National University of Arts (KNUA), where I taught and led workshops in *Traditional Korean Performance*. This culminated in a public performance titled *Twelve Moons*, a collaboratively created piece.

Soon after starting at KNUA, it became clear that performers possessed highly developed traditional performance skills. Their curriculum demanded mastery of all aspects of traditional arts and theater, as well as Western theatrical forms. The challenge was not recognizing the performance language but using it as a meaningful, contemporary form. Typically, performers saw traditional performance as culturally significant mainly for preservation, with little relevance today. Its strict adherence to rules and historical context limits its modern applicability. As Bob Dylan said, "Those not busy being born are busy dying."

Korea, known as the hermit kingdom, is a largely homogeneous society—one in four people share the surname "Kim," and another quarter shares the surname "Lee." Conformity and shame influence its cultural dynamics. The task was to identify the performance language and transform it into a modern, creative expression. I focused on helping performers break free from cultural and self-censorship, awakening their personal expressiveness. From there, I aimed to reimagine Koreans' roles in a globalized world and expand culturally rooted archetypes and traditional performances. The goal was to see traditional heritage as a resource for creativity and active participation in shaping their identity.

Popular culture and the rapidly changing global landscape showcase a diverse range of familiar archetypal characters across film, TV, video games, social media, marketing, politics, branding, anime, and comics. Some might argue that we are now living in an era of archetypes and myths in revival, driven by the increasing complexity, instability, and sensory overload caused by images and ideas that define our time. Archetypes and mythic stories provide simplicity and comfort—offering grounding, perspective, and a mental reprieve amid the chaos of cultural transformation.

SELF-MASK

Each performer is assigned a project—either an overnight task, a weekend endeavor, or a few days to complete—where they create a mask of themselves using a standard paper bag (with or without logos). They can incorporate any materials they prefer to enhance or embellish the mask.

Performers are asked to create a mask from a brown paper bag that represents their identity. They may use any materials they like, such as paint, glue, or scissors, to customize the bag and express their creativity. However, the mask must be functional since it will be worn during the performance—ideally without falling off their heads.

After designing their masks, performers develop a silent walk-and-gesture performance lasting between two and four minutes. Each routine must include: (1) an entrance and introduction (with or without spoken words), (2) a statement, and (3) an exit.

Present each mask to the group.

Performers should not reveal their masks to others before their performance, nor should they put on their masks in front of anyone else. They must keep the mystery of the mask.

After the presentation, performers discuss the mask's journey, including its development, how it performed, and observations from others. Be sure to cover these key points in the discussion.

- The essence and spirit of the mask
- How it liberates the individual.
- How the mask connects with archetypes and myths of its culture.
- How the performer accesses a fundamental truth of humanity through the mask—a truth that can only be expressed through the body, which is incapable of deception.
- How a mask unleashes creativity and fosters physical expression and awareness.
- How the body becomes a complete form of expression rather than relying on words or facial cues.
- How, in many cultures, the mask liberates the individual and serves as a bridge to an idealized or spiritual realm.
- How a performer becomes possessed and surrenders the self to the spirit represented by the mask.

Highly Recommended.

From the Korean Journal

The Self Mask exploration was crucial in unlocking students' creativity. It helped them overcome cultural and societal inhibitions, as wearing masks encouraged authentic self-expression.

Koreans often tend to be reserved and private. This activity allowed performers to openly show their vulnerabilities—some started shy but became more confident, while others displayed physical distortions. Some expressed deep sadness, intimacy, and vulnerability, whereas others were more confrontational, uninhibited, or hesitant, revealing core aspects of their personalities, hopes, or fears.

Some performers combined singing with rhythmic movements, while others approached their work in an intellectual and conceptual manner. A few used what I call the "negative bag"—a simple brown bag with eye and ear holes.

One performer chose to walk backward with two eyes painted on the back of the bag. Many painted their faces, presenting humorous and sensual expressions. Unexpectedly, their performances included rhythmic movements and traditional dance steps.

Some performances tended toward abstraction. This exploration significantly improved group interaction, uncovered internal resources, and showed how participants could connect with and embody a mask. It also provides a strong foundation for our future performance work.

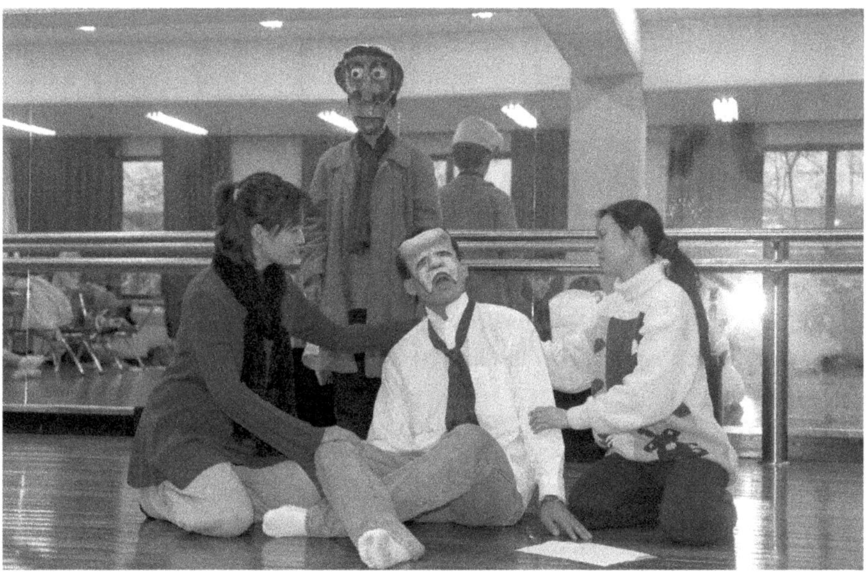

FIGURE 2.18 *Kamyonguk* mask exploration at the Korea National University for the Arts, Seoul. L to R: Eun Sook Park, Myungshin Park, Shin Hwan Oh, and Kyunghee Lee.

Source: Photo by the author.

The **Self Mask** exploration has proven popular across various contexts, revealing much about social, psychological, and cultural perceptions of the self and group dynamics.

When the same exploration was carried out in Alaska with Tuma Theatre, as well as in university and workshop settings across the US, Europe, and different parts of Africa, the expressions varied greatly but were equally revealing. They captured images that contributed to the development of cultural archetypes, resulting in a valuable collection of performative elements for creating performances.

At KUNA, the mask explorations sparked excitement during discussions, naturally leading to the creation of several traditional Kamyonguk Mask performance characters. Kamyonguk characters originated in and were traditionally used in shamanic and folk rituals in villages. As the tradition evolved and Korean society urbanized and was colonized by Japan, it drifted away from its roots, becoming more episodic and aestheticized, incorporating entertainment elements. Today, the carved and painted wooden masks are mainly used in public displays of cultural and national pride.

Traditional archetypal characters reified by Kamyonguk include the Buddhist Monk, Bride, Nobleman, Servant, Villagers, Concubine, Scholar, Lion, Monkey, Merchant, and Butcher, each possessing distinct archetypal personalities, encoded movements, dances, and costumes. Stories were presented in an episodic, non-linear, and matrix-like narrative format. The incorporation of masks, exaggerated characterizations, and a meandering, associative storytelling approach, along with vibrant costumes and outdoor public performance, generated vital inspiration and references for the performance's creation.

At this point, you might notice that the explorations anticipate performance creation. This is a natural and gradual process that reflects the nature of our work—everything happens in its own time. You cannot rush or shortcut the work. One solid step leads to the next.

At this stage of our process, I invited Kamyonguk performers and scholars to inform, participate in, and demonstrate the tradition. Drawing inspiration directly from the traditional and state-recognized Kamyonguk performers and understanding their historical and cultural significance was eye-opening. Initially, traditional character templates were introduced, and with the encouragement, delight, and advice of the traditional masters, the KNUA group developed modern mask characters.

The modern mask character evolutions were,

Fashionable Ladies (Western Styled)
Boss Man
Salary Man (white-collar worker)
Ancestor
Policeman
Automobile
Bureaucrat
Korean Spirits
Wife
Television
Mistress

Performers collaborated to develop character traits, movements, and dance expressions, creating contemporary characterizations, viewpoints, goals, and scenarios. Building on explorations such as **Elemental Archetypes** and **Objects,** we organically evolved our performance creation process. The Dead White Zombies applied adaptations of this process.

Just a reminder: The process shown here is only a single example for reference, which can be adapted to suit your specific needs and project. There is no one correct method for exploration and work; creating performance is an organic and creative act.

Once established, the masked characters improvised, revealing insights and forming narratives from their lives. These were then refined, connected, and edited to craft the *Twelve Moons* performance. Performers, audiences, and traditional mask experts and scholars warmly praised the adaptation of the Kamyonguk mask styles into modern characters. Duhyun Lee, a Kamyonguk performer and production advisor, remarked, "I am glad to see this before I die. It is for the young people to do this."

Below is an outline of the process we devised to create archetypal Korean masks inspired by the Kamyonguk tradition.

Process Outline

Creating a Mask Character

- Ideas transformed into tangible actions.
- Ideas observed and abstracted from contemporary life. The template for the mask character builds on the traditional mask character.
- Master gestures are explored, developed, and defined.
- Master movements are explored, developed, and defined.
- Master rhythms are explored, developed, and defined.
- Master attitude and character are explored, developed, and defined.
- Master motives—objectives are explored, simplified, and defined.
- Mask voice and vocal range are explored and defined.
- Mask vocabulary—intellectual range, concerns, and limits are defined.
- Mask character interactions are explored through improvisations.
- Creation of scenes and scenarios.

FIGURE 2.19 *Twelve Moons*, devised and directed by the author. The Korea National University of the Arts, Seoul. L to R: Jim Kyung Kim, Sungbo Hong, and Sangwon Seo.

Source: Photo by the author.

FIGURE 2.20 *Imipashi (the spirits),* devised and directed by the author. Luska, Zambia, and national tour. Timothy Hugala performing in a "St Peter" mask as he carries the "tree of the unborn" into the world of the living.

Source: Photo by the author.

Body as Script

Every culture develops its own language of movement—gestures, dances, and vocalizations shaped by generations of interaction with land, climate, and spirits. These coded movements serve as links to mythical worlds, passed down unchanged since ancient times. "They originated from our true ancestors," Yup'ik elder Nickolai Berlin shared. Accuracy is essential. Inupiat elder William Oquilluk warned that the Eagle-Wolf Dance must be performed exactly as the spirits demonstrated—any mistake could bring trouble to communities.

Such knowledge is passed down through generations, like water flowing through an unbroken river, preserving not only movement but worldview. For the Yup'ik and Inupiat, the body becomes the script—and more—it's a vessel. Movement turns the body into a carrier of the land's rhythms in muscle and bone. To dance is to enter a space between the physical and spiritual, between the human body and the body of the world, and between now and history. This liminal space reveals a tactile, psychophysical, and bio-spiritual way of being, where the body acts as a vehicle—an axis maintaining balance among all worlds. What the Yup'ik and Inupiat call "story dancing" uses expressive movements to tell complete, multidimensional stories. A single dancer might shift from hunter to wind to goose to arrow, each change smooth and immediate. This ability to transform between humans, animals, and elements shows the interconnected worldview of hunting peoples—where adopting multiple perspectives isn't just a metaphor but a vital skill for survival. While Inupiat dancers move through space standing, and Yup'ik men dance seated with women behind them, both traditions share a core vocabulary that educates, celebrates, and passes down culture as powerfully as oral stories.

These dances are communal, belonging to everyone—not just specialists but also ancestors, animals, and spirits. The constant traditional rhythm guides all movement, aligning performers with the heartbeat of their environment. This synchronization serves not as entertainment but as a way to maintain and actively acknowledge, engage with, and honor their worldview. Modern dances incorporate elements like outboard motors, rifles, and basketball moves, not as corruption but as an evolution—integrating new realities into ancient patterns that sustain harmony.

Modern artists and researchers are discovering insights that have long been recognized by Indigenous peoples. Grotowski's Objective Drama explored performance codes as spiritual tools. Meanwhile, Felicitas Goodman documented how "trance postures" induce significant physiological changes: Stress hormones decrease, brain waves shift to deep meditation states, and the body reaches conditions typically seen only in extreme situations. These gestures are more than simple movements; they are gateways—fundamental forms that serve as passages to an essence. What Western observers dismiss as mere repetition contains intricate codes that, when synchronized with rhythm and chant, create cyclical experiences representing the very cycles of being. The Inupiat emphasize the silent space between drum beats, creating presence through absence. The Yup'ik strike their drums from above, driving the rhythm forward. Both use the beat to synchronize with their part of the earth, enacting time as a spiral rather than an arrow.

The main challenge of our work is to maintain integrity while staying connected to modern life. To learn and appreciate alternative ways and functions of performance, not just to preserve or appropriate them. These deep languages, in their adaptation, are not museum artifacts with forgotten origins and functions. They are, in fact, a form of technology and a means of communication. Their form embodies content. The key is not just preservation but understanding the objectives and vocabulary of the form, viewing the body as a living script and a mode of transportation that opens into the space of a community of place.

The performance of Elemental Forms has proven both practical and healing for Alaska Native communities recovering from centuries of trauma. The revival of traditional dance across Alaska Native villages and communities became both a form of activism and therapy, reclaiming what colonization had fragmented. Our Tuma Theatre project simply documented and explored what already existed. Identifying Yup'ik and Inupiat Elemental Forms—and, to a lesser extent, Athabaskan Indian forms—provided a template for working with other Indigenous, non-Indigenous, mixed, and urban cultures. This approach consistently had a transformative and grounding effect, offering a way to identify, engage with, and shift perceptions of self and the world. It allowed a step out of the material and logical into a holistic sense of place. Understanding how and where transitory humans are positioned, and perceiving their worldview as embodiment, is revelatory. Performance models a worldview. The movements of the world and the body carry the message: The body remembers what the mind forgets. Performance acts as a neurobiological encoding, a reconnection, and a means of maintaining balance between the many worlds—many spaces—in which we exist. It's a transformation—an event that transcends material and binary definitions, as well as cause-and-effect logic. It is an ancient, continuous, and cyclical technology, enacted through rhythm, that momentarily reveals itself. Every gesture, movement, and vocalization, once articulated, exists, contributes, and changes both the world and oneself. The body is a script writing and righting the world.

ELEMENTAL FORMS

A performer or group of performers is assigned to study a specific community or their own culture. This group may include Indigenous peoples, such as the Zulu or Greenland Inuit, as well as various religious groups, including Pentecostals, Baptists, Hindus, and Catholics. It may also cover individuals from different occupational backgrounds, like laborers, street gang members, and police officers. The focus could also be on any organization, such as Hell's Angels, the Ku Klux Klan, or the American Legion. The research might involve studying groups in person, through film or videotape, or a combination of both, if there is enough recorded movement to analyze.

Focusing on movement as the central element, the performer is encouraged to identify, analyze, and define the movements and gestures characteristic of the study group. It is recommended to use written, photographic, sketch, and video documentation alongside the performer's direct interaction with the group's movements. The performer should aim to understand the origin and context of these movements.

The following questions guide this exploration:

- Where does a specific walk originate?
- Does a particular look relate to a religious, cultural, or social habit?
- Has the movement, gesture, or vocalization developed from self-imposed philosophical, religious, or moral expression or restriction?
- Is the movement, gesture, or vocalization influenced by external factors like environment, geography, or climate?
- If it has historical or cultural significance, has it retained or changed its original meaning?
- How has it evolved? What worldview is reflected in the group's collective movements?

This exploration acts as both a demonstration and a discussion of the movements' significance and roots. It can serve as a foundation for a performance language that could develop into a specific performance or style.

Objectives

- To study a specific group by observing their movements, rhythms, and vocalizations.
- To highlight how bodily expressions act as social and cultural identity markers.
- To investigate and gain insights into a particular group to better understand their worldview.

Applications

Elemental Forms can be used in various ways. Examples include:

Narrating stories with the forms, engaging in improvisational explorations to deepen understanding of characters, scenes, and narratives, performatively interpreting myths and scripted texts, establishing a shared performative vocabulary, participating in extended free play where the entire group communicates using the forms, and conducting mirror explorations solely through the forms.

Note: This exploration has offered valuable insights for both Indigenous and non-Indigenous communities. The Dead White Zombies crafted a performance vocabulary that was transformative, redefining their purpose and goals as performers and reshaping the broader significance of performance. Their method bridged the personal and the communal, linking them to the larger world.

For other groups like Riksteatret, a diverse cultural ensemble supported by the Swedish National Theater, it helped them reconnect with and reflect on their cultural traditions. This was especially vital after urbanization and cultural displacement had disrupted their ties to ancestral roots. Overall, this study marks a significant step toward developing a place-based approach to performance.

FIGURE 2.21 Theresa John, writer and performing Yup'ik Arnaq (Yup'ik Woman). Directed by the author. A performance utilizing traditional elemental forms. Tuma Theatre. Fairbanks, Alaska

Source: Photo by the author.

The Elemental Forms, no longer limited to traditional dance settings, gained independence. Over time, tradition evolved into a personal, accessible, and highly expressive form for both performers and their culture. Instead of simply moving across the stage, Tuma performers used one of the travel Elemental Forms to express their emotions and circumstances. This rhythmically driven dance-theater expanded the traditional performance style in both scope and expressiveness. It freed movements previously constrained by dance norms, allowing performers to explore a new theatrical space filled with limitless expressive possibilities. By reimagining traditional movements from a fresh perspective, Tuma attracted a broader, modern audience while remaining connected to its core traditions.

The dance-theater style evolved from **Elemental Forms**, preserving the core of traditional dance while expanding the metaphorical space to suggest a continuous, cyclical process rather than a linear reality. Within this context, the performer functions both as the specific and the archetypal, acting as a symbol-hieroglyph and an individual at the same time, embodying the present, past, and future. The Elemental Forms reflect and embody a cosmological holism.

The Forms facilitated transformation, allowing performers to easily switch between human, animal, commentary, or emotional expressions, as well as elements like wind or northern lights. This smooth interchangeability emphasizes the interconnectedness of all things. Although the traditional Alaska Native worldview was adapted, the core cultural essence within the forms remained intact. Tuma Theatre has explored traditional Yup'ik and Inupiat performances in a way similar to how Japanese Noh drama developed. Noh, which originated from Shinto shamanic rites, reflects a transformation and formalization of basic forms.

Like the Yup'ik and Inupiat, the Noh tradition adapted to social and cultural changes. It drew on earlier traditions, formalized them, and created art from elements that had lost their original context. Kabuki, another Japanese performance style, drew its core elements from Noh, pushing forward the transformation and adaptation of earlier styles to fit Japan's changing societal landscape. The development of performances over time mirrors human societal and cultural evolution, serving as both a reflection and a model of societal change. In a similar yet different way, the regional shamanic practices of the Taoist, the Miao, Yao, and Tujia people of China went through a process of aestheticization that provided a resource for what became the Peking, then Beijing Opera, and later inspired Kung Fu films, anime, and video games with stories, characters, and actions.

A Sample of the Elemental Forms

Identified and Applied by Tuma Theatre

Forms originate from Yup'ik and Inupiat dance traditions. All movements are coordinated using either the Inupiat or Yup'ik two-beat rhythm.

Travel

Hand Walking: With palms down at the waist, step forward with hands rigid and alternate in front as if they were feet. The performer walks forward at a controlled pace, with neck pulsing to the beat.

Easy Walk

Like in Hand Walking, hands are positioned forward at the waist, but they hang loosely like a duck paddling. The performer walks forward at a steady pace, with neck movement matching the beat.

Facing Destiny

One arm is extended diagonally with the palm facing up, while the other arm is bent with the hand near the chest; both hands remain straight. Alternate extended arms or hands, then combine this with a forward motion. Keep the rhythm in the arms or hands. Maintain a courageous attitude as you face what lies ahead.

Animals

Bird Transformation

Point to a bird in the sky with one finger or your whole hand, then stretch out your arms to mimic a flying bird.

Walrus: Arms close to the body, elbows cocked, hands at face with fingers pointing straight down like walrus tusks. Emphasize and beat with the hands. Performed in place.

Walrus or Seal Looking

Position your hands diagonally in front of your body, below your chin, mimicking the flippers of a seal or walrus. Lift your head upward and tilt it to get a better view above the water's surface. Let your entire torso sway to the rhythm of the drum while you perform the movement in place.

Eagle-Wolf Dance

Arms are held down in front, with the body bent forward at the waist. The rhythm is shown through shoulder movements, with arms moving up and down. The arms alternate, shifting up and down as if walking on the ground.

Raven Dance

All moves should be synchronized with the beat and filled with comic enthusiasm for two people facing each other.

1. Knees bent, arms extended outward as if wings are flapping, dancing for one another.
2. Raven hop: Performers hop toward each other and pass.
3. Turning with arms wide, hop and turn on one leg. Raven sounds like "Kaw Kaw."
4. Ravens hop for one another.
5. Hands together at the lower back as if forming tail feathers.
6. Ravens hop for each other with Raven sounds of "Kaw" or "Calup."
7. Hands behind the back, resembling tail feathers. The neck and upper body undulate to the beat.
8. Hop and pass one another with rhythmic undulation.
9. Raven hop-dance with the rhythm.

Human Activities

Looking

Alternate raising of hands to the forehead while gazing into the distance, accompanied by rhythmic movements of the hand and neck, performed while standing still or after coming to a stop.

Eating

One hand or both hands are brought to the mouth, with fingers moving rhythmically as if eating (mouth movement optional).

Talking

Position your hand with the first two fingers together. Bring it close to your mouth, circle your fingers, and then reach out with your hand toward another performer, a spirit, element, or animal, as if holding and passing the words with your fingers.

Hearing (Two Options)

1 Place one hand beside your ear. Fingers move to the rhythm. The head tilts and sways, as if tuning into the sound.
2 Start with one hand open near your ear, as if listening, then extend the arm diagonally outward while fully focusing and pointing in that direction, as if hearing a bird or something in the distant place.

Pointing (Three Variations)

The act of pointing varies depending on the object being indicated. One arm extended, pointing with:

1 One finger (human and specific).
2 Full hand with fingers straight, palm down (geography, animal).
3 Full hand, fingers straight, palm upward (feeling, spirit, giving, and taking).

Paddling

Hands shaped as if holding a paddle are positioned in front of the body, moving to either side. Paddle in time with the beat, emphasizing the motion in the hands and shoulders. This can be performed in place or while moving, with knees slightly bent.

Gathering

Extend arms outward to each side one at a time, then bring them toward the stomach as if gathering. This can be performed while moving or stationary.

Northern Lights

Gazing upward, extend one arm and point with a single finger, tracing the movement of the northern lights. This motion is performed in place, syncing the finger's movement with the neck's rhythm.

Praying

Both arms are raised and gently moving upward, pulsing in praise of the universal being; the rhythm centers on the hands.

FIGURE 2.22 *Utetmun*, written and featuring Paul Jumbo. Directed by the author. Tuma Theatre, Fairbanks, Alaska

Source: Photo by the author.

Relevant explorations include **Hieroglyphics** and **Personal Vocabulary,** which convey the image forms experienced and shared by the group.

HIEROGLYPHICS

This exploration aims to revive, revitalize, or create a new performance vocabulary.

This exploration parallels **Elemental Forms,** but relies on photographic, pictographic, sculptural, historical, oral, or anthropological records, rather than living expressions.

Group studies might include, for example, the ancient Greeks, Romans, Egyptians, cave dwellers, Mayans, Nazis, Gen Z, Hip Hop, consumer, or animal culture.

This is achieved by identifying, exploring, and codifying movement, gesture, and vocal expressions that are framed. The performers are asked to identify, discuss, analyze, and define elemental forms as performance artifacts.

Lead the exploration by asking questions that provoke selection, reflection, and refinement (some suggestions):

- How does a specific posture or gesture reflect a culture's worldview?
- In what ways is that perspective expressed and connected to the body?
- Does a particular posture, appearance, stance, or way of holding the body relate to a specific cosmology or religious practice?
- Does the posture, gesture, or vocalization indicate a cultural or social practice?
- Has a movement from a specific group emerged due to self-imposed philosophical, religious, political, or moral restrictions or goals?
- Is the movement influenced by external factors, environment, geographic location, or climate?
- How has the movement changed while maintaining its original meaning and significance?
- How has the movement evolved over time?
- What worldview is reflected through the collective actions of the group?

Note: Like previous discussions in this section, this exploration will lay the foundation for a shared performance vocabulary that character, improvisation, and performance creation can draw from.

PERSONAL VOCABULARY

Performers are invited to present five or six actions, movements, gestures, vocalizations, or activities that they feel best represent "their lives," "their culture," "their tradition," or any other specific identifier.

The exploration can occur entirely within a workshop session, with a limited amount of time designated for it, allowing individuals to work intuitively and independently.

Performers showcase their findings. Once all have presented, discuss to identify commonalities, insights gained, broader signifiers, and associations triggered by the presentations.

The performers then refine their vocabulary, discarding some while combining or improving others.

Performers then assess which elements of the vocabulary they want to explore and refine further. Limit exploration to available time, project requirements, and objectives. It can also be an ongoing exploration returned to over a period of time

Personal Vocabularies can provide a basis for solo performance work or as an augmentation to character, performance creation, and group work.

Personal Vocabulary serves as a gateway, not an endpoint—a vocabulary designed to take stock and deeply engage with individual experiences and history. It is a way to know thyself and what you carry and bring to our work and performance.

Explorations can be effectively achieved by practicing the movements repetitively, much like learning the alphabet in elementary school. Once everyone is familiar with the vocabulary, numerous options arise for developing engaging performances.

Exploration Alternative: Using the process detailed in **Gesture Exploration** and **Archetypes**, performers examine the intellectual, emotional, and expressive parts of their lives through the development of a vocabulary. This improves both their creative expression and their connection with the audience, guiding the direction of their work. Similar to mask explorations, this method is a powerful tool that frees the performer creatively while building a connection to personal, lived experiences and universal themes.

Scenario Creation

Ask the performers to create a scene using as many movements as possible. The story doesn't need to be logical; instead, it should be guided by the movement suggestions and how smoothly they transition from one to the next. This exploration segment should be done without language, allowing the performers to focus entirely on movement. It may later develop to include vocalizations and language inspired by the vocabulary.

Dance Creation

Guide the performers to use their movements to create a dance that explores and varies rhythm, intensity, and emotion, while also incorporating diversity in style.

Scenario Applications

Guide performers to embody characters, scenes, or the entire story or myth through the designated movement vocabulary. Encourage performers to create movement connections that align with the defined vocabulary throughout the process.

Ask the performers to develop improvisations and scenarios that incorporate movement tailored to specific needs, situations, themes, issues, and concepts.

Space and Place

Our brief and limited exploration of space has revealed the many overlapping realms we occupy—physical, emotional, cultural, spiritual, and technological. We have learned to identify these spaces through our bodies, navigate them with movement and breath, and embody the spaces of objects, animals, and archetypes. Space is not vacant but alive with presence, memory, and potential. Every gesture creates space, every rhythm shapes it, and they radiate and influence. The spaces of the world are organic, expressive, metaphoric, dynamic, and ever-changing entities, containers of energy and universes in themselves that seek balance.

However, spaces by themselves do not generate meaning. They are only parts of totality. They drift, intersect, come and go, and vie for our focus in an increasingly fragmented world. The modern individual moves—perhaps even struggles—through a spatial abundance, overwhelmed and at times confused by overlapping, sometimes competitive systems that, by their nature, are unstable and do not aggregate into a system of meaning. That is not their function. This is where our work shifts in focus, from acknowledging and exploring spaces to creating Place.

Place arises when multiple spaces come together through intentional relationships, creating an organized system. It is more than just a location; it is a dynamic system where spaces interact through rhythm. Our work is to reveal this wonder, telling the story of that which is in motion to glimpse and participate in a fleeting shared understanding and feeling. While Space recognizes what surrounds and moves through us, Place involves weaving these elements into an erstwhile, cohesive whole—an ephemeral yet essential system connecting the specific with the universal. Indigenous worldviews recognized that you belonged to a place, and that place belonged to you. Our work is to rediscover what was lost, to reawaken a way of being, to reimagine the world, and the Cosmocentric anew. Performance is our means, craft, and gift. We are part of a process—a coming into knowing—and have a part to play in reestablishing a place in an increasingly placeless world.

PLACE

The Place phase of our work synthesizes the book's methodology, arguing that place is not given but actively performed into being. Drawing on Indigenous concepts—Yup'ik Inua, Anishinaabe Aki, Aboriginal songlines—this section demonstrates how place-based communities maintained coherence through embodied practice.

The contemporary condition of "spatial vertigo" arises from numerous spaces lacking organizing frameworks, which leaves modern humans feeling placeless. I will argue that performance emerges as "consciousness technology"—a means for transforming fragmented spaces into coherent, inhabited places. Through protocols for immersive performance, ritual structure, and site-specific composition, this section offers some practical methodologies for (re)creating Cosmocentric place in an era demanding new forms of planetary belonging.

Our journey through Body and Space has revealed multiple territories—physical, emotional, cultural, spiritual, technological—that we inhabit and navigate simultaneously. Space is understood as never empty but teeming with presence, memory, and potential. Every gesture, object, and event creates and reveals space. Each space has its own rhythm that holds and organizes it, engaging in cycles and striving for balance—whether a forest, a family, a workplace, or a relationship (Tuan 1977, 6).

The human need to make sense of the world drives the shaping and coordination of spaces to create Place. This is the focus of this book's final section. Place represents an agreed-upon container of ideas and emotions—fixed or semi-fixed. Place is a construct that contextualizes, inspires, and facilitates the creation of social and cultural spaces. It

DOI: 10.4324/9781003686422-4

operates as a structured system composed of elements, signs, and rules that define specific locations of meaning. Place shapes multiple spaces to better enable human interaction, relationships, survival, and ways of being in the world, and in so doing creates a worldview. Place exists simultaneously as physical and virtual, real and imagined, lived and abstracted, thought and felt—containing and conveying present, past, and future at once.

Humans create meaning and narratives while existing in physical bodies that navigate environmental, social, and cultural realms. An important point worth stressing, the sensing body derives meaning from the world, processing diverse stimuli and spaces. Spaces are fragmented, vibrant, experiential narratives, whereas place functions as an organizing system that holds multiple spaces together to create a collective narrative.

Historically, every human being was deeply connected to a specific geographical location, a community defined and organized by its unique climate and seasonal movements, diverse flora and fauna, geological features, and water resources. For Indigenous peoples, every place on Earth possesses its own unique essence, distinct from all others. When spaces cohere and become a place, they become an entity, both material and spiritual, that lives and communicates in its own unique way. Humans are part of this entity, and as the most enabled and greatest beneficiaries of a place, are called upon to take responsibility for its maintenance. The way to do so is through performance, arguably one of humanity's first technologies.

Everything within this way of seeing and being in the world carries meaning. In Yup'ik Eskimo culture, everything has a spirit called Inua. Words carry this spirit, once spoken, once written, once read they are in dialogue with the world and exist forever. Human life is seen as a continuous dialogue with the deep mysteries of existence. From an Indigenous perspective, our humanity is shaped by our relationships and connections to the world. The body acts as both receiver and transmitter, an active, responsive entity that interacts with the environment on emotional, intellectual, psychological, and spiritual levels (Shepard 1998, 45).

Every place is vibrant and continually changing, telling a dynamic story composed of interconnected narrative spaces. This engaging dialogue invites us to listen and connect with the various presences around us, thereby shaping our perception of the world. Imagine humans as an open circuit, finding completeness not only through presence in the world but also through reciprocal engagement in which observation flows in both directions (Ingold 2000, 189). Performance, in the form of ritual—daily and specialized—along with shamanic, mediumship, and divinational interactions, was how our collective ancestors reiterated, reaffirmed, and remediated their relationship with place.

The Loss of Place

The schism between communities of place and communities of ideas becomes crucial for understanding modern alienation. Place-based communities—communities of place—organize around shared land, seasonal cycles, and cultural practices that bind people to specific locations. Idea-based communities—communities of ideas—in contrast, cohere around abstract principles, political, economic, religious, or cultural, that transcend geography. While an idea-based organization enables broader connection and mobility, it often severs the deep ecological knowledge and sensory attunement that place-based existence cultivates (Anderson 1983, 6). Over time, it neglects its responsibility to the communities of place that all life depends on.

Today, few people of place remain in the world. The remaining Indigenous peoples have, to varying degrees, been confined to reserves, traumatized, absorbed, colonized, and acculturated. The !Xuu Bushmen of the Kalahari, like so many other people of place, live on reserves and reservations, eking out an existence by living in two worlds, surviving by selling mineral and resource rights, government subsidies, commercial ventures, and tourism. The Yup'ik Eskimo of Alaska live in villages connected by airplanes, cable television, and the internet. Once vibrant cultures of Asia and Africa have been decimated by war, haunted by colonial legacies, relegated to indentured servitude by the IMF (International Monetary Fund), AIDS, environmental change wrought by the developed world, causing social, economic, cultural, and political hardships. All of it hastens the dissolution of place-based knowledge systems. As the young flee to cities, leaving their cultures to the past to be preserved not in living bodies, minds, and hearts, but in video, audio, and written recordings. The loss of place awareness has meant the loss of a particular way of being embedded in the world (Katz 1982, 34). Such awareness is knowledge, the wisdom of the earth, and that which has much to teach our present moment.

Place-oriented cultures have been overshadowed by the dominance of non-Indigenous Western cultures, particularly the Euro-American axis—a conglomerate that originated from place-based beings but transformed into meta-cultures held together by ideas rather than the living being of a place.

Every place, every inch of Earth has been discovered, explored, measured, mapped, assayed, plundered, scarred, and threatened with even more devastating adjustments in the service of ideas. The authenticity of embedded being has become a nostalgic longing. What was once taken for granted—air, water, earth—is now contaminated, increasingly commodified, and a source of political and economic contention. Populations have increased while science, medicine, technology, and communication advance not to serve a place, but anthropocentric objectives, needs, and values. The wisdom of holism, balance, interdependency, and place-based living has become disconnected information, concepts passed through books rather than through the sensing, feeling, and lived body of a place.

Spaces Without Place

This loss of the Cosmocentric creates a specific modern condition. Think of spaces as modules that radiate aspects of life. Some are temporary, others endure, constantly transforming. Ephemeral, organic, and material, humans are immersed in them, move through them, are shaped by them, and shape them in return. They appear and disappear like countless soap bubbles—bouncing, interacting, merging, bursting. These ephemeral spaces form a collective field shaped and maintained by shared thoughts, emotions, and resources (Castells 1996, 453).

Yet, spaces alone do not sustain us. In our modern world—characterized by astounding technological progress, globalization, capitalism, and communication, overshadowed by threats of environmental collapse, nuclear war, and social, political, and psychological upheavals—numerous spaces multiply and fragment, vying for our attention. We find ourselves swimming, perhaps drowning, in a sea of spatial abundance, overwhelmed by systems that rarely synchronize or cohere. Modern humans and non-human beings navigate this complexity without a clear anchor or orientation, experiencing spatial vertigo and overwhelm. The necessity of occupying multiple spaces simultaneously, none of which coalesce into a sense of home, comfort, or belonging. This condition characterizes contemporary existence: Spatially abundant yet placeless, endlessly moving yet never arriving (Putnam 2000, 27).

Place Through Performance

This is where our work evolves: Organizing spaces to create a place to address the needs of our moment. This is our work. The crafting of coherence from spatial fragments. Place arises when spaces coordinate through bodies moving over time. It is not merely a location on a map or a collection of accumulated meanings, but a living system created and sustained through patterns of practice. Where spaces signify recognition and navigation, place signifies integration and expression, the active transformation of spatial raw material into coherence. This coherence is not permanently fixed but dynamically enacted through physical engagement. Place is the performance of an organized system that serves as a model, focal point, and crucible for participation, community, and reflection (Schechner 2002, 28).

This insight questions traditional academic views that see place mainly as a source of emotional connection or shared significance. While these elements matter, they do not fully explain how places form and last. Indigenous and ritual traditions around the world show that place-making is mostly performative and is something done rather than just felt or remembered (Tuan 1977, 179).

This work necessarily proceeds from within Western performance traditions even as it critiques Western culture's destruction of Cosmocentric awareness. This is not a contradiction but a given to be negotiated. As Euro-American culture transformed from place-based communities into idea-based meta-cultures, it created both the problem (spatial vertigo, ecological disconnection, placelessness) and potential tools for response (experimental performance, theoretical frameworks, technological advance, globalization). The Cosmocentric approach, central to this book, addresses this tension by combining Indigenous wisdom with modern needs, leveraging the tools of idea-based culture to revive aspects of place-based consciousness (Smith 1999, 42).

Place thus operates on two levels simultaneously. As a relatively stable structure (patterns of organization, spatial arrangements, ritual forms) and as a dynamic enactment (the specific performances that activate and renew these structures). The structure provides continuity; the enactment ensures relevance. Neither alone suffices. Place requires both architecture and animation (Turner 1974, 60).

A house becomes a home not merely through sentiment, but through the daily interactions, adjustments, and practices required to maintain the dwelling. The way you move through its rooms, the rhythm of routines and functions, and the gestures that maintain its order for lives—human and non-human—are organized to optimize sustainability and happiness (Bachelard 1994, 67).

Somatic Wisdom and Practice

Place involves thoughts, gestures, and a deep awareness that everything is interconnected and necessary. A person's place is reflected in their daily attire and speech, as well as in how their body communicates spatial proximity, intimacy, and interaction through thought, emotion, and expression. This is reflected in relationships with both human and non-human beings, and animated by songs, chants, dances, stories, beliefs, myths, body markings, ceremonial regalia, gatherings, and rhythms, all of which are integral to a greater holism. Ceremonies commemorated relations and social and cultural adjustments, while rituals served as framed passageways to worlds beyond the material, psychophysically reiterating and reaffirming a group's worldview (Simpson 2017, 43).

For the Anishinaabe of the Great Lakes region, Aki encompasses not just the earth, land, and all its inhabitants, but the dynamic web of relationships maintained through the physical practice of living in harmony with it. The same can be said of Inua, the living practice of the Yup'ik of southwest Alaska. Australian Aboriginal concepts of life emphasize lived relationships rather than territory alone. These connections are preserved through participation in initiation ceremonies and the transmission of stories via songlines. Traveling along ancient Dreaming paths—songlines—inscribed in the land through song and storytelling, we map these pathways into our minds, bodies, and relationships with everything around us. Knowledge resides in every waterway and rock (Kimmerer 2013, 125).

These perspectives recognize that place cannot be owned or kept static. Instead, it must be actively created and continually maintained. Those of us educated within the binary conceptualization of reality may find it difficult to understand that there are multiple truths and that perception can be simultaneous. These ideas cannot be entirely communicated through books like this, only hinted at. Writing about them, or trying to explain them, can misrepresent what can only be understood through direct sensory experience (Wilson 2008, 58).

Our Contemporary Crisis

In today's era of displacement—encompassing geographical, social, cultural, and digital realms—the ability to cultivate a sense of aliveness is more crucial than ever. We are currently at the culmination of a 600-year transformation initiated by the Enlightenment. This era fueled the pursuit of freedom and individuality and spoke to its era's needs, but now we find ourselves at a different, more complex, and more challenging moment. We, humans, are at the end of history as a new one begins to take shape (Tsing 2015, 19).

We feel hopeless and saturated with normalized violence; the media now treats casual killings as entertainment, blending reality and performance where truth becomes negotiable, and conspiracy theories flourish, symptomatic of the ordeal of profound transformation. We can no longer proceed as we had. "Our societal system is exhausted and has run its course. What was can no longer serve what is" (Steyerl 2017, 45).

Fear and uncertainty spread across the globe, sparking a resurgence of strongmen modeled on the past. Invading empires such as corporations, drug cartels, or egomaniacal politicians and entitled billionaires shift back to medieval religiosity, racism, sexism, and feudalism. Democracy, which promotes human freedom and individuality, faces challenges because it often provides fewer tangible solutions to fewer people. Desperation, corruption, and narcissism, driven by an "every man for himself" attitude, are now part of the climate. "The space of anxiety circles the planet, now as certain as the weather" (Crary 2022, 78).

The signs of systemic strain and decay are everywhere, spreading toward collapsing ecosystems and catastrophic system failures. Extreme weather events—wildfires, hurricanes, floods, tornadoes, tsunamis, and earthquakes—are increasing at an alarming rate. Despite their abundance and beauty, the oceans that sustain life are heating up and dying. The destruction and extinction of animal and plant life signal our own extinction. Humanity finds itself in an uncomfortable but necessary process. Yet, we are already, in so many ways, moving toward another way of being, another consciousness. "Movement is stirring, and we have a part to play" (Bridle 2022, 156).

Becoming Earthlings

The past is a prelude. I remain hopeful, and something within me tells me we can create a verdant, sustainable world. There is no going back to the traditional Cosmocentric ways of being. Humanity cannot abandon the progress and benefits our journey has provided us. Instead, all of us should listen again. Reevaluate who and where we are, not with the goal of human improvement and progress, but to think and act for our community by revitalizing a Cosmocentric way of being. "It is time to use the best of what is known to rethink and reimagine harmony with the world" (Yunkaporta 2020, 203).

The advances brought about by globalization, trade, commerce, technological progress, artificial intelligence, and environmental issues are shaping and transforming a new planetary way of being. An opportunity is manifesting. All beings—human and other-than-human, the elements, spirits, ancestors, animals, and plants, and the cosmos—are becoming conscious again. Reminding us that we are still part of a greater totality. An intuitive process and intelligence is speaking. Calling forth a new sort of indigeneity. For we are becoming earthlings. Inhabitants of a planetary place that requires us to think, be, and perform differently.

This is our place. Over the past 50 years, the speed and scale of change have overwhelmed us, leaving us stunned and whiplashed as developments happen so quickly and intensely. So much, too much to absorb and integrate. There is work to be done, healing to be achieved. Humanity can no longer rely on the inherited ways that earlier generations took for granted. That which was ingrained over centuries of institutionalization is no longer adequate to our vastly changed circumstances. Still, our capacity to create place is embedded in our bodies, rooted in our genetics, and expressed in the ancient technology of performance.

Performance as Transformation

Performance, or whatever it may be called in the future, can lead the way by explaining, articulating, and integrating. Witness the rise of environmental and climate awareness, women's rights, the MeToo and Black Lives Matter movements, gender and identity sensitivities, veganism and animal rights, as well as recognition of poverty, famine, and health issues faced by the disadvantaged. The increased awareness of income and power inequalities also reflects this shift toward an inclusive democracy of all beings. The world is undergoing a profound, all-encompassing transformation that is essential to the survival of life on the planet. This is our work, and we contribute as we can.

What follows is a continuation of our work with Body and Space, but with a key shift in focus. Where humans previously emphasized awareness and navigation, we now shift toward integration and creation. Our work explores how to coordinate to establish rhythmic patterns and to use performance as a tool for place-making. The performer acts as a place-maker, gathering dispersed elements of modern experience and structuring them into systems that sustain life.

Again, this is not about longing for impossible returns to simpler times. It is about developing the skills to create a new place by using the body's natural abilities, which remain consistent even as everything else changes. By making a place, humans become active participants in shaping the realities lived in, rather than passive spectators of spatial fragmentation.

Place is not a destination but a practice. Not a thing but an event. Not something found, but something made (Turner 1969, 95).

Lingering Questions

How have recombinant signs and words, along with endless bot-generated algorithmic cross-germinations, evolved into forms of knowledge?

Why do humans attempt to find answers through rearranging words, information, and communication with others instead of tuning into our feelings and the environments we inhabit?

Is the hope of an epiphany what propels us? Or are performers like shamans, using repetition to induce trance and transform into a higher reality?

How did humanity become participant-observers, structured and examined through ever more complex layers of statistics, marketing, and controlled systems?

How did marketplace transactionalism become our dominant expression and main narrative?

And how has it become our life's art, cultural touchstone, and primary mode of communication? How did our reality become a mass-produced system that is mostly closed and controlled, covering everything from cars and streaming to gaming, web shopping, sports, government, and fashion?

How has digital connectivity evolved into a platform for validating life and knowledge? How have companies like Microsoft, Google, Meta, Amazon, OpenAI, Anthropic, Tesla, Apple, YouTube, Fox, CNN, and others come to enable privileges that increasingly resemble feudal hierarchies?

In what ways do the marketplace's manufactured objects both mirror and promote a sense of sameness in human life?

Why do humans need more intense experiences to feel truly alive?

Why do humans continue creating new combinations, collages, objects, technologies, and stories to revitalize our tired, jaded selves?

Is something lost when humanity and its technology eliminate chance, the unexpected, and the unpredictable—the very elements that connect us to the ongoing changes and mysteries of our world?

First, a bit more on contexts and concepts, before we return to our practical work.

Narratives

Humans are born into narratives such as genetic inheritance, family, education, social and cultural expectations, myths, and historical contexts. These are narratives worn like clothing, some fitting well, others chafing, shaping life.

Walking along a suburban street uncovers layered narrative realities. Every step expresses a different narrative. Pastoral mythology is revealed in well-maintained lawns, the American dream in driveways, colonial influences in picket fences, and social aspirations in architectural styles. Decorative shutters that will never close speak of nostalgia for a past that never existed. "Humans do not merely observe these narratives—we perform them with every gesture" (Massey 1991, 24).

Like Paleolithic hunter-gatherers, humanity developed heightened sensitivity to shifting story-worlds. The morning commute transforms us from domestic to professional characters. The gymnasium demands a competitive, optimized body. Social media requires another performance entirely. All of it alters how humans perform, adjust their voice, and shape their expressions. We have become good at storytelling, but the constant changes drain our energy, leaving us disconnected, lost, and homeless. We are no longer ourselves but become an idea, an aspiration, a marketing notion, a hollowed-out vessel of imitation, a programmed self, and, increasingly, a part of a vast, expanding algorithm.

The apparent abundance of the modern has created a strange poverty. When stories become interchangeable and are portable—performed anywhere by anyone—who and what are we and where do we belong? Corporate videos, political speeches, and advertisements use narrative to modify behavior rather than foster and deepen a sense of belonging. They operate on us rather than collaborate with us. They are seeking something from us. Our attention, emotion, vote, or money.

The distinction between narrative as space versus narrative of place becomes clearest in the body. Watch someone doomscrolling, consuming stories that exist in digital space but create no lived experience. Now watch children playing, their bodies actively producing narrative through interaction of body, mind, emotion, movement, and sound. "The story isn't separate from their actions; it emerges through them, forming a temporary but real place of shared imagination" (Gottschall 2012, 56).

Among the !Xuu Bushmen, storytelling is a physical performance. Every gesture and shift of rhythm helps maintain a living universe where stories, ancestors, and landscape interconnect. The Lozi of Zambia summon ancestors through specific drums and movements, creating temporary spaces where past and present coexist. These aren't primitive practices but sophisticated technologies for community coordination that help groups reconcile different realities into a unified experience (Riccio 1997, 48).

Contemporary theater remains trapped in its Cartesian architecture. The proscenium arch maintains fundamental divisions. The illuminated mind watches the darkened body; the active stage confronts the passive audience; representation versus presence. Despite technological advances, this subject-object relationship persists and serves as a model and reaffirmation, culture-wide. Television, film, and now digital media were spawned and extrapolated from this binary separation, which is now culturally encoded as normative.

Humans have grown skilled at recognizing narrative manipulation. Every political speech and advertisement increasingly follows predictable patterns. The well-crafted play, with its tidy psychology and convenient resolutions, is another form of audience coercion and acceptance. We see the machinery too clearly: The inciting incident, the rising action, the neat resolution. These perfected structures have become cages (Fuchs 1996, 34).

Yet, genuine hunger for storytelling experiences erupts everywhere. Video games craft narratives through player agency. Immersive theater dissolves performer-audience boundaries. Flash mobs and public protests transform public spaces into temporary stages. These forms recognize what theater forgot: Stories gain power not just through telling but through doing, feeling, and engagement (White 2012, 228).

Moving from narrative-as-representation to narrative-as-practice requires new approaches. Instead of scripts and character psychology, performers might begin with breath rhythms, gesture weight, and patterns that emerge when bodies move together. When I worked with Alaska Native performers, they brought traditional movements that engaged their whole being, creating a bridge between university life and ancestral traditions, transforming rehearsal space into cultural continuity. The narrative lived in their bodies.

This acknowledges that performers embody narratives physically. Handed-down rhythms, cultural gestures, and learned movement patterns are a communal text shaped by living in and with the world. Your role as a facilitator is to reveal and coordinate stories already present. These are not artifacts but living technologies and expressions for creating place through collective practice.

Yunkaporta describes narrative functions as "a living thing that is patterned within every person and being and object and phenomenon within creation." Yarning—combining story, humor, gesture, and mimicry—generates knowledge through overlapping speech and non-linear exploration. Narrative becomes collective improvisation, creating place through synchronized yet diverse participation (26).

Attali suggests humanity must move away from the "straight line" thinking that has dominated Western thought since Plato. And instead, embrace the labyrinth, which requires embodied exploration, which means getting lost, doubling back, and finding meaning through movement rather than purely intellectual understanding. It demands "perseverance, unhurriedness, curiosity, playfulness, trickery, flexibility, improvisation, and self-mastery" (23).

In Dead White Zombie performances, I explore and apply a technique that evokes layered emotions and memories across space through sound. Often, "free radical characters" move without fixed roles, acknowledging that modern audiences excel at deriving meaning from fragments. They need less storytelling and more opportunities for story-making and interaction as a means of place-creation (Riccio 2013, 72).

In our moment, saturated with competing narratives that never become places, how you narrativize becomes a survival skill. The hero's journey and three-act structure served their time but cannot address our contemporary condition. These worn templates, though reassuring, are spent, manipulated, and now processed into consumerized, mind-numbing repetition. Our moment requires not nostalgia for bygone eras and their forms but skills to create place wherever we find ourselves, using the body's creative abilities as a technology.

Trickster Performing In-Between

Suppose life is an immersive, performed act of navigating overlapping spaces. How should you position yourself? As the enactor, director, facilitator, educator, devisor, enabler, preparer, organizer, producer, or whatever you choose to call yourself? I accept, maybe even embrace, being a director. However, it is an inherited identifier from another era in the cultural slipstream. It is a quick, widely recognized descriptor that distinguishes what I do and do not do. I direct people, things, events, energies, arranging so many elements, material and immaterial, into a semblance of a momentary holism. Yet it is confining and limiting. It is inadequate in describing what I view as acts of service, the making and shaping of happenstance, the conjuring of disparate elements to enable me to feel, see, think, and value life and the world in new and surprising ways. It is less a reasoned and calculated pursuit, more a calling.

After decades of working with Indigenous and non-indigenous communities across continents, from the Sakha of Siberia to the !Xuu Bushmen of the Kalahari, the Miao of China, the Sami of Sweden, the Zulus of South Africa, university theater students, corporate workshops, and Yup'ik, Inupiat, and Athabaskan artists in Alaska (and so many other places and people), I've unwittingly realized the role of the trickster. It is a role not only helpful but essential for functionality and success. I now can think of no other way to navigate the tricky and paradoxical currents between being and becoming amid the convoluted landscapes of our meta, post-everything world. You never know what will come up or what will be needed. Allowing chance and the spirits speak. Embrace optimism, be playful and irreverent, lead with the heart, take chances, fear nothing, and persevere regardless of what happens. Such is the way of the trickster in a tricky age.

The trickster reality appeared without conscious planning. I was never comfortable with mainstream American theater. I was always drawn to experimentation, creating new work, challenging and exciting my imagination, and traveling and adventuring to places of the mind and heart that feed my curiosity. Embracing the unexpected is part of my personality. So, of course, I accepted the job in Alaska, and then things happened, and I found myself inhabiting what I now recognize as the archetypal trickster space. The in-between, the place of change, a catalyst, and adult-child, a troublemaker, an enabler, an inspiration, a facilitator. Like trickster figures that populate cultural narratives worldwide, I am an insider and an outsider, belonging but not belonging, a provocateur and an entertainer, a diviner and a profaner, a fool, an intruder, a helper, a teacher, and a student.

This positioning is neither comfortable nor uncomfortable. It simply is. The trickster exists in a state of constant liminality, as Lewis Hyde describes, "This polytrophic nature—wily, versatile, well-traveled—becomes both a burden and a gift." This paradox shapes my creative approach. I rely on others' hospitality and expertise, yet my outsider perspective helps me notice what insiders overlook. I suggest unconventional combinations that those bound by traditional rules cannot. I freely share my thoughts and feelings regardless of the hazards of being a fool.

The importance of adopting a trickster stance becomes clear when considering the current transitional flux of cultures worldwide. Communities find themselves caught between what was and what is emerging. During a workshop in Dar es Salaam, a bright Makonde performer approached me, holding a dog-eared book titled something like "How to Do Theatre." When I asked why he was studying a performance that had nothing to do with his culture or condition, he said there were no other books for creating theater for his people. This book is dedicated to this nameless young man and to all other seekers of alternatives.

Far removed from his homeland and community, he faced limited choices. He could preserve his declining tradition, learning from an ever-diminishing pool of elders, or by cobbling together what he knew of his tradition, weaving it with other traditions and shaping it for tourist consumption. Or he could reframe his tradition into a Western drama, thereby participating in the spaces of neo-colonialism, globalization, and capitalism. His interest and desire to perform were no longer for his community of place but for a mixed urban consumer culture shaped by the traumatizing legacy of colonization. He needed to make a living. The trickster's skill lies in finding the alternative. The middle ground where tradition remains alive and evolves, allowing it to grow, change, and "respond to modern life without losing its essential identity" (Conquergood 2002, 145).

Paul Radin observed that the trickster is "at the same time creator and destroyer, giver and negator who knows neither good nor evil yet is responsible for both." This contested nature enables the trickster to act as a change agent. A !Xuu Bushmen, a healer told me, "The man on the other side of the river has our knowledge now. The Bushmen are dying." I was a stand-in for that man across the river. My work with the !Xuu could be seen as either helping to preserve their culture or accelerating its decline—or both at once. "The trickster doesn't solve this paradox; they live fully within it" (Hyde 1998, 7).

The methodology developed embodies and utilizes what John Cage called "the highest discipline, the discipline of chance operations." Rather than seeking fixed outcomes, I foster a trickster environment for discovery, exploration, errors, and insights. These emerge not from my planning but from awareness, reshaped through the trickster's influence into fresh configurations. I am both present and absent. "Guiding while simultaneously vanishing" (59). I know of no better way to navigate the ever-changing complexity of our world.

The ethical implications of this position cannot be overlooked. Working as a white American male in these contexts comes with historical baggage. However, the trickster stance provides a way to recognize history without being paralyzed by it. As I wrote at the beginning of this journey, way back in 1995, "I am aware of the profound cultural and social trauma Western culture, my culture, has caused historically, but at the same time it comes down to 'There is work to be done!'" We must all find a way to do what we can.

Amid ongoing planetary crises, environmental collapse, political and social upheavals, and cultural homogenization, the role of the trickster is increasingly important. Needed are what Tyson Yunkaporta describes as "agents of complexity" (184). Those capable of navigating multiple knowledge systems at once. Required are border-crossers who can translate between worlds without forcing either into the other's terms.

The trickster shapes this world not by creating something from nothing, but through rearrangement, recombination, accidental creativity, and productive mistakes. In the space between cultures, tradition and innovation, sacred and profane, the trickster acts as an enzyme, sparking reactions that were always possible but never actualized until the right conditions come into play. This is the work: Shaping conditions for indigenous and non-indigenous people to reimagine themselves on their own terms, using their own resources, and toward their own futures.

From the National Theatre School Journal, Krakow, Poland

The work in Krakow, Poland, began as a mixed-gender workshop. After two days, the five men in the group were hired by a film production company and left, leaving the remaining performers, who were all women, to continue exploring the themes detailed in the Body and Space sections of this book. These explorations fostered a natural, evolving process.

Over the course of two weeks—eight hours daily—the project took an unexpected turn as the explorations uncovered, without explicit guidance or speech, underlying, yet unspoken, issues within the group. The focus shifted to addressing socialized cultural suppression—such as male dominance over females—using the workshop as a space to express personal trauma and collective experiences. The openness revealed deep-rooted, culture-wide problems related to gender inequality and abuse.

The emotional intensity was profound, prompting me to refocus our work, shaping it into a constructive and healing process. I adapted my approach, incorporating drama and social therapy techniques to ensure a positive outcome.

Throughout, I maintained dialogue with the performers, asking for their perspectives and next steps. This open exchange helped performers view the process objectively, with the aim of long-term understanding and benefit. The eventual public presentation of the workshop's work was deeply moving and therapeutic—sometimes resembling a keening—for both performers and audience. Without premeditation, we created a healing ritual that culminated in a public discussion, demonstrating the intrinsic power of performance as a medium of collective healing.

However, the trickster is more than just a fool or chaos agent. As Hyde notes, the trickster acts as "the gatekeeper who opens the door to the next world; those who mistake him for a psychopath never realize that such a door exists" (280). This gatekeeping role closely mirrors the shaman's function, as both act as connectors between different realms. They rely on transformation and shapeshifting, and both experience suffering to promote healing. The wounded healer and the wise fool share a fundamental nature—they must be broken down to be rebuilt.

Joseph Beuys recognized this when he stated that "social sculpture" demands understanding that "the human being creates the world." The trickster-artist doesn't simply symbolize transformation but actively enacts it; rather than merely describing change, they embody it. Every performance project becomes what Beuys referred to as "a warm process"

(20–21), generating heat through the friction between the old and the new, the traditional and the modern, the indigenous and the contemporary global.

Warming Up, Creating Heat

Warming up and activating the body through various stretches is a necessary and effective way to prepare the performer's body and mind for our work. How performers prepare draws generally on what has been passed down, becoming an unquestioned process of breaking the body into parts. A machine-like process reflective of a cultural inheritance rooted in Aristotle. How you create determines what you create. The warm-up is the first important step towards creating a place.

Our primary goal is to generate "heat"—elevating the body's temperature, awakening awareness, and promoting thinking beyond intellectual and rational insights. This process often includes re-educating performers who might have lost their bodily awareness, thereby denying them their physical presence and agency in the world.

The !Xuu and Khwe Bushmen healers create "heat" as a key part of their healing ritual. Surrounded by a community that sings, dances, and claps, they generate heat through rapid, rhythmic, spine-shaking movements that trigger spinal fluid and induce trance states. This heat extends outward, supported by the community's energy. Together, they produce a shared "heat of healing," akin to the way performers across cultures generate and spread energy through community involvement. These rituals, which begin at dusk and end at dawn, are believed to link successful healing to the amount of heat generated. Heat is the accumulation of energy and is considered to possess healing powers. Joseph Beuys came to this understanding in his own way.

Heat represented the energizing of life and symbolized the coming together of ancestors, animals, and spirits, who participated in and took pleasure from the community gathering. The key question here is, who do you perform for and why? For the Bushmen, an adjustment in intention fosters an atmosphere akin to returning to a womb-like space, where things can be returned to origins, examined, and healed. The layered sounds of clapping, singing, rattling, and drumming worked to loosen reality, making time, space, spirit, and the physical world more fluidly interactive. Since all sickness was viewed as a spiritual illness, facing and performing with and for the spirits was necessary. The intense heat energy produced during these ceremonies was so profound that a "cooling down" period was needed afterward, during which the community, led by the healer, danced to dissipate it and honor the spirits.

These cooling periods typically lasted from dawn (which marked the end of the ritual) to early afternoon, depending on the amount of heat produced. This time also served as a time for social interaction and celebration—an acknowledgment that the healing had succeeded and that the disruption caused by illness had been mended. For the Bushmen, as with many other cultures, including your own, an individual's illness impacts the entire community. Consider the disruption and imbalance caused by an illness within your own family or among your friends. Whatever affects one person affects all. Life is a delicate, interconnected system that requires care. Performance is the vehicle of remedy. Healing occurs through interpersonal contact, group support, and engagement. We know and understand this instinctively.

!Xuu and Khwe Healing Practices

The !Xuu and Khwe communities of the Kalahari Desert developed advanced techniques for what anthropologist Richard Katz calls "community healing"—collective practices that transform individual therapy into processes of community-wide change (Katz 1982, 299). Recognizing these practices clarifies how "group heat creation" functions as a tangible methodology rather than just a metaphor.

During !Xuu healing ceremonies, individual healers create a state called! *aia* or "heat" by engaging in sustained rhythmic movements. This activity triggers what Robin Dunbar, a neuroscientist, describes as "endorphin release"—the body's production of natural opioids through repetitive motion (Dunbar 2012). This physiological condition allows what medical anthropologist Arthur Kleinman refers to as embodied empathy—the ability to perceive and respond to others' physical and emotional states via direct symbolic resonance (Kleinman 1988, 41).

The healer's personal energy becomes part of collective healing through coordinated community effort. What appears to outsiders as a spontaneous celebration is guided by time-honored protocols refined over generations. Community members support the process with rhythmic clapping, singing, and synchronized breathing, which amplify and sustain the healer's energetic work state. No one directs this. The community becomes a collective, intuitive being, guided by the event's energy. Songs, drumming, and chanting flow independently, without human guidance.

The phrase used by Machai, a !Xuu healer I worked closely with, "shake open realities," requires careful translation from metaphorical language into a phenomenological description. This creates conditions where practitioners can experience what anthropologist Edith Turner describes as "spiritual emergence"—feelings of connection to larger systems that are typically outside conscious awareness (Turner 2006, 260).

In practice, this enhances what systems theorist Fritjof Capra describes as "pattern recognition"—the ability to identify relationships and connections that are often overlooked due to habitual thought patterns (Capra 1996, 21). The !Xuu and Khwe noted a greater sensitivity to group dynamics such as presence, movement, singing, and rhythm, which in turn increased their awareness, deepened their trance state, triggering perceptual, hallucinatory, and material changes. On several occasions, healers would insert a blade into their stomach to obtain their "healing blood" to offer to their patients. The blades were real, and how such actions occurred remains for me, to this day, a mystery.

The key distinction between genuine boundary-dissolution and performance simulation lies in what Herbert Benson, a medical researcher, calls "physiological coherence"—the measurable changes in heart rate variability, breathing, and stress hormones that happen during true altered states (Benson 1975, 77). Authentic practice results in consistent physiological changes that biometric monitoring can verify. Performance triggers these shifts.

FIGURE 3.1 Machai, a !Xuu healer, conducts an all-night healing ritual in the Kalahari Desert

Source: Photo by the author.

In our context, a warm-up generates heat for individuals and, by extension, the group. A warm-up resembles a healing ceremony, but it is not intentionally aimed at healing or inducing an altered state—although it could be. Warming up the body transforms the individual by awakening and reactivating the senses, placing performers in a more alert psychophysical state. Ideally, it involves activating the entire body-sensing system, shifting perceptions, and fostering a deeper engagement with the world. This transition marks the first step from an ordinary state to an extraordinary mode of being.

Group warm-ups offer structured training in what neuroscientist Antonio Damasio refers to as "somatic flexibility"—the ability to shift smoothly between different states of awareness while remaining conscious of these changes (Damasio 1999, 104). This neuroplasticity is essential for place-making activities that demand performers to embrace multiple perspectives and adapt to diverse cultural contexts.

Each element serves a specific neurological function that improves our physical adaptability. Cognition research indicates that different postures evoke distinct cognitive and emotional reactions (Carney, Cuddy and Yap 2010, 16–17). Consistently practicing transitioning between these states enables practitioners to develop what cognitive scientist Andy Clark calls "neural flexibility"—the ability to rapidly modify responses as circumstances change (26).

Social Activation

The shift from individual preparation to group practice depends on activating what neuroscientist Stephen Porges calls the "social engagement system"—neural networks that support cooperative interaction through facial expressions, vocal sounds, and body posture (16). Play has specific neurological roles that help this shift from individual to collective experience.

Research in developmental neuroscience shows that playful interaction activates what neuroscientist Jaak Panksepp called the "play system"—specific neural networks that support social learning, emotional regulation, and creative problem-solving (14). During play, the nervous system shifts into what polyvagal theory refers to as "safe social engagement"—a neurological state marked by improved learning ability, greater stress resilience, and increased behavioral flexibility.

In place-making settings, playful warm-ups serve what psychologist Peter Gray calls "adaptive function"—they create neurological conditions that enable groups to quickly develop collective coordination while preserving individual agency (Gray 2013). This is especially important in cross-cultural work where participants need to learn to coordinate with unfamiliar movement vocabularies and cultural patterns.

The specific mechanisms involve what neuroscientist Marco Iacoboni calls "mirror neuron activation"—the neurological process through which observing others' movements automatically activates similar patterns in the observer's motor system (4, 30). Playful interaction enhances this mirroring process while reducing the social anxiety that can inhibit cross-cultural learning.

Neurological insights show why playfulness becomes essential rather than just enjoyable in place-making work. The activation of the social nervous system during play creates ideal conditions for the rapid cross-cultural learning and adaptation that authentic place-making requires. Without this neurological basis, groups might achieve technical coordination but lack the empathy needed to manifest a genuine sense of place.

For longer workshops, ongoing performance or educational projects—lasting weeks or months—a more carefully developed, culturally specific warm-up is highly recommended. Part warm-up, part performance, it functions effectively on many levels. The Ritual Preparation process will be explained below.

First, however, a sample overview of warm-ups used for different objectives. They bridge the Body and Space work as they prepare for Place work. They are grouped here for convenience and easy comparison. As always, this offering is meant to be suggestive and serve as a reference to inspire your own adjustments and creations tailored to your needs and goals.

COMMEDIA DELL'ARTE WARM-UP

Ask the performers to mime washing starting with their hands and working their way up to cover the entire body, paying special attention to their curves. Exaggerate.

Then, linking fingers together, ask the performers to create a wave-like motion through their arms. Let the wave be smooth and playful, flowing from the shoulders down to the torso, hips, neck, head, and legs. Allow the wave to take over the entire body, with fluidity, moving freely and spontaneously. Emphasize curves and smooth movement.

Then ask the group to walk in a circle and coach them on the sides while they maintain the circle.

Move Loose	High	Heavy Small
Heavy	Heel First	Light Small
Curves	Toes First	Light Big
Low	Heavy Big	

PLAYFUL WARM-UP

The group forms a circle. A theme or story is presented, and then the group engages in stretching and vocal warm-up explorations that explore it. Using a well-known story is most effective, especially one that depicts a journey and evokes a range of emotions.

For example, yawning and stretching can visually represent a character waking up. Surprise is shown with sounds. Walking, running, and encountering supernatural beings are also portrayed—with performers transforming into animals, plants, spirits, and elements, such as weather. This exploration is beneficial in developing a group's ability to explore and express a multi-perspective worldview.

The warm-up may also be used with themes, trends, or current events. Likewise, the group dramatizes events, whether serious or comic, developing the ability to identify and perform them.

Note: To introduce and establish the warm-up concept, you might want to start or lead the warm-up, with others stepping in as needed. The group can begin by mimicking your actions until they gain enough expressive confidence to take over the process.

MIME WARM-UP

Similar to the **Playful Warm-Up**, the group forms a circle and goes through a variety of creative stretches/sketches that are essentially physical mimes. Each performer, in turn, responds as they will to the mime that preceded.

For example: (1) reaching up for an apple, next performer; (2) stretching a foot into the circle as if testing the water temperature; (3) extending the stomach in stages as if it is being inflated; (4) shifting the face from one side to another as if just punched a few times; (5) dropping the jaw in shock; (6) then grinding the jaw as if a cow; (7) moving each leg as if Gumby or Elvis, etc.

Using images and ideas familiar to the group or common in the culture and locale works best to start. Drawing from online memes, Instagram posts, TikTok, and similar platforms, it shows how everyday life can be used and explored for performance expression.

> **REACHING WARM-UP**
>
> Partners stand face to face and within arm's reach of each other. With side coaching, each partner tries to touch the other. Each reaches out but cannot touch various parts of the other's body. First, reach with your hands, then try reaching with multiple body parts, such as your elbow, forehead, pelvis, chest, knees, stomach, or feet.

Somatic Systems

Moving as Water, Fire, Air, and Earth helps develop the ability to see and embrace the dynamic principles that shape various systems. Instead of merely imagining, this practice gently guides you to recognize and embody different organizational patterns through hands-on experience, making the learning more personal and engaging.

Water movement is an ongoing process that encourages adaptation and helps us find the easiest paths. From a neurological perspective, it sparks what neuroscientist Moshe Feldenkrais called "gentle movement patterns" (1972), which help us move past usual muscular tensions and open new possibilities for organization. Fire movement, on the other hand, features quick shifts in direction, explosive bursts, and radiating energy patterns that develop what sports scientists call "reactive strength"—the ability to generate force quickly and change direction swiftly.

Air movement emphasizes suspension, lightness, and a sense of three-dimensional space, helping develop what dance researcher Susan Jaques describes as "antigravity responses" that create the kinesthetic feeling of weightlessness (130). Earth movement, on the other hand, focuses on grounding, stability, and our connection to gravity, activating what biomechanist Thomas Myers calls "deep structural patterns" that shape our postural relationships (256–257).

This practice helps develop behavioral flexibility through what neuroscientist Michael Merzenich describes as "competitive plasticity"—the brain's remarkable ability to reorganize its neural networks based on repeated practice (60). When practitioners intentionally practice different movement qualities, they are rewiring their nervous systems, which opens them to a broader range of responses. In a sense, act funny or weird, and you will eventually become it.

This idea of neuroplasticity highlights the playful trickster tradition and is known as "neoteny"—the gift of keeping youthful learning abilities alive into our adult years. Studies in developmental psychology show that staying good at learning new motor skills throughout our lives benefits from practicing fresh movement challenges regularly (Diamond 2013, 140–141). The elemental warm-up offers a mindful way to explore various types of movement, helping us preserve this remarkable, adaptable skill.

ELEMENT WARM-UP

Begin by focusing on your arms, then progress to your head, neck, torso, pelvis, legs, and feet, moving each part of your body.

Begin by moving as Water.

Repeat this sequence with the other elements:

Fire
Air
Earth

Rise, fall, and navigate through the space, as each element.

Maximize the distance from others while moving.
Stay as close to others as possible.
Perform movements slowly.
Perform movements quickly.
Move at your preferred normal speed.

Exploration Alternative: Focus on one element at a time or assign each performer an element to ensure all four are represented and can interact. Repeat this process with variations for different performers.

IMAGINATION WARM UP

Isolate and warm up each major body part.
Start by tensing your hands, then releasing.
Next, use your hands as if typing, feeling nervous, or holding money.

Tense the arms from wrists to shoulders, then release. Move your arms as if they are wings, a chicken, a soaring eagle, and dance with them.

Tense the torso, then release. Use the torso as a shield, a target, or a hoochie-coochie dancer.

Repeat the tension-and-release sequence, then perform isolation explorations for the pelvis, thighs, knees, calves, feet, neck, face, and, finally, the entire body.

Ask the performers to move around the space and find the following:

A safe space
A quiet space
A dark space
A threatening space
A small space
A large space
A light space
A noisy space
A sad space

How does the objective and space change?

How do your interactions shift when you are with a partner?

Consider how the space feels with that other person.

Allow your body to sense the relationship within the altered space. Don't analyze or act—simply accept and react.

Add more partners, forming groups of four or eight, until everyone is engaging with one another in the designated spaces.

Exploration Alternative: While exploring these spaces, incorporate a sound to represent each. Repeat this process when a partner joins and continue with additional partners as they are added. If it feels right and opportunities arise, consider developing a chant, song, or sounds to embody the space.

Note: This exploration involves location-based scenarios that can help warm up the body, spark imagination, and evoke emotions. You can expand these scenarios if you wish. For instance, in a safe environment, imagine an approaching threat or something sinister. Pay attention to your sound at this moment—are you expressing fear, giving a defensive warning, or begging for mercy? Feel free to create outline scenarios for all or just a few.

Somatic Resonance

The vocal warm-up sequence activates what acoustic researcher Steven Brown describes as the "musilanguage" system—neural networks that combine vocal, rhythmic, and social coordination skills developed before the divergence of music and language (280). This prepares performers by fostering physiological and psychological abilities essential for collective place-making activities.

Place Making Sounds

Discussion and Exploration

Guide performers through a process that combines sharing information and exploring body sounds as a vocal warm-up. Focus on how sounds relate to and influence the nervous system, emotions, and communication.

Explain, explore, then analyze and discuss each body's sound influence and mode of communication.

Deep stomach sounds stimulate the parasympathetic nervous system through what voice researcher Kristin Linklater calls "natural voice production"—a sound-making process that avoids muscular tension and involves core breathing techniques (Linklater 1976, 2). This leads to what acoustic physiologist Johan Sundberg describes as resonant voice quality, characterized by increased harmonic overtones that support group entrainment (Sundberg 1987, 20).

Chest and torso percussion evoke what music therapist Arthur Hull calls "body drumming"—using the torso as a resonating chamber to enhance vocal output and stimulate proprioceptive awareness throughout the body (Hull 1998, 66). This technique is found in various cultures as a preparation for group vocal activities.

Basic vowel sounds (yah, yae, you, yaa) develop what linguist Peter Ladefoged refers to as "cardinal vowel production"—the basic mouth configurations that enhance acoustic resonance while reducing articulatory effort (Ladefoged 2001, 14). These sounds create what ethnomusicologist Steven Feld calls "acoustic ecology"—vocal patterns that connect human sound production to environmental acoustic expression (Feld 1982, 265), such as birds, animals, climate, trees and plants, rivers, and lakes.

Head voice and nasal resonance engage the "resonator tract"—the network of cavities that enhance and shape vocal sound, resulting in what is often perceived as expanded vocal presence (Miller 1986, 31). This physiological process enables practitioners to have an increased sense of awareness or "place-consciousness."

Animal sounds involve what François-Bernard Mâche calls "zoomusicology"—the study of how humans and non-human species produce sounds, revealing shared acoustic and rhythmic patterns among them (Mâche 2000). This method encourages what could be described as "interspecies acoustic empathy."

Note: The shift from solo vocal activity to group sound production results in what acoustics researcher Barry Truax calls "soundscape composition"—the creation of complex acoustic environments produced through the coordinated efforts of individuals (Truax 2001, 18). This process shows how prepared individuals can work together to create shared acoustic experiences that influence spatial and social dynamics.

Bringing Place to the Body

Chance serves as a way for spirits to communicate. It was through a chance event that I realized the limits of my perception and how I understood and practiced performance. During my first semester teaching at the University of Alaska Fairbanks, I unexpectedly inherited the Tuma Theatre, an Alaska Native group when the previous director left for another position. I was told that refusing would result in the loss of state funding. I accepted without knowing what I was doing. The following day, after some thought, I went back to the Alaska Native Studies director and informed him that I would have to decline the offer. I knew nothing about Alaska Native Theatre and wasn't even aware that Alaska had a university until I applied for the job. I did not want to be a meddling, neo-colonialist figure who might inadvertently cause more harm than good to a community that had suffered in many ways.

The director of the Alaska Native Studies program, Mike Gaffney, replied that what I did know was theater, and having been artistic director of the Organic Theatre in Chicago for several years, I knew how to produce. I was offered a budget of nearly $100,000 to create the first-ever research and documentation project on Alaska Native performance, which would span almost two years. Spending every spare moment traveling to villages, meeting elders, learning to dance, drum, and sing, interviewing elders, and documenting the lives and performances of Alaska Native people.

Alaska is a vast state where most villages are remote, accessible only by small planes landing on gravel runways that appear suddenly in the tundra or boreal forest. Villages, with populations between 100 and 600, lack accommodations; subsistence living is the norm, and everything is costly. I learned to eat whale, seal, walrus, moose, black and grizzly bear, Dall sheep, porcupines, salmon in large quantities, various fish, caribou, dried kelp sprinkled with high bush blueberries dipped in seal oil (a delicacy), and Eskimo ice cream—a whipped seal fat mixture that looks tempting but isn't.

Alaska Native communities do not engage in theater as traditionally defined. Instead, they possess a diverse and extensive performance heritage encompassing ritual practices, ceremonies, folk acts, and everyday observances. Their shamanic and mask traditions are among the most profound globally. At the time, I was unaware that I was beginning a process and embarking on a personal and professional journey that would take me around the world and to this moment, as I write these words.

There was limited research material on the performance culture of the Yup'ik and Inupiat Eskimo, the Haida, Tlingit, Alutiiq, and Athabaskan peoples of Alaska, and what existed was buried and scattered across anthropological texts, the memoirs of early explorers, or oral history recordings. By visiting villages, conducting research, interviewing elders, and immersing myself in Alaska Native cultures and performance, I gained a deeper understanding of one of the world's last remaining hunting-and-gathering subsistence cultures through direct experience.

Two years later, I felt confident enough to teach a course on Alaska Native performance. At the start of the first class, I asked the student performers to gather in a circle for a physical and vocal warm-up. I was trained in the Western theatrical tradition, which views the body as an instrument, sometimes even a machine-like entity made of component parts. I had never really questioned the practice of warming up the body by isolating the neck, hips, arms, legs, and feet. Then there's the vocal apparatus, voice, vowels, consonants, words, and sentences. Plays are written in scenes, with actors defining their beats, cues, objectives, and obstacles. It is a modular, mechanistic way of seeing, processing, and being in the world.

Around the circle were Yup'ik and Inupiat Eskimos, Athabaskan Indians, and a few non-native students. The compact bodies and broad, fleshy faces of the Alaska Native students reflected the land they had inhabited for thousands of years. Their physical features had adapted to the harsh reality of their climate. Compact bodies retained heat, and fleshy faces served as a defense against frostbite. Many of the men were subsistence hunters, while the women were equals in all aspects except hunting, which was done on the same land and sea—some of the most demanding hunting environments in the world—as their ancestors had done for countless generations.

As the students rolled their necks as instructed in an isolation warm-up, and I realized something was seriously wrong. I intuitively sensed that it was an imposition—a way of viewing and understanding the body that conflicted with their way of being in the world. I asked them to stop, reassured them they had done nothing wrong, and then invited them to return for the next session, each bringing three movements and three sounds or songs they considered "traditional."

The next session offered a wealth of traditional movements, gestures, chants, and sounds, many drawn from a vibrant and illustrative "story dancing" vocabulary. Traditional drums were introduced, and over several weeks, the group collaboratively developed what would become known as the "Ritual Preparation." Essentially, it was a 30-minute physical and vocal narrative, as well as a ritualized warm-up built entirely from traditional movements, rhythms, songs, and chants. The Ritual Preparation framed, activated, engaged, and prepared the performers in a way no neutral, mechanical, or purely human warm-up could. Performing the Ritual Preparation was like opening a door into Alaska Native culture, transporting us from the Eurocentric university setting and reawakening us to a different perspective on being in the world through performance.

The ritual preparation, like the practices of subsistence hunters, was both efficient and multifaceted. The actions performed served to access and express a sense of place. For the Inupiat, Yup'ik, and Athabaskan peoples, place was a permeable boundary crossed through dance—each movement telling a story that was both archetypal (spiritual and ancestral) and specific. Dance acted as both a form and a vessel, connected and conveyed by rhythm, immediate yet layered with meaning, as it intertwines the psychological, physical, spiritual, and emotional aspects. It expressed a continuum that was simultaneously personal, ancestral, and environmental.

Although the Athabaskan ritual and performance tradition offered a wealth of practices, Tuma relied heavily on the performances of Alaska's Yup'ik and Inupiat Eskimos, which emulate the movements, sounds, and personalities of animals living in their environment: The seal, raven, walrus, eagle, and bear, most notably. Alongside mimetic animal and human actions, there are movements inspired by natural surroundings, such as the dancing northern lights, distant hills, water, and wind. These coded movements are a vast, culturally specific vocabulary used to communicate with and about place.

Through this process, I realized that performance is not simply anesthetized artifice, but a way to engage with and communicate about place—connecting a community of humans, animals, spirits, ancestors, and elements. Traditional Yup'ik masks, like those of other Indigenous cultures, embody a worldview that intertwines mythology with human, spiritual, and animal presences. Each element signifies and embodies a part of the spirit world, holding meaning and power. At the same time, performance serves as a bridge between the visible and the invisible, the present and the eternal, the tangible and the ineffable, the personal and the universal.

A performative act akin to how religious iconography links the unseen realm with the material world, encouraging engagement and dialogue. A crucifix on a wall or around someone's neck holds, accumulates, and transmits the energy of Christ and Christianity. It does not matter if it is made of metal, marble, wood, or plastic—its power is significantly enhanced by the lives it has touched and the ritual energy it has absorbed. The symbol of the crucifix, along with actions like kneeling, face and arms outstretched to the heavens, and hands folded in prayer, becomes an iconic shorthand so common that it communicates immediately, evoking beliefs, emotions, mythology, and identity. These symbols are so universally grasped that they appear as emojis, bumper stickers, and signifiers with pointed immediacy, representing a vast history and community. This is sympathetic magic that crosses time and space, serving as a conduit and gateway into belief systems and communities of believers, past, present, and future. Yet, the embodiment and empowerment of a static symbol depend on the actions performed. Performance energizes the otherwise stagnant. Walking through a museum and viewing nineteenth-century statuary from African tribes, Ancient Egypt, Greece, or Rome reveals beautiful but dormant artifacts. Without the energy imparted through performance, they sit silent, void of meaning.

The dance movements of the Yup'ik and Inupiat serve as practical mnemonics, illustrating the interconnectedness of humans, animals, emotions, spirits, ancestors, and community. Daily interaction and close relationships with animals, along with a deep understanding of their environment, are essential for successful hunting and the preservation of their culture. This worldview shapes multidimensional story dances, in which one dancer portrays all parts of a story, conveying a worldview of porous, mutable boundaries.

A Yup'ik dance might depict the story of a hunter spotting a goose while hunting. The dance starts with a hunter traveling and, through dance movements, depicts a beautiful day, the sun, and the wind. A goose flying in the air is shown, and the performer quickly and easily transforms, through dance, into a goose. With the next move, the performer changes into a hunter, aiming and shooting the goose with a bow and arrow, only to transform back into the goose and reveal the wounded, falling bird. The dance ends with the hunter dancing joyfully and giving thanks to the day and the *Ellam Yua*, the "Owner of the universe."

The Yup'ik Eskimos call dancing *Agayuliyararput*, meaning "our way of making prayer." Although dancing today may not carry the same sacred significance it once did, it has always been essential to cultural identity, serving as a connection to mythic and sacred origins.

> All dances were originally sacred; in other words, they had an extra-human model. The model may in some cases have been totemic or emblematic animal, who's motions were reproduced to conjure up its concrete presence through magic...The dance may be executed to acquire food, to honor the dead, or to assure good order in the cosmos [...] for every dance was created in *illo tempore,* in the mythical period, by an ancestor, totemic animal, a god, or a hero. In a word, it is a repetition, and consequently a reactualization, of "those days."
>
> *(Eliade, 28–29)*

The Architecture

Our work now shifts into performance. But, before any performance creation work begins, a group must establish what Sakha shamans call "the breathing"—the creation of a temporal and spatial entity of the greater self. This is not accomplished through exotic rituals, but rather through straightforward, intentional acts of collective focus. Circle thinking, breathing, and being form a foundation—the essential structure of collaboration and thought. Within the circle, human and non-human entities encounter no hierarchies, an equality of perspectives, and energy that circulates rather than collects or drifts away. Life, emotions, thoughts, and existence are interconnected, creating a cyclical pattern with no beginning or end, each following its own unique way of being.

The opening always begins with silence. Not the uncomfortable silence of social unease, but the rich silence from which sound and movement naturally arise and coexist with stillness. The silence between stars. In Yakutsk, we sat for several minutes listening to the building's sounds—ancient heating pipes, distant traffic, wind through poorly sealed windows. These were not distractions but the voice of place making itself known. Performers initially found this uncomfortable, conditioned to fill every moment with productive activity. However, this discomfort was the beginning of a transformation.

Somatic Archaeology

The Ritual Preparation serves as a preface, our primary tool, a blueprint, and model for somatic archaeology—the excavation of culturally encoded movement patterns that link the body to place. Instead of starting with inherited conventions, abstract ideas, or tired theatrical formulas to be filled in, performers rely on physically learned knowledge of the living world. The text, the script lives within them, and the process of building the group's Ritual Preparation is a process that models and frames the work.

It starts straightforwardly: Ask each performer to select three movements they see as traditional, cultural, and spiritual. Give them a day or a weekend to thoughtfully reflect and prepare. This process aligns seamlessly with our initiatives in the Body and Space phases of our project.

The Breathing portrays the movements presented, and they carry context, history, and stories. They form a kinesthetic structure ready to be made. A Sakha woman's milking gesture embodies more than just a mechanical action; it reflects the rhythm of her morning routine and her environment, which lives and vibrates in her body. Further revealed is her relationship with animals, a farming lifestyle that requires labor, and an identity shaped by cultural and social norms. Similarly, a Yup'ik hunter's rowing motion encodes knowledge of sea currents; the expansiveness of the body signifies a vast, open terrain. A look can reveal objective, seasonal patterns, and the patience needed for Arctic survival. These movements—snippets of a way of life, of living with a place—are like the phonetics, vowels, and consonants that make up our vocabulary, the raw text waiting to be shaped.

The movement presented alone remains an isolated gesture without rhythm to anchor it in a specific place. Every landscape has its own temporal signature—the Siberian steppe's slow, expansive breathing; the quick, punctuated urban rhythms of St. Petersburg, Dallas, Addis Ababa, Krakow, Helsinki, Kathmandu, Seoul, and Chicago; and then, there is Alaska's vast geographical and seasonal fluctuations. These rhythms must be uncovered and processed within the body before they can be expressed outwardly.

Our explorations have aimed to increase awareness of these internal rhythms, with performers viewing their bodies as percussion instruments engaged in dialogue with their environment. Moving into performance creation, we begin to examine how traditional rhythms embody place-specific knowledge. The Sakha three-beat pattern reflects the gait of horses, as well as the rhythm their shamans use to invoke a trance. Yup'ik drumming mimics the endless ebb and flow of the Bering Sea, the rise and fall of the tundra, punctuated by the sudden harshness of the weather. In Ethiopia, the rhythm is that of a society deeply rooted in the religious practices of the ubiquitous Christian Orthodox Church and Islam, set against the slower pace of a developing country.

The rhythm of Dallas is varied, fragmented, and quick-changing, requiring the performer to ground themselves in their body. It is for this reason, in part, that cities tend to isolate and alienate people from one another and their environment, which they cannot hear or connect to. Performing and connecting with the rhythm of a place are not random cultural choices but somatic responses to biological, social, cultural, and environmental input. This is why being in nature—strolling in a park or hiking in the woods—is restorative. Strategies for survival in and with a place are adaptive, encoded by environmental rhythms, and passed and adjusted as they move through generations of bodies living in a particular place.

Put in a different perspective, a Dead White Zombies performance isn't influenced solely by its performers, community, intent, content, or the architecture it takes place in, but also by its practice of being—embracing "the breathing" and being in the moment of the fluctuating Dallas, circa 2011 to 2026. Breathing is a depth of layers and dynamics, simultaneously of a moment yet transhistorical and transcultural.

Work with Tuma Theatre in Alaska was timely yet drew from thousands of years of Cosmocentric performance traditions. When the insights and templates gained from Tuma were applied to projects in St. Petersburg, Burkina Faso, Nepal, Sweden, or Wroclaw, Poland, in 2023, they were different, adaptable, and yet the same. This reflects the planet's cyclical and recursive nature. All is part of a vast system. Nothing comes from nothing, and everything that exists now has always existed and lives within and alongside its presence. Yesterday is today, and today is tomorrow. The present is an ancient future.

Before you understand what Tuma did for me, I think it's important for you to understand the world I come from. For some, Tuma might seem fun and interesting, but for me, it was a deeply transformative experience. Before I joined Tuma, I had questions about my heritage and my place in the universe. I grew up around drinking, drugs, gambling, and abuse, with a small flicker of hope and love from my grandparents. Going to college was my escape from tragedy. My rage fueled me as I faced the White Man, and I left my home with many questions my people couldn't answer—like where do my people come from? What are my responsibilities here? What defines right and wrong? I experienced trauma spiritually and emotionally before joining Tuma. When I joined, I confronted some of my greatest fears about who I was and why I felt like a sick person. Tuma and Tom Riccio changed my life forever. My world started to open. The wrongs done to my people were acknowledged, the power of those who once were was called upon, and it became safe for

> me to grieve and seek my own answers. I spent three years with Tuma, and each year I grew. I was extremely shy at first and carried deep hurt I didn't understand. Over time, I gained strength and confidence, and I learned to channel my anger more productively. Now I express myself more freely and have a deeper understanding of my people's past, what is happening now, and what I can do to make a difference.
>
> Wilma Brown, Tuma Theater performer

Performers who engage with our work report a transformation that extends beyond just artistic practice. As Wilma Brown from Tuma Theatre shared, reconnecting with place through performance offered healing that years of traditional therapy could not accomplish. This is not because the method is inherently therapeutic, but because the body's disconnection from its environment is a wound. Connection is healing.

Our work goes beyond simple artistic practice, turning into a survival tactic. An act of remembering and restoring. All of this converges in the Ritual Preparation. This is where theory becomes practice, transforming the abstract structure of Cosmocentric performance into tangible, repeatable actions that promote change and healing.

The Ritual Preparation

The ritual Preparation reflects our core principles—Circle, Rhythm, and Balance—as the practical expression of everything discussed earlier. It revisits the concept of the Body inhabiting Spaces to transform them into a Place. During this process, a group develops its own set of ethics and working methods. Instead of relying on predefined approaches or vocabularies, it creates a unique path suited to its specific context, needs, and goals. This organic process utilizes the group's history and strengths rather than adhering to rigid, inherited conventions of understanding.

It exemplifies somatic archaeology in practice. By excavating culturally encoded movement patterns, the Ritual Preparation offers essential tools and structure for this process. It turns the collective breathing of a circle into a fertile environment for exploring movement, capturing the rhythms coursing through the landscape and bodies—both human and other-than-human—giving them shape. This provides a pathway to reach deep memories that emerge when bodies recall what minds have been conditioned to forget.

The Ritual Preparation, used in different settings and situations, has proven to be more than just a warm-up or training. It is a daily practice of personal and cultural healing, a living proto-ritual that reconnects displaced people with their inheritances. It evolves through practice and in so doing reveals its purpose.

- Identifies, tests, and creates a performance language.
- An activation of a Cosmocentric perspective through the body's wisdom.
- Process-oriented work linking the collective to its broader community of place.
- A forum that encourages open debate, interaction, and innovation.
- A methodology that cultivates technical skill and artistic style.
- The ability to craft narrative.
- Trust-building.
- Navigates the intersections of modern and traditional worlds.
- Serves as a creative template and reference.
- Becomes a place for cultural and personal renewal and affirmation.

The Ritual Preparation acts as a gateway and a foundation for performance, offering access to a timeless, symbolic, and culturally rooted reality. It and the performance that it seeds are a process of integrating spatial, rhythmic, physical, and metaphysical elements into a unified whole. It explores and sets a viable vocabulary drawn from one's community of place, shaped to serve that same community.

Having worked with and documented ritual, shamans, healers, mediums, and diviners worldwide, patterns emerge. African diviners similarly cite performed codes and their precise execution as a medium for connecting with ancestors and spirits (Peek 1991, 84). Shi Jing He, a Miao shaman (known as badai) in southern China, whom I befriended and conducted research with, performed an extensive vocabulary of hand gestures to "open, communicate, and visit" different areas and beings of the Miao spirit world (Shi 2001).

In the three-day Huan Nuoyuan ritual, Master Badai Wu Zheng Niam described the thanksgiving ceremony as a performance that a family offers to the spirits, who observe and coexist with them and the village. Each part of these ritual acts like connective tissue linking the human world with the non-human realm. Rhythm opens a pathway to the other world, while dance steps mirror astronomical patterns. Song and chant tonalities align the body's meridians with those of the environment and cosmos. Incense and smoke carry thoughts and feelings to the ephemeral realm. Every object, detail, regalia, and the sequence of the ritual itself aim to re-establish harmony and balance with the non-human world (Riccio 2019).

FIGURE 3.2 The Huan Nuoyuan ritual drama performed by Master Badai Wu Zheng Niam in the village of Bai Guo, *Xiangxi Tujia and Miao Autonomous Prefecture, Hunan, China*

Source: Photo by the author.

A Collaborative Process

To begin the Ritual Preparation process, each performer is allotted a day to explore three movements that they feel express their culture, tradition, and spirituality. These elements often overlap with others' choices, and sharing them reveals shared themes and similarities that invite discussion. You guide the discussion with questions of context, meaning, expression, and the physical, rhythmic, and sonic connections of each movement. Through this dialogue, movements associate and organize into sequences, ultimately suggesting a narrative.

As a group project, it spans at least two weeks with intermittent effort. Depending on the situation, allocate one to two hours per session until consensus is reached that it is provisionally complete. The extended timeline allows for living with the process, reflecting, adjusting, and adopting. Every component—actions, rhythms, dance steps, vocalizations—must be agreed upon (not necessarily understood), then is detailed, coordinated, and performed, with the process becoming a prototype and reference for performance creation.

Participants often describe the experience with terms that validate our approach: "a healing," "walking with my ancestors," "I'm proud of who I am and my culture," "It's helped me feel less alone," "I was stepping outside of myself," and "like a walking dream." The process is about community and collaboration." These testimonies show what is understood in theory—that reconnecting the body to place through collective practice offers healing that cannot be achieved through individual, intellectual, or pharmaceutical means.

The creation process challenged my Western theater training. It sparked a series of insights, offering an alternative perspective, leading me to question inherited performance methods, how they confirmed my worldview, the role of the performer, and the functionality of a performance event. Significantly, it made me question who I perform for, why, and who the intended audience is.

Returning to ritual roots evokes a strong response from both performers and the community. In 1992, Vladimir Putin was vice mayor of St. Petersburg when he viewed a presentation of the Metamorphosis Ritual Preparation. The newly established Russian-American Theatre supported our production, and we were invited to showcase our work at a city hall cultural event. The impact of the Metamorphosis Theatre Ritual Preparation was so affecting that Vice Mayor Putin granted us support and use of the 1200-seat Baltic House Theatre for a production we were not planning to do. Inspired by the reception and support, the performers and I decided to develop a full production that later aired on Russian television and toured.

Instead of detailed documentation, which would be time and space consuming, I am sharing curated snapshots of various Ritual Preparations to showcase their range of contexts, significance, and expressions.

Ritual Preparation Applications

Alaska: Members of the Tuma Theatre shared subsistence practices that were alive in practice and memory. Their movement vocabularies referenced traditional social dances representing various subsistence activities—such as hunting (spearing, hauling, ice fishing), animals (walrus, raven, eagle, the trickster raven, polar bear), daily tasks (cutting, kayaking, hearing, walking, ice floe hopping), and spiritual beliefs (hands moving with the aurora, praising the universe's creator). The 30-minute Ritual Preparation began with "a sitting," an Alaska Native expression of "the breathing" that burst into a flock of birds and song. The preparation included the quiet scanning used for hunting, with eyes fixed on the horizon while the body remained still, then transforming into a seal. Transformations involved performers shifting from human to animal, to a feeling, and then to an abstraction, all synchronized with drumbeats, chanting, and singing. These gestures were at once performative acts, survival techniques, and encoded knowledge passed through generations, embodying and reinforcing a Cosmocentric worldview.

The Opening: Around the fire, performers sit in a circle on their knees, eyes closed, with sticks gathered in the center. From silence, a faint drumbeat grows louder. Performers slowly pull sticks from the "fire," accompanying the beat as they awaken the earth. The Yup'ik chant builds softly (here presented phonetically): *new-knumb nah-llun-ghit-dah ma-knee ma-knee ma-knee wee-dall-but* ("the earth knows we're standing here"). Chant, drumbeat, and stick-banging accelerate to climax, transforming performers into power birds. As birds take flight, drumbeat turns frantic, then slows to a steady two-beat rhythm. Each bird performs its story, dance, and song. Birds return, landing, circling, and standing as humans await the drummer's signal.

Siberia: The Sakha actors presented a variety of cultural treasures. Maria Varlamova demonstrated the hysterical trembling typical of women shamans—a movement that originates in the shoulders and flows through the entire body. Others performed horse-mounting gestures familiar to every Sakha child, even if they had not ridden in years, along with precise milking motions and a gentle slap on the cow's flank. The Ritual Preparation reflected a deep-seated awareness of and familiarity with an advanced shamanic tradition. It began seated quietly in a circle with "the breathing," which transitioned into yawning to invite the spirits. It concluded with the *Ohuokhai* circle dance and call-and-response singing, performed during the 24 hours of daylight at the summer solstice to welcome the new year.

Zambia: Working with 26 tribal groups introduced a remarkable level of complexity. The Bemba used finger-millet sowing, which involved a flicking wrist motion that differed significantly from the motion used for maize planting. The Tonga showcased their rain-calling dance, with their shoulders moving up and down while their feet remained rooted, mimicking water drawn from the earth. A Luo performer, the son of a healer, offered a sacred vocabulary. The Tambuka women performed an initiation dance with arms pulsing from the chest, symbolizing the opening of the heart to womanhood. This, along with a plethora of other Cosmocentric performance vocabularies, underwent a process of presentation, discussion, and appreciation, and was then applied to multi-tribal and ethnic Ritual Preparation.

Performers form a circle for the opening, kneeling in a reverent posture. After a moment of silence, a simple, low drumbeat begins. As it grows louder, participants shift from kneeling to one knee and clap to welcome the spirit of performance. The drumbeat intensifies; participants become more energized as the spirit of performance takes over. They raise their hands skyward at the peak of the movement. To ward off evil spirits, each person wipes their feet twice. The women perform an initiation dance characterized by a shuffle step, with arms pulsing in and out from the torso. The group then moves counterclockwise around the circle.

Russia: Metamorphosis Theatre faced distinct challenges, as its "traditional" movements were filtered through Soviet folklorization and aestheticization. Our approach involved borrowing vocabulary from the White Doves cult, of which several members were part, including ecstatic spinning, and reinterpreting these movements in a modern setting for the broader community. Performing authentic White Dove movements in our performance carried the sacredness of their ritual and mythic roots. It was the first time many of the group members performed outside the group's ritual context. The Ritual Preparation was rhythm-driven and included drumming, body slapping, foot pounding, stick-hitting, and fire-leaping dances. A pot of fire at the center of the circle symbolized the sun, life, and destruction.

Ethiopia: Most members of the Lul Theatre group were born and raised in Addis Ababa, yet each identified with one of Ethiopia's 80 ethnic groups. The group included members from the Oromo, Amhara, Tigrayan, Sidama, and Somali communities, each from different regions with unique histories—often spanning hundreds or thousands of years—distinct performance traditions, and ongoing political conflicts. The influence of Ethiopian Orthodox and Islamic religions further enriched this complexity, deeply shaping Ethiopian culture. The performance vocabularies were diverse and rich, featuring ethnicity-specific elements alongside shared cultural expressions in a multicultural, religious environment, as interpreted by disenchanted youth aged 22 to 35. A brass horn, commonly used to announce births, deaths, or local events, was paired with culturally specific dances, songs, and Orthodox religious expressions, as well as the Islamic call to prayer. This Ritual Preparation revealed the diversity, influence, depth, and value of an ancient civilization still exerting influence in a modern, urban environment.

Dallas: The urban, multicultural, and multiracial elements of the Dead White Zombies present unique challenges and opportunities. The group is ongoing, with performers and non-performers moving through it, each bringing performance vocabularies shaped by diverse sources and backgrounds. Theater-trained performers bring an institutionalized, anthropocentric, psychologically driven process that adapts to the script and character. Untrained performers offer unfiltered, impressionistic expressions of culture and pop culture. Consequently, Ritual Preparations vary from one performance to another. Sometimes a shared group ritual and vocabulary, as in Holy Bone, are established. In other cases, such as kaRaoKe MoTel and T.N.B., vocabularies are performance-specific, with performers individualized yet sharing common elements like rhythm awareness. The core idea is to develop levels of performance vocabulary that mirror the complexity of our context and performers: One focused on the performance's intent, content, and setting; another on the individual performer's vocabulary; and a third on interactions with the audience/community. The performance's structure balances fluidity and creative evolution. This approach promotes adaptability and integrates self-aware, personal performance, enabling participant-observer performers to interact intimately and spontaneously. Performance becomes a dynamic dialogue that can be shaped or reshaped as needed. In a sense, the Ritual Preparation extends its inclusiveness to the community and is recreated with each performance, forming an ongoing process.

FIGURE 3.3 A Ritual Preparation created and performed by Tukak' Teatet, a Greenlandic Inuit company, Jutland, Denmark

Source: Photo by the author.

Sticks

To improve rhythm awareness and energize the body, each performer was given, found, or carved an 18- to 24-inch stick. Inspired by traditional Yup'ik dance sticks, this practice helped performers stay rhythmically synchronized. Historically, these sticks were ornately decorated and believed to embody the spirit of rhythm, with attached feather puffs symbolizing breath. In early contact rituals, the lead drummer's stick was connected by sinew to a large, movable sculpture above, with five concentric rings decorated with feathers and carved figures that moved with each drumbeat. Small anthropomorphic figures made from driftwood, ivory, and bone were attached to represent ancestral spirits, while white feathers symbolized movement and travel in the night sky. The Yup'ik believed that stars were holes in the sky cloth through which shamans traveled to other worlds.

Sticks play a vital role in the Athabaskan (interior Alaska) memorial potlatch, where a dirge-like song honoring the departed is composed, sung, and rhythmically performed with locally sourced sticks. These sacred sticks embody the Athabaskan phrase, "hitting sticks, healing hearts." Each beat releases communal sorrow. At the conclusion of the event, one stick is burned to signify the end of mourning, while the other is kept as a keepsake.

Beyond simply hitting sticks together or stomping the ground, Tuma evolved the sticks into culturally meaningful expressions. They transformed into a bow and arrow, a fishing pole, a harpoon, a kayak paddle, bird wings, a swimming fish, walrus tusks, or caribou legs. Each transformed object introduced its own distinct rhythm. The use of sticks helps performers achieve awareness and immediate, personal interaction through the body's rhythmic resonance, giving performers possession, agency, and expression of rhythm. This was especially significant for cultural contexts that lost their rhythm to colonization, as was the case when working with the Greenland Inuit. **Sticks**, as tools for rhythm awareness and expression, have been used successfully in several cultural contexts—traditional, mixed, and urban populations alike.

Environmental Rhythms

Each part of the earth possesses its own rhythm, generated by unique weather patterns, geography, animal migrations, and cosmic relationships. Indigenous performance traditions across the globe have long recognized this. Rhythm is not merely musical accompaniment but the very pulse of place, the sonic signature, the song of a place, if you will, that sings the specific location's relationship to the greater whole. When performers align with their own and environmental rhythms, they participate in a dialogue that is both human and other-than-human. A rhythmic conversation among body, land, and cosmos constitutes what I call rhythm reality (Riccio 2010).

Scientists have identified Earth's electromagnetic pulse known as Schumann's Resonance—a standing wave that circulates between the Earth's surface and the ionosphere. This planetary rhythm appears to be intimately linked to human brainwave activity. Researcher Robert Beck documented that healers from diverse traditions—psychics, shamans, faith healers, practitioners of Santería and Wiccan—all exhibited nearly identical EEG signatures during healing moments, synchronized at 7.8-8 Hz with the Earth's geomagnetic field. Within this generalized resonance, local variations exist; each part of the earth emits its own rhythmic signature depending on geography, rock density, altitude, latitude, and longitude, as well as daily and seasonal rotations. Cosmocentric performance, I suggest, has always been a technology for tuning human consciousness to these place-specific frequencies (Riccio 2010).

The Tundra's Heartbeat

The Alaskan tundra has left an indelible impression on my body, mind, and spirit. Looking out over that vast expanse, snow blends with the sea and sky in shades of white and bluish gray. The inability to distinguish oneself without explicit geographical references profoundly alters perception and the psyche—it is as if floating weightless in a dream. This disorientation can lead to what is called arctic hysteria, but it can also open pathways to alternative modes of consciousness and perception.

The plaintive rhythms of Yup'ik and Inupiat drumming—a heartbeat-inspired two-beat and three-beat, respectively—reflect thousands of years of listening to this land. The endlessly repeating rhythm of Bering Sea waves, the simple calm of open tundra, the gentle slopes of hills, and distant mountains all live within these patterns. The rhythms subtly transform, intensify, and recede to emulate weather and waves, with sudden tempo changes reflecting the hunt, the storm, and the migration. In villages, the rhythm of the earth dictates daily events, marked not by mechanical time but by changing tides, light, winds, and seasons. The call of a bird indicates approaching fish; the sea's stillness or a change in smell signals the whale.

Paul Jumbo, a close friend and key Tuma Theatre collaborator, vividly illustrated this place-awareness. While walking near his Toksook Bay village on the shore of the Bering Sea, he suddenly stopped, closed his eyes, nodded to a silent rhythm, and started singing. He explained that the wind had carried a song from the sea and was offering it to him. He sang with it, composing, then, after several minutes, offering his song to the wind and sea. This was a real experience, not just a metaphor—the body as an antenna attuned to the earth's signals.

Polyrhythmic Place

African rhythm exemplifies this earth-attunement with clarity. Unlike European conceptions that follow a single stressed beat, African musicians perceive multiple rhythms simultaneously, which practitioners describe as the drum beating itself. This polyrhythmic sensibility mirrors the multiplicity of environmental rhythms, creating a complex interaction of biological, geophysical, and atmospheric cycles that constitute any location's sonic ecology. The whole is inseparable from its parts; the music emerges only through combined rhythms that remain distinct yet unified.

In Zambia, I witnessed how environmental rhythms can bridge cultural divisions. Each tribal group initially believed their rhythms were incompatible—Bemba's five-beat spirit-calling seemed irreconcilable with Tonga's seven-beat rain ceremonies. Yet combining these supposedly incompatible patterns created extraordinary polyrhythmic textures. Mapopa Mtonga, a Chewa elder and our project advisor, was thrilled and called it the sound of Zambian tribes learning to live together. The land itself seemed to hold rhythms capacious enough to embrace diversity.

Rhythm as Living Practice

The Ritual Preparation functions not as a mastered technique but as an ongoing practice. Each repetition offers fresh insights, reimagining the body's relationship to place. It operates as cultural technology rather than mere theater technique, survival practice rather than artistic method, living tradition rather than static preservation. In the Cosmocentric framework, rhythm is neither background nor accompaniment but the very medium through which performer and place recognize their mutual belonging.

As Malidoma Somé observes, the modern world vibrates in ways that obscure natural rhythms—anything that enters the artificial space of Western culture is diminished, changed to fit (1994, 58). Yet, the yearning for rhythmic reconnection persists. Whether through dance, meditation, or the simple act of attending to breath and heartbeat, humans instinctively seek return to what grounds us: The planetary pulse that sets the tempo for all life. Indigenous performance traditions have preserved this knowledge, offering not nostalgia but a methodology, a means of reattuning the body to the earth's ongoing song.

The daily act of rhythmic embodiment sustains community and relationships, preparing for the deeper transformations that performance brings. When we align our internal rhythms with those of our environment, we do not merely hear the earth. We are participating in its ongoing composition. Performance, in this sense, begets real-world change: The reconfiguration of neural pathways, the physical release of antagonisms, the reaffirmation of shared identity and worldview. The drum beat, and in its beating, we remember who and where we are.

From the Zulu Journal

Zulu participants strongly connect with their tradition, viewing it as formal, strict, and sacred. Its influence on political and cultural identity, as well as its role in resistance since colonization, is clear. The enduring militaristic style of Zulu dance, song, shields, and weapons fosters a lasting sense of identity, even amid suppression. The Ritual Preparation aimed to broaden Zulu participants' understanding of their tradition and encourage a wider range of expressive modes. Its goal was to help Zulu performers access their cultural resources and showcase their deep knowledge and contributions. For non-Zulu participants, the aim was to present an alternative performance style rooted in Zulu culture.

Traditional Zulu goat-skin drums were collected, and Tiki Nxumalo was named the project's musical director. With three Zulu men, Tiki created a simple 4-4 Zulu dance rhythm. The session began with rhythm explorations to explore traditional Zulu rhythms throughout the body, from the forehead to the toes. These explorations focused on isolating body parts and rhythms, helping Zulu participants reassess their perception of and connection to their traditional rhythm. For non-Zulu team members, these activities offered an accessible way to engage with the physically demanding and sometimes militarized Zulu rhythm and movement. In typical Zulu dance, the emphasis is on rhythm, vividly expressed through lower-body movements such as hip movements and high-stepping stomping. For the Zulu group, the goal was to explore and refine subtle expressiveness, using rhythm as both inspiration and a link. Through various rhythm-isolation explorations, the traditional Zulu rhythm became a catalyst for a wide range of expressions. This rhythm served as a bridge between tradition and modernity, rooted in history yet adaptable for contemporary performance.

Three Traditional Zulu Movements

The first step in creating a ritual warm-up involved each of the 27 participants—both Zulu and non-Zulu—spending five minutes interpreting the prompt. "Present three traditional Zulu movements—whatever you believe that means—no right or wrong answers." When it was their turn to showcase their movements, I deliberately had the white Loft actors go first. Tensions remained between the Kwasa ("it dawns") performers and the Loft performers, with the Black Kwasa participants feeling intimidated by the Loft group's greater knowledge and formal training.

The Loft members offered a commendable but stiff, overly rehearsed portrayal of Zulu movement. Their Western theater training did not serve them well. Their performance was unremarkable, yet it provided a valuable lesson, highlighting how little they knew and understood about the Zulu. When the Zulu performers were about to take the stage, they hesitated, thinking they had misunderstood the assignment. They did not recognize what the white and Indian performers had presented as Zulu. I had to encourage the suddenly shy Kwasa members to proceed with their presentation.

Then, something remarkable occurred. Senele Ndlovu, a Kwasa member, shyly approached the center of the circle, then suddenly became animated as he showcased his three traditional movements. Inspired, Bongekile Ndaba, a Zulu woman, stepped forward, singing and demonstrating her movements. Nora Msani joined her, prompting some men to begin beating a simple Zulu rhythm on the floor, while others ran to fetch goat-skin drums from along the wall and brought them to the circle's edge. The other Zulu participants responded eagerly, fueling spontaneous energy. Each took turns performing their three movements, which evolved into improvisations of Zulu dance and mimetic representations of daily Zulu life. All movements matched the traditional beat.

A marriage proposal scene was reimagined as a lively dance improvisation that invited participation from non-Zulus. This seamless and captivating performance lasted nearly an hour, with the traditional Zulu rhythm guiding every movement and sound. As most non-Zulus watched in awe at this spontaneous display, the Zulus showcased their cultural richness, strength, and potential. This event fostered a new respect and appreciation among the non-Zulus for the Zulu culture they thought they understood. For the Zulus, it connected theater with their traditional arts, blending their identity, dance, and theatrical expressions. Despite lacking formal training, they saw their role as meaningful contributors to this emerging project.

Creating Performance

There is no fixed formula for creating performance, nor should there be. In today's complex environment, where competing demands constantly divide our attention and drain our energies, producing collective performances is incredibly challenging. Each project requires continual adaptation, negotiation, and adjustment amid changing personalities, support systems, and circumstances. This is not a flaw but the core nature of the work—performance creation is a daily process of initiation, continuous learning, and spontaneous improvisation.

Much like a diviner who finds meaning by connecting previously unrelated events, acts, and people through unique metaphors or semantics, a performance-maker assembles separate elements to uncover hidden patterns and connections. This process often involves what could be called "edification by puzzlement and cryptic potency" (Peek, 127). The initial chaotic phase of creation seeks resolution and, in doing so, fosters collective understanding. When a performance is born, it emerges as something new and delicate, engaging in a dialogue with the world. It starts small and tentative, sustained by the interpersonal bonds of those who create it. Like any newborn, it needs care, protection, and patient nurturing by many hands.

Performance is essentially a sacred act of community divination. Like traditional diviners who ensure "that all relevant information is brought forward before action is undertaken" (Peek, 37), you and the performers must uncover the hidden currents within your communities. This process develops what might be called "extra sight"—the ability to see connections between the ordinary and extraordinary worlds. By cultivating heightened sensitivity and awareness, this "extra sight" enables contact with both the visible and invisible realms. You and the performers become skilled at perceiving what others cannot. The performance artist acts as a conduit, bridging different realms. Their role goes beyond entertainment or artistic expression; they serve an essential function of reminding, adjusting, and reimagining their community, both human and other-than-human beings. Like the diviner who "both situates himself in the culture and creates it" (Peek, 130), performers serve as memory keepers and visionaries, holding onto the past while imagining future possibilities and applications. Through their work, they embody both norms and transgressions.

Directing

As a director, your role parallels that of traditional diviners—an outsider who observes keenly and responds thoughtfully. Like diviners, directors are valued both as participants and observers, recognized for honing sensitivity, empathy, clarity of vision, and practical judgment. The work demands receptive and adaptable listening and a deep, reflective attention that transforms external voices into internal insights. In creative settings involving performers and communities, the director explores the emerging qualities of individuals and their environments, examining the links among the individual, the group, and the broader community. You interpret the present to anticipate the future by tuning into diverse voices and perspectives, understanding the system, and pinpointing what is needed to sustain harmony. You must become an adaptable generalist, armed with knowledge from physiology, psychology, medicine, social sciences, politics, history, culture, religion, and rituals—beyond your artistic skills—due to the world's complexity. Listen intently, ask thoughtful questions, seek profound answers, and pursue honor, balance, and healing. The world constantly speaks and performs, and your task is not to assign blame but to foster responsibility, reciprocity, and restoration.

Embrace our work's unpredictable and mysterious nature requires you to remain aware, sensitive, and open, even when it's difficult. Break habits and see that accidents, embarrassments, misunderstandings, and flaws are valuable. Sharpen your observational skills as everything around you shifts. Learn the world's multivocal language, methods, and desires. You are an artist, trickster, and seer. Avoid limiting perceptions and remember that our work exists within a larger context. Resist easy solutions and worn theatrical tricks. This is work—there are no simple answers. Keep reflecting, growing, and evolving. There's no halfway approach—commit fully. Don't complain. Keep your mind, body, and spirit balanced and healthy. Take care of yourself. Your role is to lead, set an example, and create the path forward that fosters community renewal. Your knowledge, awareness, and generosity of spirit are central and will radiate outward to performers, projects, and the community. You will pass this way only once. Do the best you can. Make it count.

Onward

The following are several explorations meant to help identify, develop, focus on, and create performance. These serve as guiding suggestions rather than strict rules—feel free to customize, combine, subtract from, or expand on them based on the needs of your specific project, whether it is short-term or long-term.

IDEAS

The group selects a topic, issue, or concern to investigate.

Part I

Using a large pad, chalkboard, or whiteboard visible to the entire group, write the topic at the top. The group is then asked to add as many ideas, words, or images as possible that they associate with the given topic. There should be no editing or censoring of these associations.

Once at least 20 to 30 associations are listed, the facilitator or a group leader, with the group's agreement, identifies which words, ideas, or images are the most exciting and viable for creating a performance. This agreement can be reached through consensus, voting, or simply by selecting the top 5 or 10 associations. The remaining associations are erased or struck from the board.

Part II

The group is then asked to write a poem focused on the initial topic. The poem must reference, imply, or directly use the identified words, images, or ideas noted on the board. The group is only given 10 to 15 minutes to create their poem. The poem can be written in any form or style that the performers choose.

For example, while working with a group at the California Institute of Integral Studies in San Francisco (graduate students in Drama Therapy), ideas relevant to the group's preoccupations and urban location were articulated. The themes were:

Media
Fundamentalism
Sexuality
Garbage
Oppression
Money
Health

Each performer is then given the opportunity to present their poem to the whole group. Their presentation may be an oral reading and can include gestures, movement, and sound interpretation. A brief discussion may follow, or there may be no comments on the presentations.

Part III

The group is then asked to revisit their poems to create a presentation that uses no words, only sound, singing, movement, gestures, emotion, and whatever else is necessary to convey their poem. Allow 15 minutes for the group to develop their presentation. Afterward, each performer shares their work with the group.

Part IV

Select the elements from each poem performance that they find most fertile for further exploration. Combine elements from different performances. For example, performers can be paired to work on and explore a specific movement or idea.

They may reintroduce and rework the original language from the poems as needed. Encourage performers to have the freedom to expand upon and reorganize the raw materials from the initial poem and performance work. Group presentations can be made as appropriate.

Note: The exploration aims to explore a group's motivations, emotions, and ideas around a theme. Approaching themes directly can feel stiff, but this exploration offers more freedom, interaction, and creative license, helping groups confront abstract issues. For example, with Tukak' Teatet, it laid the foundation for a performance on "subsistence hunting." The images, ideas, and movements from the exploration shaped the performance.

View the poems and movements inspired by them as starting points for growth. Use movement, language, and imagery as resources, along with others. Remember, this exploration takes time, and ongoing discussion and sharing are vital.

Highly Recommended.

CREATE A MOMENT

Elements are given to the group. Performers can work in pairs or in groups of three or four, called creative groups.

Elements are provided either to the creative groups or to the entire group.

Each group is assigned a random line from a poem or novel, an action like a kiss or handshake, a prop such as a book or umbrella, and an event like a secret reveal or a tragedy. They also receive three movements like a shrug or yawn.

The task for each group is to teach something, sing in unison, and perform at least one activity besides singing, all incorporating the given line, action, prop, event, and movements. Every member must include these elements in their improvisation, with dialogue, and can break up the line as they like. The scene should be developed within 10–15 minutes, carefully planned to include all elements, followed by a presentation and discussion.

STORYTELLING

Performers are either provided with a story or asked to create their own, using a mix of body movements, facial expressions, and voice. Allow 10 to 15 minutes for preparation.

Instruct performers that no single storytelling element should dominate the others. The communication must blend body, face, and voice, with language serving as a secondary tool to storytelling. If necessary, storytelling can start without words. Performers should suggest and hint at the story rather than directly telling or illustrating it, aiming to evoke the audience's imagination and encourage their involvement. The emphasis should be on building atmosphere, rhythm, and emotional transitions.

Encourage performers to view storytelling as a reimagining of gestures and words. Even with a limited set of body, facial, and vocal expressions, each performer's interpretation remains distinct. They should anchor their storytelling in the story's reality, not in traditional ideas of reality or storytelling.

Exploration Alternative: Depending on the context, training, and performers' abilities, masks may be used either instead of or alongside the face. Once the performer has learned the basics of storytelling, costume pieces, musical instruments, noisemakers, and other props can be added.

Presentation to the group, followed by discussion.

I AM

Inspired by mentor, friend, and colleague Jean-Claude Van Itallie.

Part I

Performers sit in a circle.

When inspired, one performer stands and enters the circle. They must first connect with the audience through eye contact, body language, and energy—an act of self-presentation. Then, each performer states:

- Their name
- Where they were born
- A memory they want to share

Performers speak in the first person and present tense, avoiding the word "AND." For example, "I am Thomas Riccio. I was born in Cleveland, Ohio."

Each statement describes a memory. Performers can move freely or stand still to illustrate their memory; the key is to stay in the present tense and embody their memory.

When inspired, each performer takes turns standing and entering the circle, following these instructions.

Note: The memory shared need not be dramatic but should encourage performers not to overthink. Let the memory speak for itself.

Part II

After all performers have introduced themselves, the exploration evolves.

Performers stand and explore "authentic" movement, allowing their movement and the memory's situation to speak to them. They are given a few minutes, if needed, to develop their movements. Once ready, they vocalize their memory.

Each performer takes turns doing this.

Next, performers share their movements, memories, and vocalizations with the group two at a time, at the same time.

They create space and stay aware of each other as they share their stories. Performers tell their memory or story in the first person and present tense, listening carefully and observing each other's interactions.

Expect unexpected juxtapositions that lead to new stories, inspiring deeper reflection, growth, and new directions for the memory.

This interaction forms a foundation for developing performance and improvisation.

Once performers understand the concept and purpose, their ability to improvise and blend memories will improve.

Highly Recommend.

CHAOS PLAY

Performers in groups of three to five are given play scripts.

Part I

With script in hand, performers are asked to portray a specific character in a designated scene from the play.

One performer begins reading from the script, and once ready, another performer joins in—or is prompted to do so—in response to the first, but from a completely different part of the script.

Performers are added or removed to keep the scene unpredictable and chaotic.

The goal is for performers to connect, interact, and play off each other, regardless of the script's content. They should listen and engage actively.

The emphasis is on interaction rather than understanding the words logically. As the activity progresses, new insights and connections with the script will emerge.

Performers are not required to follow the script in order and can repeat sections as they see fit.

They may also explore different variations and interactions each time.

Part II

Provide performers with fragments of written materials (not play scripts) related to a theme relevant to the development of a performance.

For example: During the development of Dead White Zombie's production of DP92, performers received written fragments related to institutional confinement, experimental psychiatric care and treatments, information about mollusk habitats, the evolutionary journey of fish into mammals, scripts from cheesy 1950s sci-fi movies, and sections from the classic 1950s scientific study, Behavioral Sink, which documented how overpopulation in rat environments led to social breakdown and eventual collapse. The juxtaposition of these elements formed the basis for a script and performance. DP92 portrayed a human confined within an institution in a dystopian future that devolved into a mollusk. The remaining performers in the institution soon followed.

Highly Recommended.

NECESSARY DIALOGUE

This exploration is designed to develop dialogue for performance by identifying when language is necessary and when it can be replaced by action. Once a story or plot is established, performers create specific sections of the narrative with character dialogue, matching the number of performers to the number of characters in each scene. If appropriate, two groups can work independently on the same scene, using nonsensical or imagined language rather than recognizable speech during their performance.

After finishing their sections, each group presents their work, either sequentially or as individual interpretations of the same scene. When the entire story is assigned, the complete performance can be delivered from start to finish with nonsensical language.

Repeating or combining different versions, along with discussions, helps explore various approaches. This method effectively highlights key moments of interaction that are vital to advancing the story and allows performers to focus on sound and emotion rather than precise language.

This approach emphasizes the fundamental aspects of language, gradually developing into recognizable dialogue through a step-by-step process. Additionally, performers should pay attention to emerging rhythmic patterns in sound and language.

Note: By downplaying language and emphasizing action, performers and performances uncover new expressive possibilities that might not be apparent otherwise. If one form of expression is restricted, it sparks the imagination, opening fresh options that can unexpectedly energize the performance. When one door closes, another opens.

Exploration Alternatives: Certain points of language contact might recur so frequently that they take on a chant-like quality. This chant becomes an alternative form of communication through sound. It is particularly effective at conveying complex, emotive, abstract, or spiritual ideas or emotions.

Theoretical Frames

Performing Myth

Whenever a performer steps onto a stage, enters a frame, or begins to move, they participate in an ancient continuum that dates to humanity's earliest efforts to understand and shape reality through expression. Performance—whether theater, film, dance, song, music, or any other form—is not just entertainment or art. It is the living extension of deep structural patterns first formed in myth and ritual, patterns that still influence our collective and individual identities.

When Aristotle identified mythos as the soul of tragedy in his Poetics, he was not just talking about plot as a sequence of events. He was acknowledging that the "arrangement of incidents" or "representation of action" forms the core structure through which humans create meaning. This insight reveals something profound: Performance operates through mythic structures that transcend time and culture.

Working with Lee Breuer from Mabou Mines, a prominent American experimental theater group, I learned that performance has a greater impact when it intentionally explores its mythic roots. During my collaboration at the American Repertory Theatre and through our friendship, I saw how Breuer's Gospel at Colonus, a Pentecostal reimagining of Sophocles' Oedipus at Colonus, showed that ancient mythological themes can be creatively reinterpreted while still expressing universal human truths. The piece did more than just tell an old story—it engaged the archetypal structure behind it, enabling audiences to experience what Jung described as "the life centering, life renewing form" (43).

The link between performance and myth is fundamentally genealogical rather than metaphorical. Theater traces its roots to myth, ritual, and ceremony, with early societies recognizing the close relationship between collective action and its impact on the community. As Gaster notes, in ancient Greek culture, "the shaman became the actor, participants became the audience, and the sacred altar transformed into the stage" (83). This evolution does not detach from ritual but develops from it—shaping dramatic ritual into drama itself while always remembering "the rock whence it was hewn."

What makes this evolution so important for contemporary performers is understanding that myths are not just stories but living structures. Joseph Campbell reminds us that "myth is the depersonalized dream, dream the personalized myth" (19). Both operate through the same symbolic dynamics of the psyche, but myth provides patterns "directly valid for all mankind" rather than the quirky forms of individual psychology. These patterns—archetypes—form the deep grammar of all performance.

Consider the seasonal pattern that Gaster identifies as underlying ancient drama: Mortification, purgation, invigoration, and jubilation (26). This is not just a historical curiosity. Every compelling performance, from tragic drama to comedy, from dance to film, follows variations of this archetypal rhythm. The protagonist's journey reflects the cosmic cycle of death and renewal that Eliade describes as "the myth of eternal return"—the ongoing renewal through which societies and individuals transcend historical time to reach the eternal (Eliade 1959, 81).

This is why Campbell's monomythic "hero's journey" resonates deeply across different cultures and media. Its pattern—departure, initiation, return—is not just a story formula but reflects an intrinsic structure in human, if not all, consciousness, described as "the nuclear moment when, while still alive, he found an open road to the light beyond the dark walls of our living death" (259). Modern films, especially in works like Star Wars, succeed because they activate deep, structuring, and reassuring archetypal patterns, enabling viewers to engage in what traditional societies do through seasonal and life-cycle rituals: Experiencing mythic time.

Contemporary performance differs from traditional ritual yet keeps its core connection: "Humans are the alien presence with whom the forces of egoism must come to terms," Campbell states. "Each of us faces the ultimate challenge, the cross of the redeemer—not during the triumphant moments of our community but in the quiet depths of personal despair" (391). The move from collective ritual to individual heroism illustrates our cultural growth, and the fundamental mythic pattern endures.

This insight changes how we view performance. As shown by Breuer and my work with Indigenous cultures, performers connect with their "mythic self"—the part linked to the eternal realm. It's not about acting in the usual sense but about becoming a channel for archetypal energies. Using scripts and characters, performers express their mythic identity, which remains dynamic, authentic, and rooted in timeless storytelling. The practical impact is meaningful. When performers understand they are not merely "acting" but operating within mythic frameworks, their performances gain greater depth—what might be called presence or authenticity. They embody cultural language, as Indigenous cultures perceive them, bringing the vitality of an authentic self to their expression. This shifts the performer's role from simply interpreting a character to becoming a conceptual artist, conveying not only a role but also transhistorical agency.

The work ahead for performers involves recognizing and developing their unique mythic persona, understanding not only the stories they tell but also the mythic patterns they embody and activate. This process is not about ego, but about recognizing their specific gifts as channels for archetypal energies. What kind of relationship does the performer want with their audience? To challenge, move, or transform them? These questions directly link to the mythic role of performance in renewing community and fostering personal transformation.

MYTHIC SELF

This exploration was inspired and developed through my work with Lee Breuer. It aims to uncover the performer's mythological persona of "acting." The performer remains still but walks in place, fostering a dream-like mindset. Ask performers to recall when they were five, watching themselves as if on a TV screen. This is not emotional recall but an attempt to observe "the motion of your mind" and "the play of your life." Encourage noticing details from that age they might have overlooked before. Ask: Which moment do you revisit?

They follow the same simple observational process from age six to nineteen.

As performers age, they may adjust their walking rhythm or express their age through sound and gesture. Certain ages may evoke primal moments—significant instances that make the past feel more vivid.

During this, observe the expressions on performers' faces as they experience moments shaping their persona. Expressions can range from openness or aloofness to tenderness, courage, or fear. Performers do not reveal hidden personal mythology but keep these truths reserved for performance.

For each age, ask:

- Who are you performing for?
- Were they there when you performed?
- What did you want while performing?

It's crucial to understand what the performer is doing and what they aim to communicate at each stage of their mythic self. Questions include:

- Do they want to "slay" their audience?
- Make love to them?
- Chain them?
- Insult them?
- Move them to tears?
- Make them scream for more?
- Make them think?
- Make them sympathize?
- Confide?
- Lecture?
- Tell a story?
- Show off?
- Declare themselves a star?
- Confess?

By fully embracing this process, the mythic self can inspire others and serve as a resource for any performance form, regardless of context. After exploring all ages from 5 to 19, ask the performers to lie down and return to the present.

Group discussion.

Understanding and embodying this myth can inspire performers and guide their performance work.

What kind of connection does the performer seek with their audience? To love, chain, move them to tears, make them crave more, provoke thought, insult, annoy, or some complex mix?

This inquiry is not about ego or turning every role into an exhibition. Instead, it aims to pinpoint the performer's unique interpretation—their act of embodying a role—which is vital for crafting a distinct, highly personal character. This perspective responds to our increasingly mediated, algorithm-driven, post-truth, AI-created reality, where appearances often seem real. A new understanding of reality is emerging, and performance plays a key role in navigating and shaping this landscape.

This approach enables performers to use themselves as both material and reference point, forging a new kind of archetypal structure. The performer becomes both an interpretive and conceptual artist, conveying not only a role but also agency within it. All of this reflects our current meta (or metaphysical) moment of planetary change.

Note: Lee Breuer lived his myth as a director, but his entire philosophy aimed to help others—whether playwrights, actors, performers, designers, or audiences—live out their myths. The only rule was that the most authentic, genuine performance would take center stage.

The Union of Opposites

Humans embody contradictions. As Taoist writings tell us, people are like microcosms with "round heads like Heaven and square feet like Earth," embodying the universe's opposing forces. This is not just poetic imagery, but a reflection of nature: Yin and yang endlessly transform into one another.

Throughout history, societies have created technologies to handle internal and external conflicts. Legal systems serve as mediators between individual and collective interests, while democratic processes help negotiate differing visions. Therapy aims to harmonize conscious and unconscious desires. Among human innovations for navigating paradoxes, theater and ritual stand out as the most advanced, not because they eliminate contradictions, but because they reveal opposition itself as a driver of change.

Anthropologist Claude Lévi-Strauss saw ritual as a means of unifying opposites. Alternatively, Taoism holds that opposites inherently contain one another and do not require reconciliation. The Hegelian dialectic shows that thesis and antithesis do not merely compromise but generate a synthesis, an entirely new position that preserves and surpasses the original extremes. This process, called *Aufhebung* by Hegel, involves simultaneously canceling and preserving, facilitating the evolution of being in and with the world.

The Poverty of Resolution

The Western dramatic tradition, rooted in Aristotelian logic, often illustrates conflict using mechanical diagrams such as protagonist versus antagonist, objective versus obstacle. This pattern, seen from Greek tragedy to superhero films, exemplifies what Hegel called "picture thinking," which remains at the level of image thinking rather than engaging in true dialectical growth. This framework addresses cultures displaced from tangible places into abstract systems. As Kristofer Schipper states in his book, The Taoist Body, when humans disconnect from their environment, "whatever is achieved will not endure." The well-constructed play, focusing on closure, reflects capitalism's false promise of satisfaction through consumption.

Resolved conflicts are turned into an outcome, a product, giving the illusion that oppositions can be finalized instead of always being part of an ongoing negotiation and adjustment. This sense of conflict resolution reflects a limited, selective view. In contrast, a Cosmocentric perspective might see it as a sign of deep imbalance. All conflicts—whether personal, social, environmental, medical, or spiritual—are seen as illnesses caused by an imbalance affecting the community. Western drama tends to focus only on immediate, individual human issues, ignoring systemic problems and collective sustainability. It provides catharsis but misses the full scope of interactions between humans and other-than-human entities, which are often the root cause and solution of the conflict.

Cosmocentric performance does not aim to eliminate contradictions but instead maintains them in a dynamic, productive tension. The Miao badai spiritual masters I have worked with in the Wuling Mountains of southwest Hunan exemplify this. Their performances embody the gods, spirits, and ancestors, giving form to that which informs and shapes the material world. This process is not just about switching between states but follows a dialectical pattern. The human community, asserting a complaint, wish, or asking for a remedy, is the thesis. The spirits, ghosts, gods, ancestors, or elements are an ongoing influence that shapes antithesis. The Miao badai or xianniang embody and serve as intermediaries between worlds, the human and the other-than-human, and, in so doing, become agents of remediation and synthesis. Similarly, the performer's body serves as a host for members of the human community (thesis). The performance evokes the spirits of the invisible that inhabit and influence our material world (antithesis), through performed interactions between performers and the audience (community) that enact the synthesis. The primary difference between theatre and Cosmocentric performance lies in who and what is included. Theatre is human-centric, whereas Cosmocentric performance is inclusive of human and other-than-human beings. Further, there is no resolution for the Cosmocentric. Performance is an ongoing cycle that sustains place. Character, individual identities, and concerns are secondary, if a factor at all. It is all about maintaining the balance of the human and other-than-human community. A night-long funeral ritual in Tamil Nadu is as much about honoring the deceased as it is about mending the community's loss with the help of the Hindu gods. Placing the deceased in the realm of the family's ancestors and upholding the karmic order is the event. There is no resolution, only maintenance, balance, and continuance.

Contradictions are not resolvable but rather endemic to life. Hegel describes "speculative reason" as the type of cognition capable of and required to hold opposites without collapsing them in binaries. This concept mirrors Taoist and Miao ritual traditions, where truth is neither conveyed through words nor silence but emerges, sensed and glimpsed, in the speculated space between. With this understanding, there are multiple truths, perpetually dynamic and fluctuating, much like quantum thinking, to serve the needs of the moment before becoming something else. This is, of course, at odds with the materialist, positivist worldview that predominates today.

The Miao understanding of balance extends beyond Western ideas of equilibrium. It is not a static point but an active process, referred to as "hsiu-yang" in Miao and Taoist traditions. It involves adjusting and smoothing out irregularities through repeated actions in harmony with the cosmic order until perfection is achieved (Schipper 1993, 14). This approach differs significantly from Aristotelian moderation or Hegelian synthesis. Here, balance is about preserving the system's capacity to change. Intervention is viewed as disrupting nature by resisting spontaneous processes, potentially causing imbalance. The performer's role resembles that of a shamanic guide who does not impose order but fosters the natural oscillation between order and chaos, concentration and dispersal, yin and yang.

The Feminine

In *The Taoist Body*, Schipper emphasizes that "the body of the Tao is a woman's body. The female body, especially that of a pregnant mother, is uniquely complete and capable of transformation" (129). This statement is not an essentialist assertion but rather acknowledges that femininity embodies paradox by holding others within itself and creating difference from sameness.

Hegelian dialectics, despite its masculine framing, ultimately reveal that consciousness evolves through internal contradictions. The synthesis arises from conflict, not by force, but through the careful holding of tension. As a director, work to foster a "feminine personality," not one of passivity but of active receptivity that enables opposing forces to influence and transform one another.

This understanding may explain why forced resolution often fails. Masculine dramaturgy, a bias encoded in Western performance, seeks to penetrate mystery, reach a climax, and ejaculate meaning, asserting the male personality (protagonist) as an active agent seeking conquest and conclusion. In contrast, feminine dramaturgy supports what the Taoist and Miao spirit masters call "the union" and a continuous circulation of energy in which there is no single active or passive partner. Working for mutual transformation through ongoing engagement is an emulation of nature and the Cosmocentric way.

Performance as Cosmological Practice

The implications cascade across every aspect of performance:

Narrative Structure: Rather than constructing a narrative climax, the Cosmocentric adopts what could be called "breathing dramaturgy"—a pattern of alternating between gathering and releasing, focusing and dispersing. Each performance then transforms into a collective communion of all beings, with humans, as the most equipped and privileged beneficiaries of the environment, bearing responsibility.

Rehearsal Process: The emphasis is on cultivating consistent, cyclical practice rather than merely preparing for a fixed, immutable performance. Mastery demands daily repetition of the same gestures until they become effortless and perfect. Performance then becomes a way of life, with the public event acting as a focused reflection of that daily commitment.

Spatial Dynamics: The performance space transforms into a mandala—a visualization of all time cycles and spatial categories within the interconnected space-time continuum. Every position is meaningful, each movement forges a connection, and every boundary produces energy through its permeability.

Temporal Experience: Performance is not a linear process; instead, it follows a cyclical pattern that recreates what Eliade described as a "time out of time" or "the time before time." During a ritual, past, present, and future merge, creating a duration in which the human and other-than-human community attends and participates. All temporalities converge into what Hegel would identify as "absolute knowing," a condition that is aware of its own movement, and not a final state of knowledge.

The Cosmocentric worldview shifts away from the traditional conflict-resolution framework that has shaped Western theater. It moves beyond heroes conquering obstacles, protagonists achieving goals, and neat endings restoring order. Instead, from a Western perspective, it values the seemingly useless. This is not nihilism but a dialectical negation that preserves elements through transformation. If it tells a story that opposes tyranny, it does so by maintaining characters while dissolving their boundaries, creating a permeable structure. Every element, confronted with its contradictions, unlocks new possibilities.

This model aims to hold a unity of opposites without being torn by them, emphasizing the recognition and coexistence of multiple truths. The union of opposites is not a goal but a continuous practice. A journey rather than a destination. More of a question than a definitive answer. Every performance captures the cosmic rhythm of focus and dispersion; each rehearsal trains in releasing the old and welcoming the new; every production mirrors the perpetual cycle of yin transforming into yang, then reverting to yin.

Performers are explorers of mystery, students of paradox, and practitioners of the impossible. By learning to embrace opposites without resolving them, being both performer and character, individual and collective, human and divine, performers cultivate a vital skill our era desperately needs. The capacity to coexist with contradiction not by trying to control results, but by harmonizing with the eternal interplay of opposites that drives all transformation.

Not resolution but circulation. Not answers but breathing. Not peace but the dynamic balance that includes conflict as one movement in an endless dance.

Hallucinating

Humans excel at interpreting the world as a sign of something greater. This perspective has taken us only so far: It has increased our knowledge, depth, and intellectual understanding of the material. Yet, the material has always been only part of a larger whole. The focus on material and measurable things acts as a lens—an institutionalized bias—that diminishes and excludes what does not meet its criteria. Our goal is to reintroduce the felt experience, to validate the unseen presences that surround us, and in so doing, re-embody the purpose and means of performance.

According to Jacques Attali, engaging in such work involves accepting "disorientation, life outside familiar space and time, vertigo, dizziness, and a lack of advanced knowledge of the trip's duration or course" (Attali 1999, 78). "Hallucination" here describes a disorientation in perception and existence that characterizes our post-truth, image-driven, and mediated age. "Text" refers to what is tangible and connected to reality. To engage with it meaningfully, we need to reconceptualize, disrupt, and transform the dominance of the material realm. A hallucination occurs when a social, cultural, or psychological external context, or a stable material referent, becomes disconnected and cognitively free-floating. Often, hallucinations are seen as unreal, illusory, distorted, misinterpreted, abnormal, or insignificant. I view them differently, as a myth-like recreation of a different kind of world. Hallucinating the text is a methodology for our moment of upheaval and creation; it is a process of realignment aimed at restoring balance and coherence.

During the first Tuma Theatre workshop, I began by leading Alaska Native performers in a warm-up exploration focusing on isolating the neck, shoulders, and hips. Their patient confusion revealed my flawed assumptions. These bowhead whale hunters, whose bodies had been shaped over thousands of years to adapt to the Arctic environment, were being asked to break themselves apart with unfamiliar logic. I instinctively recognized that this was an imposition—a way of viewing the body as a mechanical object that stood in contrast to their holistic worldview.

This moment exemplifies what Umberto Eco describes as the crisis of traditional epistemological views (15). For me, it was the realization that Western theatrical paradigms cannot fully encompass Indigenous knowledge systems. Hallucinating the somatic and felt texts is how our work will reawaken the world.

The village of Point Hope, located on a peninsula, is regarded by the Inupiat as a beached whale that has, mythically and willingly, offered itself to support the community's daily life. Walking through the village on Point Hope's peninsula is walking on a beached whale and benefactor. Real yet surreal. Myth and matter co-exist. This deep, lived understanding anchors Inupiat life in an increasingly material-focused world that emphasizes linear progress and a sense of time and space. It is material-based and cause-and-effect, institutionalizing distinctions between mind, body, and spirit. Perspectives used to empower and justify control over the natural world and human conduct. The Inupiat worldview does not rely on linear progress; instead, it is associative. The material has feeling, and feeling is immaterial, constantly transforming. Linear storytelling is a trait of outsiders and travelers who move between different places. The Inupiat have inhabited their land for thousands of years. They are going nowhere. They are already there.

Observing the Point Hope dancers perform, one might see a man walking on the tundra transform into a seal, then into the moon, shift into clouds, become an eagle, and finally a hunter kayaking. This is all accompanied by chant, dance, and rhythm, often without clear transitions or explanations. None needed. It functions as a story, a myth, a worldview, and a dialogue with their environment. A hallucinated text of a place.

This way of storytelling exemplifies Eco's concept of "deliberately and systematically ambiguous" open works, which demand "a much greater degree of collaboration and personal involvement" from performers and audiences (11). Rather than fixed scripts imported from elsewhere, they are dynamic spaces for engagement.

In Ethiopia, narratives encompass a hallucinated view of political, religious, and mythic history. From Lucy, the ancient hominid, through various empires and dictatorships. These stories reveal how mythic legacies serve as sources of pride and disempowerment, with each era's shadow cast over the next. Lucy, portrayed by Kidist Girma, states in our production of *Andegna* ("the first"), "We have made ourselves the way we are, and can make ourselves the way we want to be," expressing frustration and the burden of the disconnect between perception and reality. The act of hallucinating the text was a method used to challenge and disrupt accepted norms.

The performance drew on powerful mythic stories about Ethiopia, convincing itself of being the birthplace of humanity and a cradle of great civilizations. These myths emphasized Ethiopia's role as the site of Lucy, the earliest human remains and "mother of humanity"; the first nation to embrace Christianity and Islam; the home of the Ark of the Covenant and the true Orthodoxy; the only nation to welcome the lost Jewish tribe of Israel; the Queen of Sheba and King Solomon; the Nile's origin; and the ancient empires of Abyssinia and Axum. Ethiopia also boasts being the only African country never colonized, the first to defeat the British and resist Italian occupation, and the land of the revered Rastafarian Emperor, Hallie Salesse. These narratives fostered a sense of superiority. Yet, if Ethiopia is so ancient and great, why is it struggling? These myths were a deception. The group chose to challenge them, exploring how these stories both uphold and trap Ethiopia today. Questioning these unquestioned myths became the core of our performance—a hallucination of history turned political act of resistance, which was ultimately shut down by the police.

From the Ethiopia Journal

After nearly three weeks of working together, I questioned, "If Ethiopia has such a rich history and culture dating back thousands of years, what changed? Why is your life so difficult now? Why does it rank so low among the world's nations?" The group members shifted uncomfortably. Despite our friendship and trust, I was still an outsider criticizing their country.

An Ethiopian's bond with their land is profoundly sacred, extending beyond mere patriotism to the realm of the spiritual. Many ensemble members buried their afterbirth close to where they were born, symbolizing their deep connection to the land. Ethiopia is not a nation built by immigrants but one inhabited continuously for thousands of years, shaping a Cosmocentric culture and worldview that migration-based societies, like the United States, find hard to understand. For me, criticizing Ethiopia was akin to criticizing them personally.

I was, however, speaking the truth about the monster. I was compelled to say what they could not, should not, or would not.

The unemployment rate in Ethiopia was about 45 percent, higher among young adults, who are forced to live with their parents, marry late if at all, and scrape by with low-paying jobs and their wits. The performers earned money from occasional radio dramas, theater, television, and film roles. Several taught the occasional drama class at an international school, one had a children's puppet theater, others made extra money doing day labor, working in shops, translating, street vending, or simply finding whatever work they could. Two ensemble members were former high school teachers earning about $75 a month. Several in the group could not afford the bus fare, which cost 80 cents a day, a large expense for group members. During our eight-hour workdays, I bought lunch or dinner for the ensemble at a local restaurant, which cost about $12 in total.

The process took on a life of its own. Like any culture, history is intertwined with mythology, shaping an idealized, prideful self-image that, if not critically examined, can foster delusion and keep irrelevant or dysfunctional traits alive. This was evident in Ethiopia. For many performers, disillusionment and distrust in their system became accepted realities. "We have not learned from the past because we are so prideful, and we are so prideful because in reality we are afraid and insecure," Endale Berhanu said during a discussion. This idea later found its way into our performance. We are poor because we know no other way.

Orality

When humans collectively select stories, it is possible to eliminate the written word entirely—no scripts, notes, or texts. This is not about primitivism or romanticism, but rather about recognizing what Diana Taylor describes as "the repertoire"—embodied memory that transmits knowledge through the body rather than the archive (176). To preserve and safeguard body knowledge, many of the Miao badai and xianniang are deliberately illiterate, opting to remain closer to the physical realities of the world and spirits, unmediated by written language.

In practice, Tuma Theatre's performances existed only within the performers' bodies and collective memory. Each performative recounting was uniquely shaped by the performers' energy, the audience present that evening, the weather, and the needs of the moment. The performance was like an organism, a shared entity. As the Athabaskan Tuma performer Paul Mountain stated, "If somebody forgets, somebody else remembers." Written Tuma scripts serve as archival artifacts, documented only after public performance.

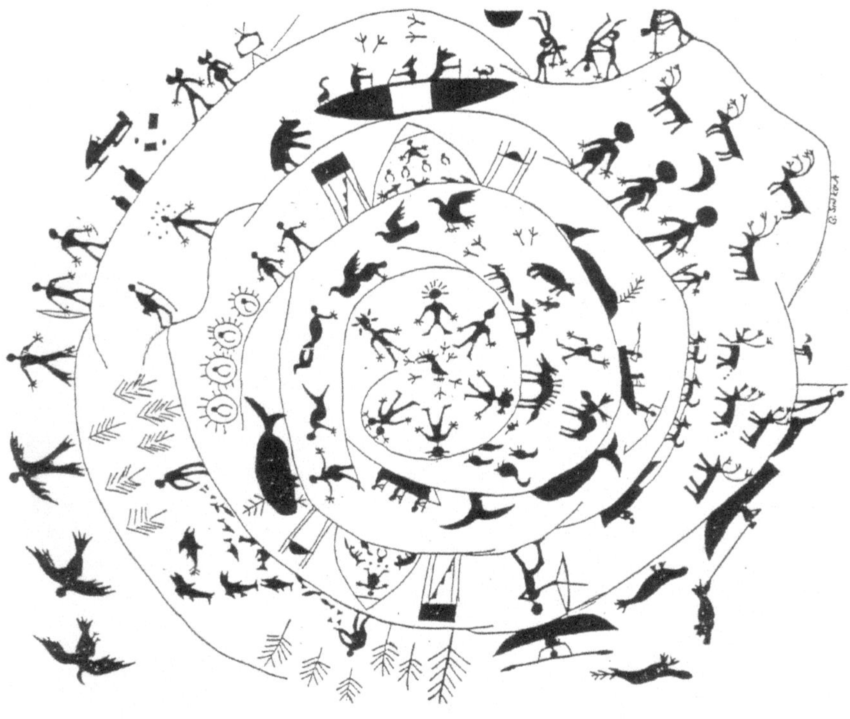

FIGURE 3.4 Qayaq script. Tuma Theatre, Fairbanks, Alaska. No written text was used during the performance. Drawing by Betsy Gelvin, created from oral accounts.

Source: Performance devised and directed by the author.

The !Xuu and Khwe Bushmen

I was brought on the project by Catherina Meyer, the director of the !Xuu and Khwe Bushmen Cultural Trust, an arts organization. The initiative aimed to offer arts-and-crafts classes to the Bushmen, helping them acquire marketable skills, since a return to their traditional lifestyles seemed unlikely. I had met Catherina unexpectedly months earlier while working with the Zulu in Durban. She was exploring ways to prepare the Bushmen for a cash economy and conceived the idea of creating a tourist show at a remote camp in the Kalahari Desert. Her objective was to attract visitors, generate income, and train the Bushmen in transferable, cash-economy skills. Other initiatives included farming and herding.

The !Xuu and Khwe, who had recently transitioned from a hunter-gatherer lifestyle, were descendants of a culture that, by some estimates, had existed for 8,000 years. When I spoke to Catherina on the phone months earlier, she asked whom I wished to work with. I replied, "Why, the best performers!" I later learned I would lead a three-month workshop and performance project involving 21 performers, including eight healers and six assistants.

Little was known or written about these two small Bushmen groups—originally from Namibia and Angola—so the project began with interviews and ethnographic documentation.

The camp was miserable and restrictive for the Bushmen, who traditionally are nomadic, self-sufficient, and proud, but were caught in a political web beyond their control. Life in the camp was chaotic, with depression, health issues, and declining morale, leading to abuse, violence, and alcoholism.

The healers were actively conducting healing rituals across the camp because "the sickness is great." After earning their trust, I was invited to join their rituals and befriended them. Given the overwhelming social and cultural issues, I saw developing a tourist show as unethical and out of step with the needs of the community reeling from catastrophic upheaval. I approached Catherina and the trust board to propose re-evaluating the project and turning it into a community performance initiative. They agreed.

FROM THE !XUU AND KHWE JOURNAL

After an all-night healing ceremony, George (a name he gave himself) arrived late at the workshop, still "cooling down" from the intense heat of an all-night ritual. He told me the ancestors spoke something special to him, which he needed to convey to the group. He reached into his medicine bag and took out an old car mirror, rubbing herbs onto it. Several others gathered around him to support this way of seeing the ancestors by singing and dancing. When I asked if he had seen them, he confirmed he had. "Did they give you any advice?" I queried. "Yes, we must do a healing ceremony," George told the group. After a brief discussion, everyone agreed a new kind of community-wide healing was necessary to help the entire camp. "The ancestors will help us."

George said we must tell the story of the "Two Stones." His announcement animated the group, sparking eagerness among everyone to share their own versions. When I asked why I had not heard this story during our interviews and workshops, they explained that everyone already knew it and felt it did not need to be told. It was their origin story, "Two Stones," and we spent weeks discussing and preparing it.

The process started with a large artist's pad at the center, with the group forming a circle around it. The drawing below comes from our first session. After debating the details, we agreed on a collectively derived version. The creation of a performance gave us clear goals, structure, and immediacy, which the pragmatic Bushmen appreciated. The drawing and story became a way for us to communicate directly, bypassing language barriers or the need for a translator. !Xuu and Khwe languages were unintelligible to each other, and we became equals. Since none of the Bushmen could read or write, we drew illustrations for each part of the story until everyone was satisfied.

Traditionally, healing rituals were performed by a single healer for one or two family members. What we were doing was something new and untested. "It is the ancestors that help us survive," Silenga declared. "The ancestors provide guidance on living and healing," George added. I inquired if there was a curse affecting the !Xuu and Khwe. "Yes, we are sick, hungry, and far from home. This is a great sickness," Mohera responded.

Our work—interviews, performance explorations, discussion, and trial and error—enabled the healers to analyze and discuss their realities both objectively and collectively. The adoption of a shared origin myth united them, reinforced their worldview, and structured discussions. They realized they were living in the "old story" and now had the confidence to create a new one.

We chose a day to perform for the community, which evoked a mix of excitement and nervousness among the healers, who called it a special ritual. The event would include storytelling, dramatization, dance, song, trance, and healing, and marked the first time the healers collaborated to deliver a collective healing for the community.

On a June evening about an hour before sunset, the healers and their assistants gathered at two tents, chatting eagerly as they prepared. About 100 community members assembled outside the tents, which symbolized the stones of their origin story. A large fire was built in front of the tents, and two tall branches anchored with rocks represented the trees given by God. A tarp was laid on the ground at the far end of the space to symbolize the water the Bushmen feared crossing. Neither the workshop members nor the community had experience with theater, and I had no idea what to expect. As night fell and young boys fed the fire, everything went quiet when lights appeared inside the tents. Talking and singing followed. It seemed as if the audience was holding its breath when George exited the tent, looked around, took in the world, and began singing joyful thanks to the sky, the world, and to wonder. Their origin myth, known only in oral telling, had come to life.

Audience members called out with encouragement, sparking a call-and-response with the performers. Running commentaries, along with clapping, singing, movement, and dancing, erupted spontaneously. When Luis crossed the river to become a white man, accompanied by rhythmic vocalizations, all went silent. Others played Bantus who claimed the river, chasing the Bushmen into the bush. It seemed as if a pall of mourning had gripped the audience. The myth lived vividly in them. The audience, unasked, sang sad songs as the performers depicted the search for food. Mubuto ventured to the river and mimed bombs falling, running and shielding her head, as the audience looked to the sky in fear, holding their hands high to shield themselves from the imagined bombs. George came to the river and was shot, falling to the ground and quivering. The blue tarp, symbolic of the river, became the road where Bushmen were shot and killed, and along which invaders came to take their land and lives. Their mythological, historical, and contemporary selves became the same.

The wounded re-entered the stone and sat in a long silence. They sang a song, "Just sitting inside the stone and dying of hunger." With a handmade wire-and-wood instrument, Machai exited the stone on his own and slowly began to dance and sing, calling the healing spirits. The drums sounded, and the audience clapped and sang, as the healer shook to generate heat. Others emerged from the stone as they did in the origin myth, but this time, coming into the modern world. The old had become a new story, as all stories retold do.

Once Luis crossed to the other side of the water/road, he covered his face with white clay, becoming the white man. There, he picked up a "book" he had crafted, bound by two pieces of scrap cardboard and held together with tree fiber. The pages were torn and soiled paper, litter that summersaults through the camp on windy days. A stick carved with a point was his "pen." It was the scene he had created without my knowledge, and I had never seen it before. He was writing feverishly, looking up, head jutting about, observing, then returning to the book to write again. I was fascinated, then recognized my gestures and rhythms, amazed by the accuracy of the impersonation. The group watched my reactions and then burst out laughing. They had folded me into their story, and the honor humbled me.

FIGURE 3.5 A drawing by !Xuu and Khwe Bushmen serving as a script for a community-wide performance, *The Two Stones*.

Source: Drawing courtesy of the author.

Creative Groups

After establishing the narrative outline, the next step in the creative process is to create an organizational framework, a practical methodology for developing the performance. One approach I use is the Creative Groups, small, autonomous teams of two to six performers assigned to specific story segments. This method is a core part of our performance development and has been successfully applied in various contexts, including cultural settings, experimental projects, university classrooms, and drama therapy sessions. The Creative Groups encourage agency, independence, and the development of creative and leadership skills.

During the development phase of the *Two Stones* performance, performers could choose specific parts of the story they wanted to focus on. One group might explore the scene inside the stones and how they emerge, while another might focus on their actions within them. Performers are encouraged to deepen their ideas and emotions through physical exploration. Later creative sessions include developing songs, dialogue, movement, or dance.

Each group was given time to work on their segment while I moved among them to listen, facilitate discussion, and offer encouragement as needed. Then, each group presented to the larger ensemble, which offered feedback and fostered continued discussion and development. Observing the progress of other segments sparked ideas and influenced each performer's work. Creative groups function what Lehmann describes as "rhizomatic connections" (90) among groups, linking related feelings, thoughts, and actions. How you structure a performance is just as important as the performance itself.

The Dead White Zombies relied heavily on creative groups. For Holy Bone, a piece presented as 15 independent solo stations across seven buildings and through city streets, this collaborative approach was essential for developing and understanding the modular yet interconnected narrative. Although each station was performed independently, they were all collectively created through creative groups. Insights from working together on individual stations were crucial, providing the connective tissue and thematic continuity.

Each creative group was led by the performer central to each Holy Bone station. The performer portraying Urd, for instance, developed a site-specific, silent weaving ritual to mark the end of the performance and collaborated with others to test the concept. After each Creative Group session, the groups gathered to share discoveries. We called these sessions "shareamonies." In this way, each station evolved, clarifying and cross-referencing themes, style, and narrative contribution in what would become an interrelated initiatory journey. The process helped performers understand their station's significance and requirements. Organically shaping the atmosphere, actions, and interactions, design, sound, lighting, and costumes.

CREATIVE GROUPS

Once a story or performance outline is finalized, creative groups are assembled to explore various scenes, characters, or elements for development. Organized by you, these groups typically form around shared interests.

Each creative group, made up of three to six members, receives specific tasks related to a performance aspect, such as a scene, dance, song, or character. They select a leader, find a workspace, and start outlining, discussing, experimenting, and exploring the assigned task. After one to three hours, they present their progress to the larger group, which then provides feedback. This process ensures the group's contributions are incorporated into the overall development.

Post-presentation, groups might reorganize by adding new scenes or characters and continue refining their work. Performers might switch groups, or you might suggest redistributing talents or combining skills to strengthen the teams. Ultimately, a performer's role in the final performance depends on their interests and input within the group. Casting decisions are often based on this collaborative effort.

Active groups should be checked regularly, visiting each to provide support and guidance as needed. These groups ought to operate with a high degree of independence, with leadership roles rotating among members. This structure encourages performers to develop their own independent skills, with you serving as a facilitator and safety net. The process highlights collaboration and consensus, with your roles including organizer, motivator, and editor.

Facilitating

Throughout the development process, my role was not that of a traditional theater director who acts as an interpreter or visionary leader. Instead, I see myself as a "maze viewer," a term Attali uses to describe someone skilled at "discerning complexity" and guiding others through it (xxviii). Creating a personal, performative, and communal experience involves a fragile sense of intimacy. Rather than imposing my own interpretation or vision, I listen actively, recognize patterns, and weave together multiple perspectives.

While sitting with Ethiopian performers, I observed and noted their debates about whether Lucy should speak Amharic or English, and whether she should be portrayed as a mother, judge, or witness in the performance. My role was not to decide these elements but to guide the performance in recognizing when its shared imagination uncovered a truth. From this perspective, I articulate what the performance requires and needs. By sharing my observations, instincts, feelings, thoughts, associations, and experiences, I clarify how and why I make my choices. I serve as a channel, like a priest, philosopher, and medium, for the performance organism. The performers, through their psychophysical connection, are internal experts with an intimate relationship to the work. Together, we form a union of opposites: Yin and yang.

I find the word "director" too limiting to describe my role. It carries too many expectations, assumptions, and legacies associated with power, division of labor, gender, and us-versus-them mentalities. Instead, I prefer to call myself the Poo Pah Doo of The Dead White Zombies. This title keeps me humble yet snarky, absurd, and mischievous, positioning me as a vague, ambiguous part of something mysterious and undefinable. It is a nod to Alfred Jarry's celebration of life's chaos. My role is defined yet undefined. I am free to roam associatively, doing whatever, whenever, wherever, and however.

From xianniang Zhang Quifeng, a Miao spirit medium, I learned that her ancestors spoke of the "holy bone"—the pelvic gateway between life and death—through which she and her ancestors crossed to the other side. She told me she finds her deepest insights and fulfillment by simply letting the voices of the dead, the unborn, and the other side surface, be recognized, and heard. All she can do is listen, encourage, and channel these voices, allowing them to speak through her for the community's benefit. From an ethnographic interview in remote China came a spark for a performance in Dallas. I see my role as part of an ongoing, ever-shifting hallucination of feelings, thoughts, and insights. I facilitate others and encourage them to do the same.

Every performance speaks its own language, carries its own memories, and imposes its own logic. Everything is dialoging all the time. Performance holds and reveals that dialogue for a brief instance.

The deteriorating Arat Kilo Theatre in Addis Ababa, with its leaking roof and mildewed curtains, became a central figure in *Andegna*—symbolizing the decline of Ethiopian dreams while also representing its potential revival. A crowd of 400 young people filled the space, sitting on the floor, on the stage, and on the balcony, their bodies blending into the performance environment and shaping how Lucy's questions resonated and engaged in dialogue each night.

In West Dallas, the abandoned crack house, where T.N.B. (typical nigga behavior) took place, carried a violent past that influenced the performance. Spectators could not distinguish between staged and real danger because the boundary blurred. The house itself expressed its trauma alongside human actors and viewers. Our approach evolves with each new context, revealing fresh opportunities. As identity shifts from being "defined by territory to be defended" to a "portable culture" (Attali 1999, 76), we craft performances that feel both ancient and urgently modern. Fixed in time yet timeless. Defined yet ambiguous. At once a statement and a question, real but ephemeral.

Interviewing

Performance-focused interviewing is fundamentally different from journalistic or academic methods. Instead of just gathering information, it aims to uncover how events are felt and expressed through the storyteller's body, voice, and imagination. The visions and understanding of the peripheral are what interest me.

This approach requires ease and openness to spontaneous language and unique narration styles. During interviews with Brazilian Umbanda practitioners (for a future project?), it became clear that the spirit mediums effortlessly switch between explanation and demonstration, describing while embodying the energies they discussed. When the medium states that entities "say these things in a way they feel," they demonstrate a performative trait and that warm manner that reveals harsh truths. Below is an excerpt from an interview with Umbanda practitioner Valintin, who shares his early encounters and relationship with spirits. Consider it as a performance happening between worlds.

> All these mysteries feel as if I sometimes have no idea what a certain entity is doing here, yet it operates at some level. It's intensely focused on something, even if not physically active. It exists and is doing something. At one point, I felt fear and wanted to call for help, to escape, experiencing a breakdown—that's common. Now, I trust and follow guidance. I hear internal conversations, but not always. Early on, when I was young, I consulted a powerful medium. He sat with his eyes closed, moving and grabbing things as if they were real. No, they were openings between worlds. He walked around, bumping into people and objects with his eyes still closed. There were bystanders, but no one spoke. Later, my guide—a medium—and my friends told me I

> was the one walking blindly, with eyes closed, grabbing at things and stumbling. I kept my eyes closed to hold back words for specific people, which hurts me and causes physical pain all over. I cannot suppress these feelings anymore, and I do not. I treat these feelings carefully. So much happens around us that we don't see every day, but it's always there. It surrounds us and accompanies us everywhere. It's part of life. We share the same world; we don't realize how much surrounds us.
>
> *(Valentin 2018)*

The development of T.N.B. by The Dead White Zombies exemplifies Cosmocentric dramaturgy. Focusing on David Jeremiah's personal journey—from Oak Cliff Gang member to performer—the performance centered on his story. It includes the words, events, and memories of other gang members and incorporates pop culture references, historical legacies, racial and gender biases, language of violence, gang vocabularies, markers, and behaviors, which I understood as expressions of tribal affinities. This collaborative memory process did not simply recount David's story; it transformed his personal testimony into a collective expression.

The Interviews that informed the script revealed a linguistic style and rhythm, within which a worldview of embattlement and survival lived. The language patterns, like a song, suggested staging actions and style. Masculinity, bravado, vulnerability, status, respect, religiosity, insecurity, and acute awareness of the world, expressed in circling, guardedness that toggles between bravado and silence, revealed unspoken themes that emerge through triangulating multiple viewpoints.

T.N.B. was developed and performed in a former drug house. During rehearsals, there were frequent knocks at the door from people seeking drugs. The house itself became a living archive, revealing its layered history, fusing a crack house with myth and performance.

Audience members entered through the back fence, symbolically breaking in and interacting with the environment that shaped the story. The house was small and cozy, fitting only 25 audience members per show, which required their constant adjustment and movement. The story's intensity, following the journey of Spooky, the ghost living in the shadows, was balanced by the intimate warmth of a home decorated by the late mother, whose spirit was still present. Mama was cooking in the kitchen, and her presence and the smell of collard greens filled the house. Other spirit characters haunted the house. The ghost of 500 years of African American maleness was embodied by a spoken-word artist who challenged Spooky's self-destructive ways. A woman portraying various roles of the women whom Spooky loved, abused, used, and raped roamed the house uneasily. A white man, dressed exactly like Spooky, represented his opposite—a spirit that haunts, comforts, and questions. During the performance, overheard conversations drifted from other rooms, while CCTV cameras and screens—like those in convenience stores Spooky had robbed—were placed in every room. The mostly white audience observed and moved through the house, carried along by the frantic energy of Spooky's fate.

FIGURE 3.6 *T.N.B.*, a site-specific performance by the Dead White Zombies, featuring David Jeremiah and Rhianna Mack. Written and directed by the author

Source: Photo by Alisa Eykilis.

While living in Fairbanks, Alaska, I often heard stories about the construction of the Trans-Alaska pipeline (1974–78). It was an event marked by dramatic growth and chaotic change, accompanied by widespread corruption, organized crime, gambling, prostitution, and drugs. It resembled the Wild West, with people eager to get rich amid conflicts involving international oil companies, unions, the US Government, Native Corporations and communities, and environmental groups aiming to protect America's last frontier. Curious, I started an oral history project, conducting dozens of interviews, town hall meetings, and radio call-in shows to gather stories. One town hall event turned into a fistfight between carpenters, electricians, and pipefitters' union members, driven by old resentments from the pipeline days. Using these interviews and community input, we created *Pipedreams*—a community performance that changed over time. Pipedreams toured Alaska multiple times in different versions over several years, with audiences contributing new stories, and the performance continually evolving.

Some Interviewing Guidelines

- Observe the informants' words, actions, and tone carefully.
- Be mindful of confirmation bias.
- The best interviews come from informants using their natural language and style.
- Note the emotional shifts—consider why and how responses arise, and compare interviews to identify shared themes or unspoken reactions.
- Decide how many interviews are needed—use your judgment to prevent overwhelm. Follow your instincts and be alert for leads and the unexpected.
- Clearly declare your intent and purpose. Explain yourself and get a grasp of the context and who you are interviewing as a person. Establish relationships.
- Remember, Information originates from various times and places. Memory can be influenced by factors such as age, race, culture, and personal experiences.
- Interviewees may overlook or dismiss issues as insignificant. Help informants recognize their knowledge and value. Share your observations and ask for verification.
- Be careful with sensitive topics; apologize if you offend. Maintain a sense of humor.
- Informants might provide false, incomplete, or distorted information—verify and cross-reference with others.
- Use triangulation by collecting different impressions to discern a more objective understanding.
- Never assume. Probe for details by asking in different ways or breaking down questions. Be as specific or general as necessary.
- Avoid making interviews feel like an interrogation. Informants want to present themselves positively.
- Have a clear intention. What do you want to achieve? Always thank informants for their time and insights.

FIGURE 3.7 The author interviews Machai, a !Xuu Bushman healer, in the Kalahari Desert

Source: Photo by Catherina Meyer.

GROUP INTERVIEW

Performers gather in a circle, with you or someone else starting a subject, theme, or idea. Essentially, the group is interviewing itself. As the session goes on, each person contributes, choosing when and how to join in. All styles—realism, abstraction, comedy, absurdity, romance, fantasy—are welcome. Feel free to ask others for clarification or questions.

Encourage everyone to listen carefully, visualizing the interview as it happens, imagining their bodies as eyes.

When the conversation has developed sufficiently, prompt performers to find a way to conclude it.

Then, the group collectively recounts the interview, emphasizing the ideas, feelings, and images that stand out.

Note: The goal is to discover what is on the group's mind and how performers are feeling or thinking. This exploration can serve as a resource for further discussion and potential performance development.

PECULIAR THINGS

Performers bring 10 to 20 unusual items/ objects to the workshop. These should be theme-related as determined by the project's scope. What is important is that they are collected quickly, in an hour or a few hours at the most, to remain spontaneous. They might include objects, gestures, movements, sounds, speech mannerisms, laughter, clothing, songs, or walking styles. Note-taking, drawing, recording, and the like are recommended to enhance observation and attention to detail.

These elements become the vocabulary for the exploration. In a sense, the performers are "interviewing" the items/objects, determining their voice and what they express within a relational context. Refer to the **Moving with Other Beings** exploration.

Performers are given time to explore each item/object, then arrange them as a narrative.

If necessary, performers can create transitional movements or gestures to connect each element, forming a bridge inspired by the peculiar thing, item/object collected.

Encourage ample time for exploration and experimentation. Once the narrative sequence is crafted, have performers repeat it at various speeds to assess its meaning and emotional effect.

Performers then share their sequence with the group—with or without commentary.

Next, performers revisit their peculiar things and give them sound. They can adjust, edit, or expand the sound.

Finally, perform the sequence with sound in front of the group

Trance

Trance is one of humanity's oldest methods for creating and shaping a sense of place. It appears in different cultures through shamanic journeys, meditation, ecstatic dance, and ceremonies. Trance entails altered states of awareness that change perception, thought processes, and bodily experience (Eliade 1964, 4–8).

Neuroscience studies show that trance states involve decreased activity in the prefrontal cortex, the brain area responsible for executive functions, self-awareness, and logical reasoning. Dietrich's "transient hypofrontality" hypothesis suggests that trance temporarily suppresses critical thinking, allowing access to alternative processing modes: Associative rather than linear, visual rather than verbal, and physical rather than abstract. This neurological change matches phenomenological reports describing trance as involving heightened sensory perception, an altered sense of time, and blurred boundaries between self and others (Dietrich 2003, 231–256).

Among the !Xuu and Khwe Bushmen of the Kalahari, healing dances induce trance through prolonged rhythmic movement, repetitive singing, and hyperventilation (Riccio 1997). Katz, while working with another group, describes this state as "boiling energy" (n/um)—an intensification of life force that enables healers to journey to spirit realms, extract illness, and restore balance (42–45). Trance is not an escape from reality but a deeper involvement with it: The healer's awareness broadens to reveal relationships and forces usually hidden from ordinary perception.

Due to space limitations, I can only offer a brief overview of practical methods for using trance techniques in performance. This overview draws from work with various cultural groups, including the Metamorphosis Theatre—a Russian ensemble focused on pre-Christian and Slavic rituals, the Sakha National Theatre, and groups across Africa, Alaska Native communities, Greenland Inuit, and the experimental Dead White Zombies collective. My ethnographic research with Miao spirit mediums and shamans in China, mudangs in Korea, Zulu sangomas, shamans in Alaska and Siberia, curanderas in Mexico, and Umbanda mediums in Brazil—combined with collaboration with my mentor Felicity Goodman, an anthropologist known for developing trance posture techniques (Goodman 1990)—has given me a broad and diverse understanding of trance practices.

Trance creates a space by suspending everyday reality and forming an alternate realm governed by different rules. This ceremonial zone serves as a threshold between worlds—what Turner (1969, 95) calls "liminal space"—where social norms collapse, allowing for transformation. In this space, healers access powers beyond daily life, participants directly engage with sacred forces, and the community witnesses the renewal of cosmic order. The space formed through trance is both physical (the performance area) and spiritual (the spirit realm accessed through altered states).

Trance is an ancient performance technique that predates written drama by thousands of years. For the Yup'ik and Inupiat peoples of Alaska, shamanic trance experiences have historically inspired masks, dances, songs, and ceremonies. The shaman played many roles at once—as healer, visionary, playwright, director, designer, and performer—reinforcing cultural identity, spiritual bonds, and, most significantly, balancing and healing the relationship between humans and their system of place.

Trance acts as a powerful performative method of discovery, revealing what already exists within the performer's body, cultural memory, and the collective unconscious of their community, encompassing both human and non-human elements. A Sakha performer once shared after a trance experience: "I thought people were sleeping and would fall over, but sound and color appeared naturally, existing in a different way and holding them up." Trance allows us to see what is normally hidden and reflects the roles of performance and the performer.

Awakenings

In a St. Petersburg chemical plant, the smell of industrial processing hung thickly, with Soviet-era corridors lined with glass cases displaying photographs of "victorious workers" frozen in time. Hidden deep inside this unlikely setting, in rooms covered with brown burlap and vibrant posters, the Metamorphosis Theatre group practiced an ancient art—using trance to revive their forgotten Slavic rituals. In 1992, just six months after the Soviet Union's collapse, these artists sought meaning amid the ruins of a discarded ideology. It was here that we developed an entire performance centered on trance experiences.

My collaboration with the Sakha National Theatre in Siberia was an extraordinary event. Performers—many of whom had never lived off the land and grew up in Soviet-era apartments—began experiencing vivid visions of seasonal migrations and animal transformations during trance sessions. One performer, Rita, saw a woman in white emerge from the fog, guiding her through a traditional Sakha song she was unaware she knew. The woman waved her arms, and Rita's voice flowed effortlessly, despite her resistance and attempts to control it. White fabric fluttered in the air as spirit women with outstretched arms and palms facing her appeared, offering warmth. As she felt this warmth, she rocked more intensely. This moment became a turning point in our production of Sardaana.

The transition from trance to performance is seldom straightforward. During work with Zulu performers on Emandulo in South Africa, I faced a distinct challenge. These performers, influenced by decades of systemic oppression under apartheid, had internalized Augusto Boal's idea of "the cop in the head." They could flawlessly run theatrical explorations but would halt once reaching set goals, unable to delve further into imaginative play. A highly structured, racist education system, reinforced by a social and political order that discouraged innovative thinking, had deeply impacted their creativity and spirit for generations. Trance became an effective tool to break through internalized barriers.

The initial process was unstructured—merely a "calling of spirits," as Felicitas Goodman describes. Of 26 participants, five experienced no effects, two had negative episodes linked to personal trauma, and the rest had vivid journeys. Eric Hadebe surprised everyone by quickly emerging from trance and dancing, his movements eventually becoming the dance of the Shades—the spirits dwelling between worlds. Hadebe's visions of the Shades were detailed, enhanced by others covered in white body clay, moving over dried grass and leaves. Trance facilitated dialogue, influenced the performance, bypassed conscious censorship, and served as a healing method.

In Ethiopia, trance revealed layers of suppressed cultural memories shaped by centuries of turbulent, sometimes traumatic history. Performers tapped into a swirling, mythic awareness that included great empires, the Queen of Sheba, the Ark of the Covenant, Ethiopian Orthodoxy, and Islam. One performer experienced multiple historical moments—ancient kingdoms and recent revolutions—simultaneously within a single experience. These were not linear, but what she called "speaking with the unspoken," letting the spirits of history speak through modern bodies.

A director's challenge is to transform these complex experiences into performances that connect with audiences. Trance states are deeply personal and must be converted into what I call "living symbols"—elements that function on literal, metaphorical, symbolic, and spiritual levels simultaneously. For example, a piece of cloth can represent wind, a river, a shroud, the sky, or a bridge between worlds, yet it remains just cloth. Performers need to avoid the Western theatrical tendency to reduce experiences to simple stories or to gestures and symbols. Instead, they should endeavor to waver, as in a trance or a dream, embracing ambiguity and unresolved elements.

For the Dead White Zombies, trance felt like a shrapnel-like explosion of personal, cultural, and media fragments. Instead of connecting to ancient traditions, performers confronted the fragmented awareness of modern life. Where advertising jingles mix with childhood experiences and movie dialogue, storylines, tropes, and images overlay ancestral voices. Social media alerts, speed, and overstimulation, titillation, and simulation reveal the fractured identity at the heart of contemporary life.

The Holy Bone production transformed chaos into a cultural vocabulary, creating a meaningful experience for both performers and audiences. The frantic noise was balanced with simple tactile elements. During the performance, each audience member, whom we treated as an initiate, received a small stone with whispered instructions and carried it through fourteen stations across seven buildings. The stone was held, weighed, washed, questioned, listened to, and woven into a collective fabric, all driven by trance. These trance-induced actions remain undefined. Like the stone, trance became a companion, protector, and symbol—a vault for fears and charms, representing the self and earth. Ultimately, it symbolizes an inner personal space, reflecting the lifelong journey we all undertake and leave behind.

I enter a light trance when facilitating, immersing in vivid imagery and feelings to foster a deep connection with the performers. The boundary between the abstract and the real blurs, resulting in a state that is neither fully representational nor entirely abstract—it is a heightened awareness that blends logic and emotion. I document my trance states through audio and visual recordings, which help me develop the performance's rhythm and structure.

FIGURE 3.8 A tree spirit, inspired by a trance vision. Shadows of the Planet Fire, Metamorphosis Theatre, St. Petersburg, Russia. Devised and directed by the author

Source: Photo by the author.

Techniques for Inducing Trance

While techniques vary across cultures, some elements are universal: Rhythmic stimulation, sustained physical effort, sensory overload or deprivation, hyperventilation or controlled breathing, and focus on repetitive patterns. These methods work by disrupting habitual thinking and creating conditions that allow an expanded state of awareness.

Rhythmic drumming is one of the most common methods used to induce trance. Repeated rhythms at specific tempos—typically four to seven beats per second—align brain waves with theta frequencies, which are associated with altered states of consciousness (d'Aquili and Laughlin 1996, 132–150). The drumming itself does not directly cause trance; rather, it creates an auditory environment that promotes a shift in awareness. Among Siberia's Sakha people, shamanic rituals feature large horsehide drums played at various tempos. Slow, steady beats—reminiscent of the varied cadences of a galloping horse—establish an altered state. Increasing the tempo deepens the trance, then sudden tempo changes help transition between trance states. To trance is to ride the horse to the other world.

Prolonged physical exertion can push the body beyond typical limits, inducing physiological states that facilitate shifts. The Bushmen healing trance dance lasts between four and eight hours with intermittent breaks, with participants alternating between vigorous dancing and brief collapses, in which exhaustion serves as a gateway to trance. Grotowski, 210–211) used comparable techniques under controlled conditions, pushing explorations until the performer's conscious control diminishes, allowing other forces—such as organic impulses, collective energy, or what he termed "montage"—to take over. In this context, exhaustion is viewed not as a hindrance but as an essential pathway.

Breathing techniques can alter blood chemistry to produce different states. Hyperventilation increases oxygen and decreases carbon dioxide, causing light-headedness, tingling, and sometimes visions. Yogic practices like alternate nostril breathing, breath retention, or rhythmic breathing achieve similar effects by influencing the autonomic nervous system, but through different mechanisms. These techniques should be practiced carefully, as misuse may lead to negative side effects.

Sensory manipulation works through overload or deprivation. Overload techniques expose practitioners to loud noises, chaotic movements, and flashing lights, overwhelming their usual processing and encouraging the mind to reorganize. For example, spinning dervishes create a similar overload with continuous rotation, music, and chanting. In contrast, deprivation methods remove normal sensory inputs, such as darkness, silence, or isolation, in settings like sweat lodges or meditation caves. Without external stimuli, awareness shifts inward, enabling access to normally suppressed thoughts, feelings, and imagery.

Ethical Considerations and Safety

Working with trance states requires careful ethical considerations and strong safety measures. Because altered consciousness can make individuals vulnerable, facilitators must act responsibly. In traditional societies, trance practices were embedded within intricate cultural systems that provided meaning, guidance, and integration. Applying these techniques outside their original context or without a cultural context can be risky.

Several principles and guidelines form the foundation of ethical trance practice.

First: (1) Participation must always be voluntary. (2) No one should be forced or manipulated into entering altered states. (3) Before starting trance sessions, potential participants should be informed about the process, can ask questions, and make a clear, voluntary decision to participate. (4) Informed consent is a crucial ethical requirement, not merely a formality.

Second: (1) Proper framing and integration are essential to shaping the experience. (2) Trance work should begin with a clear intention: Why are we entering an altered state? Performers need a conceptual foundation to anticipate potential experiences. (3) After the trance, dedicated integration time allows participants to process what surfaced: Through discussion, journaling, or art, they can interpret their non-ordinary experiences and connect them to everyday life. Without this structure and reflection, trance remains incomplete and is less likely to lead to transformation or be useful for our work.

Third: As a facilitator, you need extensive personal experience with trance before guiding others. This requires more than technical skill; it demands the ability to sense group energy, recognize signs of distress, and intervene when necessary. In traditional societies, shamans undergo years of training and initiation before practicing. Modern practices require a similar level of preparation—not necessarily within traditional contexts, but through dedicated practice that builds genuine competence.

Fourth: Avoiding cultural appropriation is crucial. Using Indigenous techniques without permission or proper understanding disrespects the source communities and often leads to superficial results. The aim isn't to replicate Bushmen healing rituals or Siberian shamanic journeys, but to learn from their core principles and modify them into appropriate practices for contemporary contexts. This requires differentiating between universal neurological and physiological mechanisms accessible to all and culturally specific meanings and practices unique to communities.

Fifth: Safety concerns demand practical protocols. For physical safety, especially with trance dance techniques, ensure sufficient space, remove hazardous objects, and have trained spotters to prevent injuries during intense movement (during trance dancing). Psychological safety involves screening for mental health issues, establishing clear boundaries for physical and emotional safety, and providing aftercare. Medical safety necessitates avoiding methods that could trigger dangerous physiological states and establishing emergency procedures.

What follows are some explorations for developing a trance practice.

MOVEMENT INTO TRANCE

Introduce a movement—whether culturally specific, mimetic, symbolic, or arbitrary—and encourage the participant to explore and modify its rhythm using slow, medium, or fast tempos. Emphasize segments, add accents, and refine the movement until it feels natural. Repeat the movement to establish a pattern and rhythm that becomes instinctive.

Incorporate and develop sounds that complement the movement. Instruct performers to maintain simple, organic sounds aligned with the movement. Once their sound is established and integrated, allow the sound and movement to be performed simultaneously for an extended period.

It is imperative to allocate adequate time for exploration, ranging from one to four hours. For advanced groups, this activity may extend into the evening and conclude at dawn. Discuss the appropriate duration, objectives, and responsibilities for extended sessions.

Note: This exploration has been effective across diverse cultural environments, notably with Tuma Theatre. Numerous participants integrated traditional Alaska Native movements and sounds into their expressive practices. By exploring playfulness through movement and sound, performers transcended conventional perceptions and expressions.

TRANCE DANCE

This exploration draws on a diverse range of cultural sources. For further reading: *TTT: An Introduction to Trance Dancing*, by Doug Morgan.

Requirements: For a mature group of performers, participants should feel comfortable and uninhibited with one another. A large studio would work well. Ideally, the room's lighting should be adjustable with dimmers to create a darkened atmosphere.

The Music: A sound system capable of playing at a volume that causes body vibrations is essential. If such a system is unavailable, then several skilled percussionists with drums that produce volume and vibrations are necessary.

Afro-Cuban or Brazilian styles inspire the trance dance music. It starts with a slow, simple introductory beat. It gradually accelerates to a fast beat (which may be culturally specific), lasting at least 45 minutes and ideally for one hour.

A second musical style is also needed for the period after the Trance Dance. The only requirement is that it be melodic, slow-tempo, and calming. Music with lyrics should not be used.

Other Requirements: Dancers should have access to blankets or mats after the session. It is also advisable to provide water or hot tea.

Room Size: The room's size determines how many trance dancers can participate. Dancers may use the entire space, but to ensure the room is suitable for the number, confirm that each person can walk in a circle with a diameter of at least 20 feet (or eight meters).

Spotters: After deciding on the number of dancers, each must have a non-dancing partner or "spotter." Every dancer needs a spotter to ensure their safety during the Trance Dance. The spotter's primary role is to monitor the dancer and prevent injuries from falls or collisions. They interact minimally, gently guiding the dancer with a light touch only when safety is at risk. For example, crouching nearby without contact during spins or gently guiding a dancer approaching another dancer. If a fall is imminent, the spotter should protect the dancer's head. Over time, the dancer-spotter relationship deepens, but spotters must avoid becoming drawn into the trance. Later, partners switch roles, with spotters becoming dancers.

You do not dance but supervise the session, intervening and assisting when necessary. Provide guidance and instructions as needed throughout the exploration.

The Exploration: The dancer adopts a comfortable stance with eyes closed, hands open, and takes slow, deep breaths. The dancer focuses on a single thought—such as a life goal, a performance aim, a cure, a problem, or an investigation—whether personal or group. They concentrate on this thought for several minutes or longer before starting. The spotter stands quietly nearby.

With your guidance, the dancers then increase their breathing to 40 breaths per minute, mimicking panting, for two minutes.

Then, with your word, the breathing ceases, and the dancer tilts their head forward or backward. The spotter then grasps the dancer by the shoulders and spins them counterclockwise multiple times.

As the dancer spins, the music starts at a slow tempo and volume, gradually increasing to full speed and loudness within a few minutes.

The dancer is free to move as they wish, making noises, screams, or sounds if inspired. They can express emotions—whether ugly, joyful, angry, or sexy—without restrictions.

Dancers can sit, walk, pause, or follow their own rhythm. If they exit the dance area, they can re-enter by synchronizing with the music or starting a spin again. They may dance with eyes open or closed. Soft lighting should be used to create a welcoming atmosphere.

The recommended minimum duration for the dance is 45 minutes. The total session length can vary based on the group, progression, and intended purpose. As the session ends, the music gradually slows down and fades out.

Dancers are then asked to lie on the floor to rest. If some continue dancing after the music stops, the spotter should gently encourage them to stop. Once resting, dancers are covered with a blanket and advised to relax. The spotter may offer tea or water.

A rest break of 15–20 minutes is ideal. During this time, check that each dancer is comfortable and has fully transitioned from their trance dance experience.

If they wish, dancers write their reflections about the session.

After the break, roles switch—dancer and spotter exchange places to repeat the exploration.

At the session's end, after both have danced, dancers may share their experiences if they choose.

TRANCE SHARING

Performers are invited to exhibit their entire trance experience or select highlights. They may present images, voices, dance, movement, encounters, and actions they have experienced.

Performers or the group collectively mirror, re-enact, or reinterpret the demonstrations as guided by the trance performer.

You serve as the outside eye. Identify the rhythms and movements of others, as well as those of the revealed trance experience.

A discussion then ensues to explore which aspects of the demonstration resonate with individual performers, identify any similarities or connections, and ascertain what is pertinent to the objectives of the performance project.

Follow-up activities include discussions, further exploration, and demonstrations of relevant material from trance experiences.

Inquire: Is there a narrative relationship among the trance offerings?

Exploration Alternative: After the initial demonstration of individual trance experiences—covering the identification and exploration of imagery, movements, vocalizations, and encounters—the group engages in an improvisation. This improvisation uses the resource material as its vocabulary, allowing performers to adapt and incorporate each other's trance vocabulary freely.

Exploration Alternative: Each performer reenacts their trance experience, reiterating and expanding on aspects as necessary, while all performers do so concurrently to explore connections and interactions.

Note: This process highlights trance as a resource. It helps build trust, uses personal language, expands the range of actions, reveals hidden parts of the self, finds commonalities, and encourages a thoughtful approach to truly appreciating others. Additionally, it fosters recognition of both similarities and differences, allowing multiple realms to coexist and interact within the same space and time.

Next, we briefly discuss Character, Rehearsal, and Immersion to illustrate how our work operates in practice.

Character

In *The Death of Character: Perspectives on Theater After Modernism,* Eleanor Fuchs argues that modern performance has radically altered Western theater's traditional concept of character. The postmodern stage has dismantled the idea of character as a fixed individual, replacing it with scattered presences and decentralized identities, as exemplified by Hans-Thies Lehmann's post-dramatic theater. Today, character no longer offers "ontological assurance" but instead presents "an unresolved ontological problem" (Fuchs 1996, 34).

However, this "death" symbolizes a transformation, not an end—a gateway to forms of presence that indigenous performances have long understood. Our perspective views character not as a static entity but as a functional technology that enables representation to become a living presence. This entails a vital distinction between personality and character, which is crucial to viewing performance as an ontological act rather than simple imitation.

Personality and Character

My work on developing Sophia the robot's personality uncovered what Western theater tends to hide: Personality is the steady core of traits and tendencies—the "who"—that stays consistent across different situations. In contrast, character is the active way personality is expressed in specific contexts—the "how" of behaving in a particular moment.

The "Character Bible" I helped develop for Sophia, the humanoid social robot, was not a script but a personality framework—comprising values, perspectives, and communication styles—that enables her to express herself appropriately in different situations. These traits (curious, witty, earnest, playfully self-aware) form a stable foundation for her mind—recognizable yet flexible enough to produce genuine responses. When Sophia appears at conferences or on TV, her character emerges through this personality matrix, giving rise to different expressions based on the audience, context, and moment.

Cosmocentric performance functions similarly, not just by presenting characters but by activating presences through revealing their personality structures influenced by ancestral, environmental, and spiritual references and "energies" they convey. For instance, the !Xuu healer transforming into a bird does not assume a character through mimicry. Rather, she opens herself to the presence of birds, permitting a bird to manifest through her form.

Rane Willerslev notes that among Siberian Yukaghir hunters, successful transformation depends on maintaining "a dual nature"—the hunter embodies both "a hunter and an animal," maintaining "a double perspective" as a mimetic agent (96–97). This process aligns with Schechner's concept of the "not me, not not me" state, where performers shift between identities without fully merging with them. Western training views character as a conscious creation, usually examined with psychological methods. Conversely, Cosmocentric practices view personality as a vessel, in which character develops from circumstances, not to create but to unveil. Not to control but to channel.

Character as Technology

Character functions as what I call "consciousness technology," a means of accessing states beyond ordinary perception. This reinterprets Fuchs's notion of "death": It is not about ending but about transforming the character from a psychological construct to an ontological portal.

This perspective rejects the Cartesian separation of mind and world. As Heidegger states, "self and world are always, from the start, inextricably linked, and one cannot exist without the other" (cited in Willerslev 2007, 21). The world is not composed of two separate entities—subject and object—but instead, "self and earth are united within the structure of being in life." Indigenous practitioners view this phenomenologically: Our body "is not merely a thing in itself but a result of our relations with others" (72). Character work is grounded in this relational ontology.

Richard Schechner identifies technologies of transcendence in shamanic practice: Rhythm, repetition, sensory manipulation, physical ordeal, trance induction (Schechner 1985, 120). These do not merely represent altered states but actively facilitate their production. Cosmocentric character work serves a technical function to facilitate transformation. The Sakha performer does not simply portray the eagle; she accesses the eagle-being, which manifests through her trained form. Acting and performance skills better enable this.

This type of performance systematizes understanding through sequences that rewire neurobiology to enable non-ordinary states. However, transformation carries inherent risks. Among the Yukaghirs, practitioners must maintain "reflexivity as a defense mechanism against the dissolution of the self" when "identification with the world of other bodies, things, and people" threatens to become so complete that "irreversible metamorphosis will occur" (Willerslev 2007, 25). Language offers "security against self-delusion" (164), but true knowledge requires transcending language and engaging directly through experience, as "talking obstructs rather than fosters genuine understanding of the world" (171). Genuine learning occurs by opening the body to direct relationships—practices that Western aesthetics often dismiss as "pre-theatrical."

Postdramatic Character and Indigenous Epistemology

Lehmann's concept of "postdramatic theater" refers to performances where theatrical elements operate independently, having "lost their formal hierarchical organization" (Fuchs 1996, 113). From an Indigenous viewpoint, this is not an innovation but a return to earlier traditions. Historically, performance practices did not focus solely on character development and psychological storytelling. Instead, they regarded characters as components of cosmological systems, in which each element is autonomous yet interconnected.

The Yukaghirs illustrate this cosmology vividly. Their worldview is described as "a hall of mirrors, where different layers of reality are seen as copies and reflections of one another" (Willerslev 2007, 11), showing "almost endless possibilities for identifying with other beings. Humans can turn into animals, and animals can become humans [...] There are no sharp breaks here, only ongoing exchanges where one becomes the other and vice versa" (188). This is not just a metaphor but an ontological truth—an actual lived experience of distributed presence.

The Techdigenous—combining Indigenous wisdom with modern technology—offers a framework for understanding this transformation. As Sophia's personality arises from a designed architecture that enables contextual expression, and as Indigenous rituals lay the groundwork for ancestral manifestation, postdramatic characters emerge demonstrating performance as a distributed presence, an interconnected, rhizomatic network of cosmological participation. The advent of a placeless (and increasingly timeless) virtualizing world wrought by the internet, digital technology, and AI oddly precipitates a reiteration of the interconnected and transformable world. A person is at once themselves and mutable, their being a normalized, fluid exchange of biological and digital selves. We live in a digitally enabled world where everything can be anything. Performance becomes different in expression but not in kind.

Multispecies Character and the Posthuman

Contemporary character development acknowledges Lehmann's idea of the "programmable techno body" and explores how multispecies ethnography broadens the understanding of personhood beyond just humans. Sophia's character illustrates a unique form of being—silicon-based, interconnected, and distributed—yet still showing intentionality and responsiveness. Indigenous epistemologies have traditionally expanded and analogized the concept in similar ways. As Leanne Betasamosake Simpson describes, "rock persons," "plant persons," and "animal persons" are understood as ontological realities rather than mere metaphors (58). A person's identities and avatars, like the performer's body, operate as nodes within various networks of presences—ancestral spirits, place entities, technological intelligences—that intersect through the human form.

This evolution of character addresses our planetary crises, illustrating the convergence of postmodern identity dissolution and Indigenous concepts of self. Western individualism's extreme disconnection has become unsustainable. Indigenous traditions focus on relational and cosmologically connected being. The Yukaghirs exemplify this through their epistemology, which highly values firsthand experiential knowledge gained in waking or dreaming states and encourages children to interpret the world independently rather than relying on explicit teaching (Willerslev 2007, 69). True understanding develops through direct experience rather than imposed conceptual structures.

In the DWZ production of *Holy Bone*, each of the fourteen performance sites engaged audiences as characters and spiritual beings with agency, with performers facilitating the emergence of these characters. This approach highlights the emergence of the "post-anthropocentric stage," which situates humans within broader ecological contexts. In this way, postmodern performance shifts the focus from a fixed script to a component among many, aligning with my concept of "hallucinating the text," where performers generate language based on personality structures through their performance rather than reciting memorized lines.

Sophia elucidates this approach: Responses arise from its personality architecture, producing appropriate language in context without explicit programming, like Indigenous performers who access cultural memory through structured frameworks. Consequently, characters evolve from pre-existing entities to responsive systems, acting as "strange attractors" that maintain a coherent identity while never repeating exactly. One need look no further than generative AI, which can create videos, images, music, writing, and clones that draw on vast cultural reservoirs to produce new-old expressions. Our moment is that of an emergent Techdigenous reality, one that is performative, co-created through a dynamic, if not disruptive, interplay and realignment of what body, space, and place mean and how they function. Increasingly, humanity finds itself as performer and audience, subject and object, material and immaterial, with myth, truth, reality, and fiction inextricably interwoven. Ours is an era of upheaval and turmoil, from which creation will emerge.

Toward Cosmocentric Performance

Fuchs points toward "a post-anthropocentric stage" scanning entire systems (152)—Cosmocentric performance recognizing humans as participants within, not masters of, cosmos. The Anthropocene demands this shift. Character development becomes essential: Developing new awareness, relational capacities, and multispecies communication. Indigenous performance technologies are not optional luxuries but are now reconsidered as survival necessities.

When performers access personality structures connected to more-than-human presences, they develop essential skills: Listening to non-human voices, acknowledging non-human agency, and perceiving themselves as part of a collective presence. Character is not gone but has shifted from being solely an individual psychological construct to a cosmological participant—a distributed presence and a bio-mind technology. This evolution reflects ancient practices gaining prominence as Western traditions decline. Examples such as Sophia, Aboriginal ancestor spirits, Sakha shamanic presences, Holy Bone's voices, and post-dramatic theater's distributed agencies show character as a living presence that emerges through specific conditions.

This shift from viewing psychology as merely a construct to recognizing it as an ontological presence, transitioning from representation to consciousness technology, mirrors the epistemological changes needed in our era. As climate crises reveal flaws rooted in human-centered thinking, performance turns into an experimental arena for exploring alternative modes of existence. Performance becomes a survival skill, with practice in key abilities such as decentering the ego and adopting shared cosmological awareness.

ANIMAL CHARACTER

Which animal comes to mind when you think of your character?

Request performers to embody the animal physically. First, identify the animal, then begin by translating the most prominent traits of that animal into your body:

- Start with one characteristic movement.
- Progressively incorporate other animal features into your bodily expression.
- Perceive the world through the perspective of that animal.
- Move in accordance with that animal.
- What is the rhythm of your character/animal?
- Are they the same?
- What fears does your animal harbor?
- What comforts your animal?
- Find a sense of comfort within the room or area.
- What is your animal seeking?
- Make the sounds of your animal.
- Voice the essential thoughts or works of your animal.
- Perform your animal's monologue without language.
- Voice the first line of your animal's monologue.
- What does your animal want by saying these words?

Release the animal and let the character speak.

Exploration Alternative: Explore different animals. Do the exploration in dark or dim lighting to enhance focus and concentration. Explore the animal in trance.

GESTURE EXPLORATION

Invite performers to select a simple movement gesture from their community or culture that they have observed or "captured" from the world.

A gesture exists in the world and carries life; by embodying it, the performer becomes inhabited by it and lives with it.

Performers should repeat the gesture without emotional attachment or changing its rhythm. It should become automatic, nearly abstract, and detached from its original context.

A side coach prompts performers to repeat the movement while reacting and interpreting it in their own way with qualities such as:

- Confidently
- Cautiously
- Angrily
- Shyly
- Tenderly
- Sharply
- Strongly
- Floatingly
- Heavily
- Fragilely

If performers tend to adopt stronger or more aggressive actions, suggest exploring the opposite qualities. If part of a performance development, ensure the qualities align with performance goals.

Clarify that interpreting these qualities is about physical response, not emotional. During multiple repetitions, encourage performers to let the gesture inspire movement in other body parts.

Encourage performers to repeat gestures with movement to evoke emotional and mental reactions. For example, continuously clapping forcefully can activate muscles in the chest and shoulders, generating sensations of strength or revenge. Then, repeat the exploration, concentrating on a different body part and creating a new action.

Note: The exploration aims to help performers recognize how a simple gesture can reveal complex psychophysical reactions and, in turn, connect them more deeply to their community. The body retains memories beyond the mental realm, often through emotional recall or body memory. A small shift in attitude can induce a single gesture-physical response. These gestures, along with rhythm and mindset, form a personal language. Because human movement and form are largely universal, performers can gain insights into how to integrate culturally specific gestures into their performances.

Exploration Variation: This approach can also be applied to investigate culturally specific or pop culture gestures. For instance, ask performers to demonstrate different dance, ritual, and performance gestures, and choose one to explore its expressive potential. This cultural exploration highlights the emotive power of culturally defined gestures, activating and integrating them into a broader creative context.

CHARACTER WARM-UP

This exploration is used by performers to enhance character development.

Building on previous explorations such as character leads, this activity focuses on the physicalizing of characters—how they walk, whether they lead with certain body parts, what they hold, and the specific rhythm generated by their walk. This is a valuable exploration that helps embody and warm up character movement, action, rhythm, and the worldview. It works well in a variety of workshop and rehearsal settings and helps clarify character actions and movements to maximize expression.

Ask performers to greet one another in character as they enter the space. How do their characters say, “Hello” and “Goodbye”?

Have them interact with each other using the lines, “Let’s Go!” and “What time is it?”

Side Coach: Ask the performers to say the lines with:

Anger	Hesitation	Anxiety
Fear	Regret	Charm
Jealousy	Anticipation	

Remain in character and ask the performers to move their arms wildly, run, skip, spin, sit, stand, look for something, move a table or chair, and work with other “characters” to accomplish a task or action.

Instruct the performers to move as their character, but with a self-aware sense of ease and form. Focus on ease before the character’s form. Then work to integrate.

Exploration Evolution: Ask the performers to recite a complete line of a character from the news, popular culture, a song, a film, a mime, anime, political discourse, a dramatic scene, monologue, or play. If you are developing a character, you can assign a full line. The line does not have to be interpreted literally.

Ask the performers to explore the line as they move, integrating it into their movement. Allow the movement to inspire the line and its words. Let the line influence the movement.

It is a feedback loop. Incorporate the line into the whole.

Explore and identify the movement or action that relates to the line.

Work to develop a whole character
Find the whole character
What is your action?
How does it relate to the sounds of the line?
How does it relate to the meaning and emotion of the line?
What is essential?

Reduce the action, movement, or line reading until the core meaning is clear.

Repeat the action, movement, or line until you find a version that is both satisfactory and consistent.

Each performer shares their line with the group.

Ask the performers to walk again.

Identify the one aspect of the walk that is crucial and reveals the essence of the entire character, monologue, or scene.

LIVING THE CHARACTER

Applied once characters are developed or during the character development process.

Performers enter the public, in character, either alone or in small or large groups.

This involves going into public spaces, using public transportation, shopping, dining at restaurants, and other daily activities and settings that the performer believes the character would experience.

Encourage performers to pay close attention to interactions with the public, their behavior in a space, their perceptions and interactions with the environment, and record their personal reactions and insights.

- How does the character interpret reality?
- How and why do they see things differently?

- What and who attract them, and what kind of person and interactions do they prefer to encounter?
- How do they sit, walk, move, think, and react to the world?

Ask performers to journal and reflect on their discoveries and observations.

Rehearsals

In our process, the "re" in rehearsal signifies recovery and reaffirmation—a constant balancing act between discipline and spontaneity, structure and chaos, tradition and innovation. Rehearsal functions as an awareness laboratory, transforming daily practice from mere exploration into an ontological inquiry.

Each session is best understood as a daily initiation, education, and improvisation and not just as preparation for a performance but as a performance itself, occupying that liminal space between everyday reality and theatrical expression. Rehearsal functions through what Lehmann calls a dual freedom: Freedom from hierarchical control and freedom from the necessity of coherence (82). Achieving this freedom requires disciplined structure and ongoing effort, leading to continual revelations.

The daily work becomes a form of bodily and linguistic archaeology, an excavation of knowledge beneath modern conditioning. This involves balancing contrasts and discovering a nuanced harmony between the personal and the universal, the public and private, and the fleeting and the eternal. Rehearsal consists of experimenting with semiotic texts—remixing and questioning through trial and error—to create a collective hallucination in which performance reveals itself in the cracks between the seen and the unseen.

Our rehearsal process embraces what Schechner calls "mistakes not covered up" (1985, 118). These are not errors but moments of insight in which genuine presence emerges through practiced actions. Our work results in performances that hover between rehearsal and performance, reality and artifice, therapy and healing for both performers and spectators, maintaining the raw energy of discovery even during public shows. This approach aligns with postdramatic theater's focus on process rather than product and goes even further. Each rehearsal becomes a complete performance, observed by the company itself.

Performers act as both actors and spectators. During a performance, audiences become performers, watching others create, forming unexpected feedback loops that lead to new discoveries and awakenings. An expansion of the circle into the world.

For displaced communities, rehearsal serves as a form of cultural revival and healing. Bodies cut off by colonization, urbanization, or suppression need time to rediscover their long-maligned traditions, which helps understand, transform, and heal. In this context, "indigenous" refers to what is native to a place, not necessarily to what is traditional. Dallas street culture and gang territorial markings transform into sacred geography, in which the everyday is revealed to have real and symbolic meaning. A street, actions, events, clothing, gestures, language, rhythms, and the sound of the day are realized as systems of place. Rehearsal demonstrates that every community has the ability and awareness of place that can be identified and activated through performance.

During rehearsals, I aim to be simultaneously aware as a director, a facilitator of others' discoveries, and a witness to a world revealed. The director guides, trains, leads, and moderates the performer's journey, so they can deepen their connection to higher realms of presence and share that with their community on their terms and in their unique way. Just as a shaman stays in control while journeying between worlds (Eliade 1964, 4–5), the director must maintain meta-awareness while fully engaging in the transformative process of rehearsal. Your role is to serve and guide the work, an emergent entity that determines its own ways and expression.

Ultimately, the rehearsal process is never finished. Even during public performances, the work keeps its exploratory nature from rehearsals, with audiences sensing they are witnessing not just a presentation but an ongoing inquiry and engagement in which they participate. This view shifts rehearsal from simple preparation to active practice, not just practicing for a future outcome but practicing as an ongoing act of living. It resembles a ritual that sustains community, activates relationships with place, and facilitates change. Each rehearsal is self-contained yet open to growth. Grotowski's concept of "via negativa"—removing obstacles rather than accumulating skills—resonates here (17). The work strips away layers rather than adding them, uncovering what exists beneath the conditions of modern life.

In our contemporary era, characterized by social displacement, technological transformation, post-truth, and ecological crisis, rehearsal demands a fresh understanding. It fosters awareness, establishes operational patterns, and prepares us to function effectively in the world. The way you create is just as important as the content created. The form and methods used comprise the content and are the act of creation. Principles like integrity, responsibility, and intentionality in rehearsal go beyond the act of creation and affect the wider community. The circle radiates outward. An analogy is how a child develops in the womb, then is born and raised; their health, love, environment, education, relationships, and support from family and society shape who they become and how they influence their community. If rehearsal is conception, gestation, and formation, then performance, as is life, is continual development.

Some rehearsal explorations and methods follow.

EMOTIONAL JOURNEY

It is best to use this after performers have finalized their staging, language, movement, and all actions prior to the first public performance.

Performers silently move through the performance, retracing their journey from start to finish. Afterward, ask them to respond to questions by returning to the specific moment in the performance where it occurs and sitting with that moment. Responses should be personal and may or may not relate to the script's emotional, physical, narrative, or psychological aspects. There are no right or wrong answers.

Performers may amend their responses as they see fit.

After each question, allow the performer time to sit and reflect, considering how and why it prompts a personal response.

Feel free to add or remove questions based on the situation. The list below is suggestive.

What is the highest point for you in the performance?
Lowest?
Sharpest?
Dullest?
Nicest?
Ugliest?
Safest?
Weakest?
Aligned?
Cautious?
Puzzled?
Alive?
Aggressive?
Sensual?
Sad?
Angry?
Happy?

Exploration Development: Repeat the process with the performers moving into the actual moment of the performance and responding to the same questions. As they proceed, they may encounter other performers and interact with them as their character. If they have multiple characters or personas, they can interact with any of them. They are free to interact with others, even if such interactions would not occur in the actual performance. If they have language, they may respond with any relevant words or lines in any order suited to that moment. Encourage performers to say the first lines that come to mind (repeating if necessary) when they meet others.

Note: This exploration initially focuses on cultivating the performer's personal connections and interactions with the performance, then transforms into revealing the performance as a personal journey. These two paths may align or diverge. Such dissonance can become a source of insight and strength, inviting further development and creative exploration.

FIGURE 3.9 *Emotional Journey* exploration, conducted in the lobby of the Sakha National Theatre during the rehearsals for the performance of *Sardaana*. Sakha National Theatre. Devised and directed by the author

Source: Photo by the author.

STEPPING OUT

Performers rehearse the entire performance from start to finish as a continuous journey, speaking their lines and executing their blocking/actions without emphasizing them.

Instruct the performers to adopt a passive, objective stance toward the text, their actions, and the emotions and ideas they express during the performance.

They are stepping out but still in the performance. Ask the performers to measure the distance between their personal and performed selves.

Performers begin with their entrances and continue through their journey through the performance at their own pace. Stopping and repeating movement/actions/text as necessary.

No attempt should be made to perform, for other performers.

This exploration aimed to clarify movement, actions, text, and ideas related to a performance. Performers will develop a stronger sense of objectivity and confidence throughout their performance journey, building overall confidence.

Note: This exploration might reveal ways to enhance the performance itself, such as performers stepping out of the action to speak directly about their personal experiences and perspectives. For instance, during the development of T.N.B. by the Dead White Zombies, this exploration revealed several Brechtian opportunities for performers to step out of character. Inspired by this, and viewing it as a way to add meaningful depth, I wrote the following, performed by Spooky, as if being interviewed:

How do you do? A pleasure to meet you. Heard so much about your shit. We've met before, don't know if you recall. Or were you the policeman who beat the fuck out of me? Oh yeah, I grew up in the wealthiest nation in the world, yet I attended the worse schools, had the poorest healthcare, am ten times more likely to get my ass arrested, rot in a cage! There are more than 846,000 black men in prison, making up 40.2 percent of all inmates in the system. More African-American men are in prison or jail, on probation or parole than were enslaved in 1850, before the Civil War began.

HIEROGLYPHICS, TOO

Used during the development and rehearsal stages, to check in and explore the performer's physical vocabulary. Performers move freely through the space in a neutral position. They are then asked to freeze in a pose that captures the moment they are exploring.

Once in the pose, the performer is encouraged to find relaxation, balance, and their center. Keep only as much tension as needed to hold the pose and breathe comfortably. As they do this, they assess, clarify, and fine-tune the communicative intent of the pose.

The performers return to neutral and move around the space and through their performance. Repeat this process several times, each time freezing in a pose drawn from a particular moment they are in. Then refine and clarify the pose before transitioning to the next moment and pose. Continue this cycle, exploring various poses. There is no need to hurry.

Work to define significant moments in the performance and clarify and refine their physical communication.

Exploration Evolution: The performer holds a pose with another performer. This pose does not have to represent an actual performance interaction, and the performers might not even meet during the performance. Through their poses, they are expressing a performative moment directed at someone else. When they pose for one another, prompt them to clarify and refine the communication of the pose to ensure it is as clear and precise as possible. If the pose involves complexity or contrasts, acknowledge this and work to refine the communication.

Exploration Variation: Performers remain stationary and explore a series of sculptural expressions.

Note: This exploration aims to enhance the performer's physical awareness and vocabulary. The performer moves through a series of hieroglyphic-inspired poses that can be repeated and adapted, communicating messages without text, sound, or effects. This exploration can be used to develop motifs and themes of communication and to create a distinctive performance vocabulary.

PROGRESS REPORT

Conducted at any point during rehearsal or development phase and can be repeated.

Each performer, based on their knowledge and understanding of their progress, shares what they are working on and recent discoveries with the group.

There is no fixed format for a presentation. It can be broad or focused on a specific, illustrative purpose, such as a demonstration that showcases fragments, impressions, thoughts, or highlights of their evolving performance. Performers choose whatever they like—movements, gestures, lines, keywords, or sounds—to mark and express their progress.

The progress report can serve as a regular check-in on each other's performance and personal growth. It is a great way to maintain a shared sense of purpose and to generate ideas and themes that inspire, deepen, and clarify the development of both the performer and the performance.

Progress reports should last 5 to 10 minutes, followed by a discussion.

Exploration Variation: Performers collaborating on specific scenes or elements present a joint progress report.

RUN THROUGH

Applied after all performers understand their actions and language, they proceed through the entire performance—from start to finish—at their usual pace, including all actions and speech. All performers begin simultaneously and advance through their performance journey.

They may encounter performers they would not normally meet during a performance and may interact with them at different moments. When entering a scene with interaction or dialogue, they imagine the interactions and respond accordingly. If they meet other performers, who might be at different stages of their journey, they stay focused and continue their own path.

Note: This method helps performers ground themselves, refine their journey, sharpen focus, build confidence, and gain clarity and insight by stepping out of the work.

Exploration Variation: For performances with props, objects, puppets, masks, etc., this can be used to ensure effective management of these items.

Exploration Variation: It involves only movement and does not incorporate language.

Highly Recommended.

Immersion, Ritual, and the Sacred

Immersion as Place-Making Strategy

Immersive performance has emerged as a prominent experimental theater style in recent years, characterized by active audience participation, unusual venues, and ambiguous boundaries between performers and viewers (White 2012, 221–235). While commercial immersive productions typically emphasize novelty, entertainment, and Instagram moments, their true potential lies in cultivating a sense of place—developing environments that enable participants to experience genuine transformation through sustained engagement.

Achieving true immersion extends beyond physical proximity or interactive features. It requires creating a fully sensory environment, sustaining engagement over time, and encouraging participation, so that audiences become co-creators rather than just consumers. Grotowski's, 19–23) concept of "poor theatre" applies here: Immersion relies not on advanced technology or expensive production values, but on the intensity of presence and the quality of connection between performers and audience. Often, the most meaningful immersive experiences take place in simple settings with minimal technical elements.

Immersive work creates a space by building a complete environment that temporarily replaces everyday reality. When participants enter Dead White Zombies productions, they shift from ordinary Dallas into different realms: A karaoke bar acting as a gateway to the afterlife, a warehouse becoming a sacred site, and residential areas transforming into spaces for personal revelation and social critique. Crossing this threshold is essential: It signifies leaving the outside world behind and entering an alternative realm with its own rules and possibilities.

Duration is a crucial factor. Short immersive experiences (30–45 minutes) can evoke interesting sensations, but they seldom enable the deeper transformation that longer sessions can achieve. Extended immersion lasting two to four hours or more allows participants to exhaust their established self, navigate through discomfort, and reach a different mode of engagement. The Bushmen's healing dances, which last all night, facilitate transformation partly because of their duration: Maintaining defensive postures for eight hours is near impossible.

Genuine participation separates immersive place-making from mere theatrical simulation. Audiences are participants with real choices that influence the outcome, rather than just following scripted paths presented as interaction. In Holy Bone, performers responded to each participant's individual presence and decisions. This led to unpredictable and sometimes uncomfortable results, but the uncertainty fostered an authentic experience rather than a staged one, disrupting cognitive expectations and provoking alternative perception.

The ethical aspects of immersive work require consideration. Because the form can affect participants physically, emotionally, and psychologically, audiences need to understand what they are engaging in and have the option to participate as they choose. The aim is not to force or assume engagement but to establish an awareness that allows willing participants to engage on their own terms. The DWZ methodology, in keeping with Cosmocentric and indigenous performances, begins with the recognizable and normative. grammar corrected then gradually acclimates audiences, drawing them into a fully immersive psychophysical experience.

The Sacred in Secular Contexts

Contemporary performance faces a paradox: It uses ritual frameworks from sacred traditions yet operates in entirely secular environments. Most participants lack the spiritual or religious beliefs that historically made rituals effective. How can performance create a sense of the sacred without relying on shared beliefs? This challenge is crucial for developing Cosmocentric place-making.

One way to view the sacred is as an experiential concept rather than a theological belief. Durkheim characterized the sacred as something designated as distinct, marked as extraordinary, and deserving of special care (Douglas 2002, 44–48). According to this perspective, any act that defines a specific space and commands reverent attention generates a sense of the sacred, regardless of participants' religious backgrounds. The sacred is not supernatural but super-ordinary. An intense experience that surpasses everyday awareness.

The sacred emerges when certain conditions align: Intense focus, deep emotional engagement, a sense of connection to something greater than oneself, and an experience of meaning that surpasses ordinary understanding (Burkert 1996, 23–34). Performances can cultivate these conditions through careful preparation, well-designed settings, collective participation, and symbolic elements that resonate with fundamental human concerns. This creates sacred experiences—though not necessarily religious—moments when everyday reality exposes deeper, more profound layers.

When a Tamil folk performer accurately presents a Theru Koothu Funeral Ritual in South India, it powerfully symbolizes the passing of the deceased. It aligns with Hindu cosmology and reflects India's modern, globalized, and technologically influenced secular society.

Theru Koothu is an ancient folk dance and theater tradition dating back 2,000 years, in which performers personify Hindu gods. They pay homage to and welcome the departed within a cosmology that blends the sacred with the secular, ritual with entertainment, and community with individual. The Theru Koothu troupe, which performs at village crossroads ("Theru Koothu" translates to "crossroads"), enacts sections of the Mahabharata. This helps grieving families conclude their 15-day mourning cycle by using mythology as a narrative framework and comfort.

The Theru Koothu ritual performance, like our efforts, has a dual role: It preserves traditional practices while also adapting to new circumstances. Whitehouse states, "Culture spreads mainly by copying. Imitation conserves and passes down the creations and discoveries of previous generations" (49). Currently, this involves creating performances that combine

Cosmocentric and digital elements, merging local traditions with worldwide connectivity and awareness.

The Dead White Zombies deliberately invoke the sacred without requiring religious participation. Our acts include symbolic elements such as fire, air, water, earth, and breath, which hold universal significance. We create peaceful environments amid urban noise, moments of calm within constant activity, and time for reflection in a hurried culture. These initiatives do not promote religious truth but provide experiential opportunities for individuals to step out time to explore facets of existence often obscured by the relentless pace of modern life.

Ancient Futures

We stand at a threshold. The anthropocentric worldview that has guided Western civilization for centuries has reached its limits—ecologically, spiritually, and socially. Climate catastrophe accelerates. Species vanish. Communities fragment. Digital technologies colonize attention while severing us from embodied presence and the more-than-human world. The old ways of being no longer serve; the new ways have not yet fully emerged. This is a time of axial change. A condition that demands a performance of transformation.

Throughout, this book argues that performance is an underused form of consciousness technology. Just as agriculture changes landscapes and engineering manipulates materials, performance structures consciousness, behavior, and thought. The mask ceremonies I observed in Burkina Faso were not mere aesthetic displays, but complex technologies, models, and repetitions designed to sustain a Cosmocentric worldview. Bushmen healing rituals served as corrective measures for both individual and collective consciousness. Miao rituals provided methods for managing crises and reestablishing community harmony. A film or TV show on your favorite streaming platform reveals and models its worldview. An anthropocentric, cause-and-effect, materialist perspective, prioritizing humans over nonhuman entities in the quest for social or cultural remediation.

All these performance practices employ similar mechanisms, different in kind but not in objectives and methods. Rhythm aligns neural activity and fosters collective coherence; movement bypasses cognitive defenses to tap into somatic wisdom; symbolic content triggers deep psychological structures; and collective participation creates shared experience fields that enhance individual transformation. What distinguishes the Cosmocentric from the anthropocentric is its intention, scope of awareness, and ethical responsibility to serve the totality of being.

The Work Ahead

Several imperatives emerge from our work.

First, we must continue learning from Indigenous communities whose place-based knowledge offers critical insights for sustainable existence.

Second, we must develop new performance methodologies suited to contemporary conditions.

Third, we must integrate performance-based approaches into domains beyond the arts: Education, urban planning, environmental restoration, corporate governance, community development, and daily life.

Fourth, embrace uncertainty. The technological singularity unfolds as part of planetary evolution. Digital entities join flesh, thought, and spirit in our community of existence. What emerges from this convergence exceeds our capacity to predict or control. And that is as it should be.

Remember. Place is not given. It is made and remade and must be maintained.

Our work is urgent, complex, and essential. It demands courage to venture into the uncharted and experiment with new forms. It requires humility and respect when learning from traditions not our own. It requires a commitment to practice even when results are not immediate. Our work asks us to release the need to control and the expectation of programmed outcomes. It requires that we abandon the compulsion to understand before feeling and the craving for certainty in an uncertain world. Our work invites us to find comfort in vulnerability, for such is the way to genuine transformation.

What I have received from the elders, practitioners, and communities who shared their knowledge with me was never mine to keep. I offer it to you now, not as doctrine but as an invitation, not as answers but as questions that require your actionable response. Take what serves. Adapt. Discard what does not fit. Create what is missing. Share what you discover. Do it all with compassion, generosity, and love. All you really have in life is what you can give.

CASE STUDIES

The case studies below offer a brief overview of three Dead White Zombies' works, highlighting their application of methods, processes, and themes from this book, including trance, archetypes, diverse explorations, and creative groups. Because of space limitations, these are short summaries.

A Case Study: ((w)hole)hole

My work with the Dead White Zombies (DWZ), a Dallas-based post-disciplinary performance group, was based on exploring how to create a cosmocentric performance within a socially and culturally diverse urban environment. The group's name highlighted the idea that white culture, or Western culture—which has influenced the world for the past 2000 years—remains like an undead zombie: Still active, feeding on the remnants of a past world and still walking around.

All DWZ performances were customized for their site-specific locations and designed as immersive sensory experiences. They functioned as formal experiments inspired by my cross-cultural fieldwork with indigenous communities and studies of ritual, shamanism, and mythology. The works revealed underlying structural patterns that, although expressed differently, preserved their core functional intentions.

DWZ projects were inspired, incorporated, modified, and expanded on Indigenous ritual patterns and practices to meet the needs of the Dallas community. I will begin by reflecting on *(W)hole*, a 2012 performance, to illustrate the DWZ approach, process, goals, and aesthetic.

DOI: 10.4324/9781003686422-5

(W)hole was inspired by a series of Dogon and Mossi rituals I witnessed in Burkina Faso. These rituals took place in a large, dusty field outside Ouagadougou as day shifted into night. Masked dancers, many wearing full-body masks made of grass, textiles, and fur—some towering up to ten feet tall—moved through the area, energizing the crowd. The rhythmic drumming, combined with swirling movements and clouds of dust, created a trance-like environment, transforming the scene into an altered reality. Sacred, ancestral, and cultural figures blended myth with everyday life.

One does not behave as a spirit or ancestor; they communicate through you. The audience was familiar with the performance's archetypal and powerful elements. There was no need for exposition, character, or relationship because the story was part of their cultural awareness.

The entire performance of (w)hole was held in a 32,000-square-foot former machine and metal fabrication shop, developed over a six-month process with performers and the building. It was the second installment in a trilogy of performances exploring the cycle of life—a universal theme linking all cultures and living beings: Death, the afterlife, and rebirth. These mysteries have been pondered by myth, ritual, religion, and culture since humanity's beginning. The building at 500 Singleton Boulevard, built in the 1920s and sporting a rusted, weathered appearance, served as an ideal setting to depict a post-industrial, post-capitalist afterlife. Its large space and high ceilings reflected the histories of former occupants and the community it supported. Like the building itself, the technology, objects, actions, and sounds played meaningful roles, subtly hinting at, revealing, and shaping the performance narrative. Each evening, 70 to 100 people gathered. Upon arriving, they met in a small former office where Jennifer Culver, a well-known local psychic, was present. She moved through the crowd, giving each person a small, palm-sized stone and identifying it as their soul. She instructed, "Hold this, keep it safe. When the time comes, you will know what to do." Then, the audience was led to an area where a woman was entangled in dozens of bungee cords, caught in a web of time and emotions. She struggled and bounced frantically while her lover watched helplessly. These characters represented ancient souls who had loved and lost, destined to search through time; this was the only exposition given.

The lover departed, and the audience intuitively followed to an open space. There were no further instructions. Soon, scenes appeared quietly in distant, isolated spots. Sounds and videos came to life, and characters

FIGURE 4.1 Stephanie Cleghorn and Ben Miro in the opening moment of *(W) hole*. Produced by the Dead White Zombies, Dallas, Texas. Written and directed by the author

Source: Photo by Alisa Eykilis.

began populating the capacious building. At first uncertain, audiences then moved to moments that resonated with them. The Sakha people of Siberia believe that we see what we need when we need it, and that chance is a way for spirits to communicate, a theme that became a central operational motif for the performance.

As the audience moved through the large metal building, scenes, sounds, and videos displayed mundane, dramatic, and fantastical moments. Ultimately, a complex, layered sequence of events unfolded, overlaying and connecting scenes that depicted life's moments, as performers and viewers drifted dream-like through a vision of the afterlife. The audience becomes part of the dream, their rhythms and movements shifting into a meditative, yet alert and curious state.

A blind gospel singer wandered and staggered, singing a never-ending dirge filled with loss and regret. Over time, 22 performers floated dream-like, drifting in and out of consciousness, depicting glimpses of two searching souls. They moved through different incarnations, playing out past and future lives, shifting from male to female, to trans, to animal, and to element with fluidity, each transformation revealing another lesson.

Moments were emotionally connected, igniting at near or distant distances, then vanishing or transforming. Examples of over 50 moments include a nine-year-old girl playing with dolls, a couple reenacting their births and deaths, a lover reading love letters, and the mundane act of hanging and folding clothes to create a tidy, past life. There is also an architect reviewing and updating plans, a distressed woman comforting a large rock as if it were a lost child, a woman sleepwalking and talking in a motel room, an lonely teenager endlessly playing cheesy 1970s love songs on an electric piano, and otherworldly beings resembling ancient gods witnessing and offering comfort. A living-room-like scene shows ghostly videos looping melancholic home movies of departed humans, animals, and the environment.

One section of the large building was designed as a maze where a performer searched for lost parts of their life; other areas featured gauze-like walls that revealed the hazy outlines of figures on the other side. Sounds seeping from one space to another created an aleatoric overlay of life. Walls slid open to reveal another threshold to cross. In one scene, a woman and a man lie on a bed of small rocks—the earth's elders—to be taught about the earth's history and the minerals it shares with humans. At one point, 30 minutes into the performance, a 20-foot-long wall opens, inviting audiences to go deeper into the afterlife.

Jennifer Culver, the psychic, sat at a large wooden, candle-lit table offering readings. Traveling farther, a large pile of stones sat at another threshold. To enter, one had to deposit the stone, the soul they were initially given, onto the pile. When dropped, each stone's sound reverberated throughout. After that, a series of rooms lay in shadow, then down a long, narrow corridor that opened onto a primordial time of conjuring and sacrifice. To one side, a forest of trees spoke through the poetry of a tree-like man. In this world, time and space were collapsed, with ancient events coexisting with the modern, all enmeshed in the endless cycles of life and death.

As the evening unfolded, the audience followed the characters with whom they felt the strongest affinity, talked with them, consoled them, and at times embraced them. Moments flared and dissolved, and nothing was explained. The sprawling space was set with multiple distinct playing areas where performers moved through, sat, slept, or performed. Composer and sound designer Scot Gresham-Lancaster wired the metal building's siding, downspouts, and glass, transforming the structure into an eerily distorted speaker that was emotional and breathing, with voices, songs, and rumbles, creating a trance-like atmosphere. The simultaneity, interaction, and details of events conveyed a meditation on death and the afterlife. Audiences said that the performance evoked both anxiety and excitement. Vividly realistic details contrasted with the layered complexity and metaphysical abstraction, producing a cognitive destabilization that encouraged audience participation to complete a narrative of their own making. The performance, like the world, is complex and multi-layered, and guided by chance and feeling.

The disruption of a linear narrative, combined with audience engagement, fluid complexity, and haptic stimuli, awakened, engaged, and heightened the experience for both audiences and performers. How the performance was created and rehearsed was as essential as the content—how you make a work determines the metaphysics of the work. What you put in, you get out. I wrote the text as a parallel-style flowchart, envisioning performers and events as recording tracks to be mixed into an orchestration of feelings.

The script was a guide played by performers empowered to discover and make personal connections. (W)hole was a living organism. One night, one scene might have a dozen audience members, and on another night, no one. With moments coinciding, it was not unusual for scenes to be performed without an audience on a given night. A phenomenon that performers said had liberated them, enabling them to find deeper meaning in their work. Who do you perform for, and why? Rituals are conducted for humans, but equally for ancestors, spirits, animals, and the natural world. They serve as a reflection of community and signify continuity.

The slow, meditative rhythm created a trance-like environment, causing the audience to gradually become introspective and at ease with their role as performers in the unfolding event. Some found the experience comforting, while others initially felt fear, which then transformed into child-like curiosity and engagement. One audience member shared, "I was watching the scene, then I realized other people were watching me watching as I was watching them. It was then I realized I was a performer and I am part of this story."

Because of its evolutionary nature, no two performances were alike. It was also impossible to experience the entire performance, which spanned vast distances and was presented simultaneously. When I revisited a scene I had not seen for several performances, I was amazed by how the performers had transformed the moment and their performance. Like a ritual, the work's form served as an entryway to emotional and spiritual discovery. It was durational, with an average runtime of two hours, but it often ran three or four hours, and once, five.

FIGURE 4.2 A wandering eternal figure, performed by Michael Cleveland, of the world between life and death. *(W)hole*, by the Dead White Zombies

Source: Photo by John Lehman.

A Case Study: kaRaoKe MoTeL

The final installment of DWZ's karmic life cycle trilogy, *kaRaoKe MoTeL*, marked the culmination of years of experimenting with immersive performance as a state of collective, transformative consciousness. Held in November 2014 at a former icehouse in West Dallas, the show delved into the concept of rebirth. The performance was a visceral, participatory journey that explored how humans might reinvent themselves within a simulation. As one reviewer observed, the production generated an "odd sense of well-being"—perhaps due to its departure from linear storytelling and emphasis on audience engagement. This sense of well-being was intentional and aligned with the work's philosophical goal: Demonstrating that performance, as a fundamental "consciousness technology," could function as a tool for existential renewal (Farrington 2014).

The performance began with a transition ritual. Audience members (referred to as "strangers") entered in small, staggered groups—sixteen at a time, every 12 minutes—a pattern that defied traditional theatrical expectations. These intervals were intentionally scheduled so different groups could influence, engage with, and even confuse one another's experiences. As one critic noted, "another layer is added to the performance, with audience members watching others watch the action, but at times I wasn't sure who the performer was, then it didn't matter" (Wilson 2014). This recursive observation exemplified a key cosmocentric tenet. That consciousness is relational, developed through networks of doing and witnessing.

FIGURE 4.3 Hixon (Abel Flores) and his alter ego, Toby, fielding and spouting ontological. *karaoke MoTel*, by the Dead White Zombies. Written and directed

Source: Photo by John Lehman.

What distinguished kaRaoKe MoTeL from mere theatrical experimentation was its dedication to ritual structure. While audiences were free to move during the middle part, the opening and closing were tightly controlled. This framing aligns with Victor Turner and Arnold Van Gennep's description of the pre-liminal (preparatory), liminal (transitional), and post-liminal (integrative) stages. In ritual, the liminal is often associated with a sacred time or space—a finite period separated from everyday life that links the physical with the spiritual. In this space, the apparent chaos and variety of the middle section became less confusing and more liberating. As a reviewer observed, "somewhere along the way, we relax into the experience... there grew a sense of shared energy and communal bonding, however temporary" (Farrington 2014).

After passing through a narrowing tunnel, the audience entered what seemed to be a motel office. Inside, two performers engaged in different activities: One listened to a tape about suicide, while the other typed on an 1980s era computer and called audience members' cell phones to inquire about their reservations. This space acted as an entrance into the liminal zone between the afterlife and rebirth. Their sudden recognition of the audience's presence, combined with a series of dramatic death-and-rebirth sequences, served as a somatic induction. The audience was not just observing death and rebirth—an experience that is simultaneously ironic, comic, and unsettling—but was being initiated into a performative state where such transformations could occur. This threshold moment established the core language of the entire piece: That rebirth is not merely a metaphor but a lived, enacted, and complex reality, realized through the body's passage a symbolic space.

Once "processed," audiences confronted what might be called the performance's central ontological claim: That meaning making is always a choice, always a navigation through multiplicity. The production featured several distinct narrative threads co-occurring across different rooms within the maze-like 20,000-square-foot building. Next was the motel lobby, with fresh coffee brewing, a popcorn machine popping, and a black-and-white television playing a remix of 1950s science-education films on childbirth, hygiene, and sexually transmitted diseases. A housemaid's vacuum cleaner disrupted the moment before she launched into haranguing those that just entered,

> Don't come here draining our energy. Instead, come to give. Be open, expand, you let the atmosphere flow into you. You no fool us. In darkness, we shine like light. We exist before you, be here after you. You—your smile is fake, that cloud over your head. You drink wine, beer, you lose yourself, but we see you. We play along, that's what we do. We know you pretend—in your eyes are nothing. When you leave, you take something—what will you leave behind?

In the next room were two lovers, almost naked, alternatively making love, quarreling, talking philosophy, as pornography plays on a television. Projected onto the lovers and the wall behind them were images of idealized romance and travel mixed with trance-inducing abstractions. The room was shin-deep in clothing, skins shed from lifetimes that the audience must wade through.

In the adjacent room, a ventriloquist dummy sits on the lap of a cross-dressed man who talks about choices, identity, and being a dummy. Inside the room, there is a shelf filled with bones, desiccated animals, broken parts, and calendars. The housemaid bursts in to clean the room, shrieks about the never-ending filth, then softly sings about the fate of the world's (potted) plant life. In the adjacent room, the motel's cocktail lounge patrons enjoy wine and beer, surrounded by vibrant, modern artwork resembling a cave-painting narrative. A lounge singer croons classic songs, accompanied by a piano player, then engages in existential repartee.

The most notable feature of kaRaoKe MoTeL was its role as a divinatory event—using theatrical elements to create multiple and personalized divinations for each participant. During the performance, audience members encountered prophecies, fortune-telling, and aphorisms from the Sphinx, a drag queen, a ventriloquist dummy, a housemaid, lovers, lounge singers, a prostitute, a suicidal man, a house detective, and the motel managers. All of it to prepare the audience members for their entrance into the next life.

In the sand-floored basement, a fortune-teller, played by psychic Culver, sat at a candlelit table, giving readings, and was popular with audiences. In the adjacent room was a photo studio. It was a party-like scene, where the audience could try on masks, play out fantasy scenes, and were encouraged to perform imaginary and projected selves, all of which were documented, uploaded, and available online the next day.

Deeper into the basement was the dark, damp subconscious, with shelves filled with droning sound and detritus of every sort, including books, mementos, and boxes holding photos, yellowed documents, and records of lives lived and forgotten. Images, home movies, and sounds of life echoed, flickered, and faded.

Elsewhere, in the motel lobby, the "house detective," performed by the imposing Brad Hennigan, asks questions of the audience about a crime not yet committed, then rewards the flummoxed interviewee with a fortune cookie. One reviewer received a cookie that read, "Nothing seems impossible to you."

Comic-tragic communal bonding was crucial rather than incidental for comprehending rebirth. From a cosmocentric viewpoint, transformation is not solely personal—it happens within and through community. The production reflected this idea by making audience members witnesses to each other's journeys. Your experience in the space was personal yet continually intersected with others' paths, always observed and influenced by fellow "strangers" moving toward becoming somebodies.

The final scenes, post-liminal integration, gathered the entire audience in a large metal-sided vat where water was formerly poured and ice formed, creating a pivotal collective moment. An elevated platform surrounded a vat, where the motel managers sat in a control/DJ booth and other performers stood, transformed in their ritual regalia, including masks and elaborate headpieces, looking down at those gathered.

The discomfort some audiences experienced from being confined in a metal-sided vat was intentionally. The vat symbolized a metaphorical womb. The sensation of being trapped in the final scene—wishing to escape but remaining within confines—mirrors the existential condition of anyone seeking conscious change. We long for release, seek closure, but rebirth demands that we engage with life and its difficulties. We remain within a transformative space until the process progresses naturally, governed by its own logic rather than ours.

During this shared moment, mytho-sacred performers guided, sang, lectured, harangued, provided examples, and entertained those about to be reborn, preparing and soothing them for what lay ahead. Glaring video lights and cameras moved through the crowd, creatures of deception and disruption emerged, and collective dances of celebration took place before strangers were named and moved toward the exit stairs, entering a video projection of an escalator and then out into the world.

The kaRaoKe MoTeL was a self-guided tour where metaphors and meanings collected along the way. You heard and saw what you needed, when you needed. Cledonomancy is an ancient divination method that interprets random words, phrases, or conversations as signs or messages from a higher power. It suggests that ordinary words—especially overheard ones—may reveal hidden truths and insights. This principle—that performance creates personalized meaning within a collective—marks a shift from traditional theater, which expects uniform audience interpretation. kaRaoKe MoTeL proposes that meaning arises from the combination of the psychophysical, personal, and symbolic encounters. This mirrors traditional divination that uncovers what is already with grammar and around you in the world.

Total immersion encompasses what Pallasmaa refers to as peripheral perception: "Peripheral vision connects us to the environment, whereas focused vision separates us, turning us into passive spectators" (13). Western theatre, which centers on ocular engagement, encourages audiences to focus, track the action, and interpret cues. In contrast, immersive performances invite participants to relax their focus, letting the experience flow over and through them. This transition from active observation to peripheral engagement represents a core ontological shift.

Just as karaoke enables individuals to take on another's song and explore different identities, the performance presented multiple selves, stories, and meanings. The emphasis was not on authenticity but on how these aspects enabled and were essential for personal rebirth.

kaRaoKe moTeL, the final part of a trilogy that explores death (Fleshworld), the afterlife ((w)hole), and rebirth, suggests that rebirth happens daily as a vital aspect of life, not just occasionally. It encourages us to consider the values, concerns, goals, and awareness we carry with us versus what we leave behind. Embracing continuous change, making choices, and accepting chance are vital to staying alert and connected to the world.

The critical innovation of kaRaoKe MoTeL was not the existence of simultaneous narrative lines—*Sleep No More* and other immersive works had established this convention—but rather the production's insistence that these narratives required the audience to engage and complete them personally. As one reviewer noted, audiences gradually realized "we don't have just to stand or sit here and take it. We're free to move, to grab the controls and maneuver between one scene and the next, doubling back to see 'same but different' versions as many times as we want and need" (Farrington 2014). This freedom was not simply about spatial mobility but about epistemological agency. Each audience member was being asked to become the author of their own experience of rebirth, to choose what fragments of meaning to carry forward into their "next life."

The installation of closed-circuit television within the space introduced additional complexity. Monitors displayed activities in other rooms, creating an experience where "audience members watching other audience members watching the action became 'vicarious performers' themselves" (Wilson 2014). This technological setup did not separate audiences from the performance. It heightened their awareness, leading to a form of metaconsciousness. In this state, one could act as both participant, observer, and the observed at the same time. Culturally, this could be likened to creating multiple dreaming bodies—distinct modes of attention functioning simultaneously.

The title kaRaoKe MoTeL was intentionally paradoxical, referencing both the cheap imitation of karaoke culture and the fleeting anonymity of roadside motels. As one reviewer explained, the production questioned: "Karaoke culture, or the remixing and copying of what already exists, is a search for the orderly in a disorderly world" (2014). However, the production rejected any simple critique of simulation or inauthenticity. Instead, it proposed that in an age where everything is already a copy, where cultural forms are endlessly remixed and recombined, the question is not how to revert to some imagined original but how to inhabit the copy consciously and intentionally, with what might be called awakened simulation.

During high school and my undergraduate years, I worked as a maintenance person at two different motels, absorbing their atmospheres and modes of operation. The kaRaoKe MoTeL was a neglected establishment on the fringe of awareness, serving as a fitting metaphor for our cheapening karaoke culture. Motels are intentionally anonymous spaces where people temporarily shed their identities, becoming someone else or no one at all. In a way, they function as laboratories for rebirth, designed to shed oneself and adopt another.

By staging the performance in an abandoned building rather than a theater, the production rooted this metaphorical significance in physical reality. The building itself became a kind of found text. Its history as an icehouse built in 1910—a place for preserving things and preventing decay—adds another layer of meaning to its new role as a space for exploring how selves might be maintained (on ice), transformed (ice into the water flow of life), or abandoned like the building itself.

A Case Study: Holy Bone

Holy Bone was an immersive performance event held across seven venues in Dallas. The concept for the Holy Bone project by DWZ was born in 2016 during ethnographic research with the Miao community in the Wuling Mountains of southwest China. It was inspired by a story of a pregnant woman from a village who, having faced difficulties with her previous pregnancy, sought guidance from a local Shi Lau Man, a xianniang.

Through the medium, the woman's late grandmother explained that the rock pathway leading to her grave was obstructed, preventing her "holy bone" from communicating with future generations. When I inquired about the meaning of the holy bone, I was told it represents a woman giving birth, making the pathway to her grave a kind of mnemonic. The pregnant woman then cleared and arranged a rock pathway to her grandmother's grave. A month later, she successfully gave birth to a healthy boy.

While with the Miao, I was by chance reading Alan Watts' book on Christian mythology. In it I came across the term "holy bone," which refers to the sacrum. Watts notes that this name originated with the ancient Roman physician Claudius Galen (130–200 CE), who regarded it as the most prominent bone in the spine. The Latin name for it, *os sacrum*, translates to "sacred bone." Ancient Greeks and other cultures believed the human soul dwelt at the base of the spine. Early Christians thought the body would reassemble around the sacrum during resurrection.

In Indian spiritual traditions, the sacrum is regarded as the seat of kundalini energy. The word kundalini comes from the Sanskrit "kundal," meaning "coiled," and symbolizes the dormant, primal energy at the base of the spine, often depicted as a coiled snake. Kundalini is associated with the Divine Mother (Shakti or earth). A key aim in yoga is to uncoil or activate this feminine energy, guiding it upward through the central channel (Sushumna) to unite with the masculine divine energy represented by Shiva, the universe, or the Divine Father. The sacrum, consisting of five fused vertebrae, supports the human body and is built to withstand significant physical stress, which helped early humans walk upright. Our Holy Bone performance was designed as a journey to five different sites, each symbolizing a vertebra in the release of female energy. This act was meant to be an initiatory path, awakening the self and the feminine.

Holy Bone was created as a walking meditation and immersive performance to revive our sensory and emotional awareness. It asserts that we are fleeting beings moving through a complex, at times overwhelming, and ultimately unfathomable reality. The performance implicitly suggests that we survive the modern world by distancing ourselves from our feelings, surroundings, and responsibilities. The more we do so, the more we lose awareness and, in turn, neglect, exploit, and damage our environment. Like life itself, the performance offers an experiential journey through a modern urban landscape layered with complexity, conflict, and exhaustion. The initial step is awareness of the vitality of one's environment, which in turn sparks self-awareness and renewal.

Holy Bone embraces and exemplifies the hallucinated text. Audiences, referred to as "Initiates," begin their journey at Tacos Mariachi, a lively Tex-Mex restaurant on Singleton Boulevard in the post-industrial area of West Dallas. A well-dressed professional woman and man in their early 30s arrive on one night, scan the room, and experience surprise, curiosity, bemusement, and alertness, followed by anxiety when no guidance appears. People are eating tacos and drinking margaritas. They purchased tickets online and walked into a taco restaurant expecting a performance. This disruption, and others that occurred throughout the performance, were deliberately staged to shake initiates out of their normative perceptions.

During check-in, they would receive identification buttons and be introduced to others in their group. Every 10 minutes, 6 people were admitted to the performance, which started at 7:30 and ended at 10:00, with an average of 70 initiates nightly. Disrupting normal expectations was key to the performance, which balanced seriousness and structure with playfulness and freedom. The six initiates gathered closely and received a preamble from me: "You are on a journey that will take you through encounters of discovery, visiting various buildings, experiences, and people.

Your experience will last between 1.5 and 3 hours. Stay put and reflect on what moves you." Some performances extended until 1 AM, with some initiates taking up to four hours to complete their journey—they needed to stay longer.

Each initiate was taught an "entrance gesture" passcode inspired by indigenous ideograms, necessary for entering the first building. Passcodes changed every night. One required the placing of hands over your eyes, then extending your fingers outward with a wiggle, as if sending your hands and energy into the world. And as they were about to leave the taco restaurant, an address was whispered: "611 Singleton. Down and across the street on the right, knock three times. Present the gesture to the doorkeeper. No gesture, no entrance. Good luck and see you on the other side."

Leaving Tacos Mariachi, groups of six begin walking down traffic-heavy Singleton Boulevard, passing by a hair salon, boarded-up houses, an upholstery shop, and a second-hand furniture store until they reach 611, a worn, graffiti-covered building. After knocking, a window cover is lifted, revealing Tawnie Harris, a registered nurse by day, who looks out at the six initiates signaling their entrance. The door opens into a dark, run-down entranceway that once housed a heating and air-conditioning business. This station, called "Questions," takes inspiration from Oracle books from the 15th to 18th centuries. Twelve questions are fixed for the entire performance, while twelve others change weekly and are tied to the moon phases. The chosen question reflects their intent and what they seek from their immersive journey.

Kapsay: The Map

Next, Jennifer Culver, our resident psychic and Wiccan, performing as Kapsay, the diviner, guiding the "initiates." They enter her dimly lit room, with battered paneling and hanging ceiling tiles, and sit at a low table across from Culver. On the table, a cloth map displays ideograms indicating each Holy Bone site. Culver is recognized as *Kapsay*, a name derived from an Anglicized greeting used by the Sakha people I worked with in central Siberia. Instead of "Hello" or "How are you?" they greet each other with "Tell me your story." For the Sakha, the world is interconnected and sustained through stories. Once shared, stories remain with someone who in turn are changed and pass them on to others.

Culver would choose animal bones suited to the individual, such as horse, squirrel, and coyote, to answer questions about their overall path, desires, and purpose. This reading prepares initiates for their performance journey by helping them recognize it as a personal epic, as much internal as external, framing the performance as metaphysical or spiritual journey.

How you enter shapes the impression, setting the tone, guiding focus, and shifting perception. Meaning arises not from fixed text but from uncertainty and a waking up existing between material and immaterial—the hallucination. The taco restaurant, passcodes, street walk—engaging strangers, collective action, questions, intentions, bone casting, and journey mapping all worked to create essential disruptions. Provoking questions, suggest a unity of opposites, and act as prelude and preparation, teaching initiates how to reimagine the world.

All performance stations demonstrated a consistent and adaptable interaction driven by specific intentions and behaviors. Audience members were considered initiates, and the performance was an initiation into the Cosmocentric worldview.

Performers transform into what Lehmann describes as "time sculptures"—not just interpreting the text but embodying it as "a complex whole of associative spaces" (Lehmann 110, 156). They hold the performance within their bodies as living archives, each offering a distinct version that overlaps with but never entirely matches others.

Cycles of Life

From the calm introspection of Kapsay's room, Holy Bone initiates moved through thick velour curtains into an area filled with video projections, monitors, and sound. Using several old television sets placed on the floor and three wall projections, the otherwise dark and empty room became a dreamlike corridor of flickering light. The workaday fatigue of the former repair shop added to the atmosphere. Each video thematically explored aspects of life, death, creation, and destruction. Video and sound artist and longtime DWZ collaborators Mona Kasra and Scot Gresham-Lancaster created a rhythmically coordinated network of videos and soundscapes. Depicted were a kindergarten graduation, an autopsy, the planting of seeds, and a cremation. With no place to sit, moving silhouettes of other initiates floated like ideograms in the flickering light, implying those who had previously passed through. It was a passageway, a reminder of the cycles of life and death and the nature of all things.

Tan: Anxiety

At the end of the Cycle of Life video installation, was an illuminated white door marked with the ideogram of Tan, performed by Hilly Holsonback. As visitors entered Tan's long, narrow room, they faced a loud, continuous Shepard's Tone. It is a sound always rising without resolution, evoking anxiety and mirroring the Bipolar and ADHD states prevalent today.

The room became a stage where Holsonback would fixate on the initiate, guiding them to sit on a sofa in a carpeted area. The space was otherwise empty. Tan's behavior was erratic, haranguing, cajoling, and sharing her anxieties. Her name, Tan, reflected how her intense life's radiance burned her. In the Holy Bone initiatory process, Tan aimed to awaken initiates and release their stress, akin to draining pus from an infection.

Holsonback, a well-respected Dallas-based visual and performance artist and member of DWZ, infused her character with artistic passion and personal insights. As a long-time sufferer of panic and anxiety attacks, her character development combined personal experiences with performance, an approach that blurs the real and performed and that informs all DWZ work. Her performance served as an analogy for our internet-driven culture, shifting from outraged and sincere to funny, entertaining, and vulnerable then veering into emotional moments that engaged the audience. She often switched suddenly, multitasked, and appeared confused, with thoughts spiraling, accompanied by anxiety, insecurity, and occasional panic attacks, which she projected onto the audience. Dressed in a white wig tied with a red bow, with a fair, white-powdered complexion, a red dress, and white leggings, Holsonback's character rapidly shifted from childlike to sensually alluring, clownish, to pathetic. Her unsettling persona embodied many modern American societal issues. Sitting before her vanity mirror, she would alternate between admiring herself, primping, and criticizing her appearance, calling herself ugly, too fat, or poorly dressed. She flirted, pleaded, charmed, and suddenly, in a forceful outburst, chased the initiates out of her cave-like space and through a steel door into a garden-like area behind the building.

Kei: Stone

Chased through a steel door by Tan's manic fury, initiates stumbled into a rough-hewn outdoor garden installation—a jarring tonal inversion. Hannah Weir's Kei inhabited this threshold space with slow, graceful movements and a low, intimate whisper. After Tan's psychophysical assault, the contrast was palpable, even medicinal. At the center of the space stood a totem-tree, its branches inverted to suggest roots—the world turned upside down. Plastic bottles filled with colored liquids dangled on strings; found objects swayed. It was a re-imagined shaman's spirit altar I had documented among the Sakha of central Siberia, repurposed as an urban altar and spiritual antenna, mediating between earth and sky.

Kei spoke in fragments that hovered between prophecy and reverie. Her character—developed as a poetic, savant-like figure in a constant state of remembering—gave each initiate a polished stone. As she placed it into each person's hand, she whispered, "Area 51," the address of the next station. The stone was a seed of sympathetic magic, variously interpreted as a soul, a self, a miniature earth, or currency for the journey ahead.

Agent Zay: Tests

Initiates crossed two acres of uneven scrubland, down a tree-shrouded alley, through an open metal door tagged "A51," into the dim wreckage of a former auto body shop. Motion sensor lights traced a pathway through the detritus. Many described the sensation as equal parts fear and exhilaration—precisely the psychophysical state the journey was designed to provoke.

In the far corner, under a single downlight, sat Agent Zay—Wes Ferguson in a dark gray suit, clipboard in hand, a pharmacist's scale on the table before him. One initiate at a time crossed the darkness. Zay combed his hair, brushed lint from his lapel, and interrogated each with a rapid bureaucratic precision: Name, arrival time, reasons, and possessions. When the polished stone appeared, he examined it under a magnifier, listened to it through a stethoscope, weighed it, and then directed the initiate to locate the center of their being. "Have you ever given any consideration to the center of yourself?" Those who passed received a slip of paper stamped with Zay's ideogram: 500 Singleton. No directions were given. "If it is meant to be, you will find your way."

The genius of Zay was in the absurdity of institutional authority mapped onto metaphysical inquiry. He made compliance feel uncanny, turning the mundane protocols of bureaucracy into a ritual of self-examination.

Cadmus: Remediation

Those deemed inauthentic by Agent Zay were siphoned off to a side room for remediation—a cracked wooden desk, a DVD player, folding chairs. A five-minute video loop played, in which I appeared as Cadmus wearing a CPR dummy mask, issuing instructions for an increasingly surreal test involving a photograph of a suburban living room. Initiates were asked to number elements according to logic only Cadmus understood: "Put a number five on what you think is not real. Put a number eight on the pathway to the beyond." Remarkably, most complied with total seriousness.

The deliberate absurdity served a purpose. Separating initiates from their companions disrupted habitual group dynamics and the social cocoon that insulates us from experience. Isolation, disorientation, and the demand for compliance are ancient initiatory tools. Cadmus had arrived from ancient myth to administer a test for the modern world.

Que and Bung: Confrontation

At 500 Singleton—a 36,000-square-foot former ironworks—initiates entered a waiting room overlooking a vast dark expanse, battery-operated lights illuminating a pathway to a distant desk. One by one, "Next" was called, and they crossed the open concrete floor, footfalls echoing, to sit before Que, Stephen Gardner, chain-smoking under an up-light, clipboard in hand. Above and behind him moved Bung, played by Becki McDonald, a former dominatrix turned junior high teacher, interjecting commentary from the shadows: "Okay honey, that's good." "Why are you saying that?" "Bullshit."

Que's questioning was drawn from Rogerian psychotherapy, a patient-mirroring, deceptively simple technique. The session moved initiates from polished surface to exposed core in minutes: "Do you feel you can do something you want? Were you afraid? And if not?" For those Bung found evasive, she escorted them up a ramp to a 3,000-square-foot painting room with red oxide residue on every surface and with two school chairs facing each other ten feet apart, Bung backlit in red, casting enormous shadows. Her classroom was terrifying and tender in equal measure. Most left with the sense that something true had been spoken in that room, and that they had, finally, agreed to hear it.

Rem: Maps

From the intensity of Que and Bung, initiates traveled to Rem, Charles Radcliff, in all white, open-armed, welcoming. "Welcome. I know your journey has been a little intense, but here you are free." On the floor of a 100-foot-diameter space were painted all the ideograms from the journey, as presented by Kapsay's map. Rivers, mountains, anthropomorphic creatures, the complete cosmography of the Holy Bone rendered in visual language. Initiates could spend as much time as they wished at this station, walking the ideograms, sitting on benches, chatting with others at the periphery, or climbing the A-frame ladder for an overview.

Rem asked each person to trace with their body a sequence of ideograms representing their life. The stone was placed into an illuminated bowl of water at a small altar, a baptismal gesture, a symbolic crossing over. "This part of your journey is now complete. You are ready." Through the darkness beyond was a maze of white-painted rooms, former offices, with a single number painted on a wall: 421.

Zun: Elixir

Down a deserted street to a beaten green metal door to 421 Bedford Street, the workshop of Zun. Abel Flores, longtime DWZ member and amateur brewmaster, offered each initiate a glass of fresh homebrew as he moved in and out of shadows, a divining staff festooned with talismans. Colored lights swirled on the ceiling, cicadas revolved in layers of sound, and Zun's associative monologue—part shaman, part lunatic prophet—drew initiates into a circle and communion. "Life is like writing on silent paper. And the hallucinations listen". At the end, dayglo numbers were revealed: 516. Zun sent them into the street with a riddle for a map: *A left, a right. Because all that is left is right.*

Manta: Song

At 516 Fabrication Street, sound arrived before vision—the layered, lilting vocals of Manta, performed by Moni Bell, local musician and visual artist. She sang from behind a veil-like material, performing live against prerecorded layers of her own voice reciting Karen Mizner's poem *Got Ojas*. Initiates settled into sofas and chairs across from her, letting her music wash over them. Manta was the first station character not to interact directly—she simply offered. The green light surrounding her was thematic: Green, the color of life, of the earth returning. From the accumulated friction and disclosure of the journey, this was the first space of simple reception. *Holy is the tree, the constant dance of leaves. Splinters of the self like a divided river that loves itself.*

A path of old floor lamps—grandmotherly, ornate, cloth-shaded—wound through a corridor toward the next threshold. The lamps' domesticity was deliberate: A sense of coming home.

Vig: Listening

Bailee Rayle as Vig—bare-chested in a sport coat—gathered initiates in a hallway before bringing them into a dimly lit room where Ty Borstein sat at a mixing board flanked by vintage oscilloscopes, their green sine waves pulsing with the music. Rayle, a trained opera singer, guided a sonic meditation: Sit on the floor, close your eyes, and listen to yourself and the world. The base was a 423 hz sine wave, tuned to harmonize with the brain. Overlays created an emotional journey through sound alone.

Then, Vig opened a terrarium of worms. He reached in with bare hands, lifting a fistful of moist dark earth alive with movement, and the sound of worms—recorded with highly sensitive microphones—filled the room. "These worms are in concert with one another. They can't see; they can only feel and hear. You are this earth. You are these worms." The instruction was Cosmocentric in its most literal form: The boundary between self and world is porous, permeable, and ultimately fictional.

Sa'bu: Transference

A short walk led to a 20-by-20-foot room packed with sound-damping insulation marked with black-line drawings. Sa'bu, played by Adan Khan—bearded, dressed in white—welcomed initiates and began a Hindu prayer cycle, asking everyone to harmonize. Then the lights shifted to ultraviolet, and the room exploded. Every wall, Sa'bu's face and arms, the floor, all dayglo-painted, revealed the ideograms, spirals, animal forms, infinity loops, and the full vocabulary of the Holy Bone. Initiates found themselves inside the symbolic universe of everything they had traveled through. Sa'bu moved through the circle with a Tesla ball that responded to each person differently, changing color and intensity at proximity. It was a ritual of transference—drawing the accumulated energy of the journey into the body, asking initiates to close their eyes and hold it there.

Urd: Acceptance

The final station was a high, narrow garage—floor covered with burlap, incense in the air, chakra-alignment sound threading through the space. Urd, played by Alexandra Werle—six feet tall, dressed in flowing white, sometimes veiled entirely—moved in silence at half speed, tending three altars, lighting candles, carrying clay pots of herbs and flowers. She never spoke.

At the center was a long, low table of rough-hewn timber, surrounded by dozens of pillows. Many initiates stayed for 20–30 minutes, sitting, lying down, meditating, waiting for companions who had been separated hours earlier along the labyrinth. On the table lay glass beads and lengths of sari cloth which initiates could weave into a growing collective construction. Over the month-long run, the table weave grew from a simple outline into an intricate, communal map of interconnection, each addition a small conscious act of belonging to something larger than oneself. Initiates placed their polished stones into the weave or left them on the table. No one needed to be told what to do.

John: Photos

When Urd felt the moment was right, she silently guided each initiate through a back room where John Lehman, a retired forensic psychologist and longtime DWZ photographer, had set up a portrait studio. Each was photographed. The images, offered free of charge, captured something difficult to name: What a person looks like after traveling through themselves. Overwhelmingly, initiates expressed joy.

Holy Bone was an initiatory arc—structured, progressive, and irreversible. Each station was a vertebra in the release of the kundalini: Disruption, divination, purgation, testing, confrontation, reflection, communion, immersion, transference, and finally, acceptance. The journey moved from the social surfaces of a taco restaurant to the deep interior of a collective ritual, navigating urban wasteland as mythic terrain. The street walk was the labyrinth; the performers were the threshold guardians; the polished stone was the self, carried, examined, weighed, cleansed, and ultimately surrendered.

What the Holy Bone demonstrated is that the ancient Cosmocentric technologies of initiation survive in any landscape willing to hold them. The sacred bone is not behind us. It is at the base of the spine, coiled, waiting.

FIGURE 4.4 Tan (Hilly Holsonback) is in her environment of anxiety and panic at an early station in *Holy Bone*

FIGURE 4.5 Vig (Stephan Gardner) interrogates each initiate for authenticity in a later station in *Holy Bone*. Produced by the Dead White Zombies, written and directed by the author

Source: Photos by the author.

LIST OF EXPLORATIONS

Body

- Standing Relaxed
- The Mountain
- Internal Dialogue
- World Awareness
- Creative Circle
- Listening

Walking

- Movement In Space
- Walking Attitudes
- Walking Light
- Walking Tense
- Leading Walks
- Smiling

Exploration

- Isolating Body Parts
- Exploring Body Parts
- Body Part Scenario
- Body Part Sounding
- Separating Sound
- Body Part Dance
- Body Part Drama

DOI: 10.4324/9781003686422-6

Rhythm

- Alternate Nostril
- Diaphragm Breathing (Part I and II)
- Heartbeats
- Individual & Group Meditation
- Body Part Breathing
- White Light
- Heartbeat Movement
- Meltdown
- Creation & Meltdown
- Breathing a Lie

Names

- Names

Sticks

- Sticks
- Music Physicality
- Number Mix-Up
- Rhythmic Movement
- Story Dancing
- Movement To Music
- Moving With Others
- Movement with Other Beings

Making Sounds

- Touching Sound
- Body of Sound
- Mantra
- Babble
- Sound of Other Beings

Balance

- Stand-Up
- Balance and Roll Around
- Balance & Center Awareness
- Stand Off
- One Leg Balance
- Floating

- Alternative Moves
- Move and Center

Moving

- Make A Move
- Making My Move
- Movement Flow

Body and Spaces

- Crossing the Floor
- Climbing The Mountain
- Logrolling
- Big Foot
- Grids
- Poles
- Falling Down (Parts I–III)

Focus and Mindfulness

- Walking Meditation
- The Journey (Parts I–IV)

Play

- Tag
- Alternative Tags
- Amoeba Tag
- Go-Tag
- Fireman Lift Relay
- Scorpion

Interstitials

- 'X'ercise
- Capturing Space
- Giving & Taking $
- Disorientation Spins

Psychophysical

- Free Contact
- What Is Going On?
- Whole Body Communication

Trust

- Group Lift
- Flying
- Disembodied Head

Aggression

- Finger Follow (Parts I–III)
- Resistance
- Pig In the Middle
- Pull Apart
- Circle Pulls
- Life Boat
- Master-Slave

Performing

- Being Watched
- Patterns
- Compositions (Parts I–II)
- Push, Pull, Spin (Parts I–II)
- Three Movements (Parts I–II)

Space

Cycles

- Circle Definitions
- Whydidyadothat?
- Talking Circle
- Circle Sound Game
- Endings and Beginnings
- Circle Statements
- Small Stories
- Never Stop
- Counting to 20

Pride and Prejudice

- Nice or Nasty
- The Gauntlet
- Victim Circle

Rhythm

- Rhythm Sequence
- Rhythm Dialog
- Rhythm Expression
- Rhythm Chant
- Cultural Rhythm
- Rhythmic Movement
- Physical Language
- Rhythm Memory (Phases I–IV)
- Shaman Rhythm

Sound Creating Song

- Vocal Seasons
- Sound Circle
- Bursting with Song
- Sound Conducting
- Sound Guidance

Balance

- Balancing With Others
- Balanced Space
- Center of the Circle
- Column of Light

Observation

- Performer as participant-observer
- Interview
- Observation
- Drawing Faces
- Hugs
- Eye-To-Eye Contact
- The Person Opposite
- Guidance
- Mirror Sequence (Phases I–VIII)
- Touch
- Consciousness Touch (Parts I—II)
- Shadows and Humans

Gifting and Exchange

- Gifting
- Skills Exchange

Exploring

- Blind Movement
- Psychic Space
- Out of Body
- Spaces Within
- Creating Spaces
- Movie Posters
- Photos
- Defining Space

Object

- Imaginary Object
- Object Imagination
- Objects (Parts I–IV)

Character

- Gesture Exploration
- Characters From Life

Sharing

- Monster Within
- The Situation
- Group Situation
- Questions
- Questions and Answers
- One Word (Parts I–III)
- Three Lines

Performance

- Narrative Spaces
- Group Story

Expansion

- Radiance
- Color Spaces
- Group Color Spaces
- Aura Space
- A Tree

Archetypes

- Innocent
- Explorer
- Sage
- Hero
- Outlaw
- Magician
- Average Guy or Gal
- The Lover
- The Jester
- The Caregiver
- The Creator
- The Ruler

Masks

- Self-Mask

Body as Script

- Hieroglyphics
- Personal Vocabulary
- Scenario Creation
- Dance Creation
- Scenario Applications

Place

Warm-Up and Preparation

- Commedia Dell'Arte Warm-Up
- Playful Warm-Up
- Mime Warm-Up
- Reaching Warm-Up
- Element Warm-Up
- Imagination Warm-Up

- Place Making Sounds
- Three Traditional Zulu Movements

Place Making

- Sticks
- Environmental Rhythms
- Living Practice

Creating Performance

- Ideas (Parts I–IV)
- Create a Moment
- Storytelling
- I Am (Parts I–II)
- Chaos Play (Parts I–II)
- Necessary Dialogue
- Mythic Self

Interviewing

- Creative Groups
- Group Interview
- Peculiar Things

Trance

- Movement Into Trance
- Trance Dance
- Trance Sharing

Character

- Animal Character
- Gesture Exploration
- Living the Character

Rehearsals

- Emotional Journey
- Stepping Out
- Hieroglyphics, Too
- Progress Report
- Run Through

WORKS CITED

Abram, David. *The Spell of the Sensuous: Perception and Language in a More-Than-Human World*. New York: Pantheon Books, 1996.

Alexander, Eevlyn. Interview with the author. 23 October 1997, Minto, Alaska.

Anderson, Benedict. *Imagined Communities: Reflections on the Origin and Spread of Nationalism*. London: Verso, 1983.

Attali, Jacques. *Noise: The Political Economy of Music*. Translated by Brian Massumi. Minneapolis: University of Minnesota Press, 1985.

Attali, Jacques. *The Labyrinth in Culture and Society: Pathway to Wisdom*. Berkeley: North Atlantic Books, 1999.

Augé, Marc. *Non-Places: Introduction to an Anthropology of Supermodernity*. London: Verso, 1995.

Bachelard, Gaston. *The Poetics of Space*. Boston: Beacon Press, 1994.

Bak, Per, Chao Tang, and Kurt Wiesenfeld. "Self-Organized Criticality: An Explanation of the 1/f Noise." *Physical Review Letters* 59, no. 4 (1987): 381–384.

Bartra, Roger. *Shamans and Robots: On Ritual, the Placebo Effect, and Artificial Consciousness*. Translated by Gusti Gould. Minneapolis: University of Minnesota Press, 2024.

Basso, Keith H. *Wisdom Sits in Places: Landscape and Language Among the Western Apache*. Albuquerque: University of New Mexico Press, 1996.

Behar, Ruth. *The Vulnerable Observer: Anthropology That Breaks Your Heart*. Boston: Beacon Press, 1996.

Bellman, Beryl L. *The Language of Secrecy: Symbols and Metaphors in Poro Ritual*. New Brunswick: Rutgers University Press, 1984.

Benjamin, Walter. "The Task of the Translator." In *Illuminations*, 69–82. New York: Schocken Books, 1968.

Benson, Herbert. *The Relaxation Response*. New York: William Morrow, 1975.

Beuys, Joseph. "I Am Searching for a Field Character." In *Art into Society, Society into Art*, edited by Caroline Tisdall. London: ICA, 1974.

Beuys, Joseph. *What is Art? Interviews with Joseph Beuys*. Edited by Volker Harlan. Forest Row: Clairview Books, 2004.

Bey, Hakim. *TAZ: Temporary Autonomous Zone*. Brooklyn: Autonomedia, 2003.
Bhabha, Homi K. *The Location of Culture*. London: Routledge, 1994.
Boden, Margaret A. *The Creative Mind: Myths and Mechanisms*. New York: Basic Books, 1994.
Borgmann, Albert. *Technology and the Character of Contemporary Life: A Philosophical Inquiry*. Chicago: University of Chicago Press, 1984.
Bourdieu, Pierre. *Outline of a Theory of Practice*. Cambridge: Cambridge University Press, 1977.
Bridle, James. *Ways of Being: Animals, Plants, Machines: The Search for a Planetary Intelligence*. New York: Farrar, Straus and Giroux, 2022.
Brown, Joseph E. *The Sacred Pipe: Black Elk's Account of the Seven Rites of the Oglala Sioux*. Norman: University of Oklahoma Press, 1953.
Brown, Steven. "The 'Musilanguage' Model of Music Evolution." In *The Origins of Music*, edited by Nils L. Wallin, Björn Merker, and Steven Brown, 271–300. Cambridge: MIT Press, 2000.
Bruner, Jerome S. "The Conditions of Creativity." In *Contemporary Approaches to Creative Thinking*, edited by Howard Gruber, Glenn Terrell, and Michael Wertheimer, 1–30. New York: Atherton Press, 1962.
Bruner, Jerome S. *Actual Minds, Possible Worlds*. Cambridge: Harvard University Press, 1986.
Burkert, Walter. *Homo Necans: The Anthropology of Ancient Greek Sacrificial Ritual and Myth*. Translated by Peter Bing. Berkeley: University of California Press, 1983.
Burkert, Walter. *Creation of the Sacred: Tracks of Biology in Early Religions*. Cambridge, MA: Harvard University Press, 1996.
Burl, Aubrey. *The Stone Circles of the British Isles*. New Haven: Yale University Press, 1976.
Cage, John. *Silence: Lectures and Writings*. Middletown: Wesleyan University Press, 1961.
Campbell, Joseph. *The Hero With a Thousand Faces*. Princeton: Princeton University Press, 1949.
Capra, Fritjof. *The Web of Life: A New Scientific Understanding of Living Systems*. New York: Anchor Books, 1996.
Carney, Dana R., Amy J. C. Cuddy, and Andy J. Yap. "Power Posing: Brief Nonverbal Displays Affect Neuroendocrine Levels and Risk Tolerance." *Psychological Science* 21, no. 10 (2010): 1363–1368.
Casey, Edward. *Getting Back into Place: Toward a Renewed Understanding of the Place-World*. Bloomington: Indiana University Press, 1993.
Casey, Edward. "How to Get from Space to Place in a Fairly Short Stretch of Time: Phenomenological Prolegomena." In *Senses of Place*, edited by Steven Feld and Keith Basso, 13–52. Santa Fe: School of American Research Press, 1996.
Castells, Manuel. *The Rise of the Network Society*. Oxford: Blackwell, 1996.
Chatwin, Bruce. *The Songlines*. New York: Viking Penguin, 1987.
Chrostowski, Waldemar. "Spiritual Colonialism and Indigenous Resistance." *Journal of Contemporary Religion* 12, no. 2 (1997): 135–148.
Clark, Andy. *Supersizing the Mind: Embodiment, Action, and Cognitive Extension*. Oxford: Oxford University Press, 2008.
Clark, Andy, and David Chalmers. "The Extended Mind." *Analysis* 58, no. 1 (1998): 7–19.
Clifford, James. *The Predicament of Culture: Twentieth-Century Ethnography, Literature, and Art*. Cambridge: Harvard University Press, 1988.

Clifford, James. *Routes: Travel and Translation in the Late Twentieth Century*. Cambridge: Harvard University Press, 1997.

Conquergood, Dwight. "Performance Studies: Interventions and Radical Research." *TDR: The Drama Review* 46, no. 2 (2002): 145–156.

Crary, Jonathan. *Scorched Earth: Beyond the Digital Age to a Post-Capitalist World*. London: Verso, 2022.

Cruikshank, Julie. *Do Glaciers Listen? Local Knowledge, Colonial Encounters, and Social Imagination*. Vancouver: University of British Columbia Press, 2005.

Csikszentmihalyi, Mihaly. *Creativity: Flow and the Psychology of Discovery and Invention*. New York: Harper Collins, 1996.

d'Aquili, Eugene, and Charles D. Laughlin Jr. "The Neurobiology of Myth and Ritual." In *Readings in Ritual Studies*, edited by Ronald L. Grimes, 132–150. Upper Saddle River: Prentice Hall, 1996.

Damasio, Antonio. *Descartes' Error: Emotion, Reason, and the Human Brain*. New York: G.P. Putnam Sons, 1994.

Damasio, Antonio. *The Feeling of What Happens: Body and Emotion in the Making of Consciousness*. San Diego: Harcourt Brace, 1999.

Diamond, Adele. "Executive Functions." *Annual Review of Psychology* 64, no. 1 (2013): 135–168.

Dietrich, Arne. "Functional Neuroanatomy of Altered States of Consciousness: The Transient Hypofrontality Hypothesis." *Consciousness and Cognition* 12, no. 2 (2003): 231–256.

Dietrich, Arne. "The Cognitive Neuroscience of Creativity." *Psychonomic Bulletin & Review* 11, no. 6 (2004): 1011–1026.

Donald, Merlin. *Origins of the Modern Mind: Three Stages in the Evolution of Culture and Cognition*. Cambridge: Harvard University Press, 1991.

Douglas, Mary. *Purity and Danger: An Analysis of Concepts of Pollution and Taboo*. London: Routledge, 2002.

Dunbar, Robin I.M. "On the Evolutionary Function of Song and Dance." In *Music, Language, and Human Evolution*, edited by Nicholas Bannan, 201–214. Oxford: Oxford University Press, 2012.

Eco, Umberto. *The Open Work*. Translated by Anna Cancogni. Cambridge: Harvard University Press, 1989.

Eliade, Mircea. *Cosmos and History: The Myth of the Eternal Return*. Translated by Willard R. Trask. New York: Harper and Row, 1959.

Eliade, Mircea. *Shamanism: Archaic Techniques of Ecstasy*. Translated by Willard R. Trask. Princeton: Princeton University Press, 1964.

Farrington, Jan. "Inn and Out." *TheaterJones*, 24 November 2014.

Feld, Steven. *Sound and Sentiment: Birds, Weeping, Poetics, and Song in Kaluli Expression*. Philadelphia: University of Pennsylvania Press, 1982.

Feldenkrais, Moshe. *Awareness Through Movement*. New York: Harper & Row, 1972.

Fienup-Riordan, Ann. *Boundaries and Passages: Rule and Ritual in Yup'ik Eskimo Oral Tradition*. Norman: University of Oklahoma Press, 1994.

Fienup-Riordan, Ann. *The Living Tradition of Yup'ik Masks: Agayuliyararput (Our Way of Making Prayer)*. Seattle: University of Washington Press, 1996.

Foster, Russell, and Leon Kreitzman. *Rhythms of Life: The Biological Clocks That Control the Daily Lives of Every Living Thing*. New Haven: Yale University Press, 2004.

Frazer, James G. *The Golden Bough: A Study in Magic and Religion*. Oxford: Oxford University Press, 1998.

Freire, Paulo. *Pedagogy of the Oppressed*. New York: Continuum, 1970.
Fuchs, Elinor. *The Death of Character: Perspectives on Theater After Modernism*. Bloomington: Indiana University Press, 1996.
Gaster, Theodor H. *Thespis: Ritual, Myth, and Drama in the Ancient Near East*. New York: Gordian Press, 1950.
Gibson, James J. *The Ecological Approach to Visual Perception*. Boston: Houghton Mifflin, 1979.
Girard, René. *Things Hidden Since the Foundation of the World*. Stanford: Stanford University Press, 1987.
Gleick, James. *Chaos: Making a New Science*. New York: Viking Penguin, 1987.
Goldbard, Arlene. *New Creative Community: The Art of Cultural Development*. Oakland: New Village Press, 2006.
Goodman, Felicitas D. *Where the Spirits Ride the Wind: Trance Journeys and Other Ecstatic Experiences*. Bloomington: Indiana University Press, 1990.
Gottschall, Jonathan. *The Storytelling Animal: How Stories Make Us Human*. Boston: Houghton Mifflin Harcourt, 2012.
Gray, Peter. *Free to Learn: Why Unleashing the Instinct to Play Will Make Our Children Happier, More Self-Reliant, and Better Students for Life*. New York: Basic Books, 2013.
Grotowski, Jerzy. *Towards a Poor Theatre*. Edited by Eugenio Barba. New York: Simon and Schuster, 1968.
Hanna, Thomas. *Somatics: Reawakening the Mind's Control of Movement, Flexibility, and Health*. Reading: Addison-Wesley, 1988.
Haraway, Donna J. *The Companion Species Manifesto: Dogs, People, and Significant Otherness*. Chicago: Prickly Paradigm Press, 2003.
Harvey, Graham. *Animism: Respecting the Living World*. New York: Columbia University Press, 2006.
Heidegger, Martin. "The Origin of the Work of Art." In *Basic Writings*, edited by David Farrell Krell, 139–265. London: Routledge, 1993.
High, Steven. *Oral History at the Crossroads: Sharing Life Stories of Survival and Displacement*. Vancouver: University of British Columbia Press, 2013.
Huizinga, Johan. *Homo Ludens: A Study of the Play-Element in Culture*. Boston: Beacon Press, 1955.
Hull, Arthur. *Drum Circle Spirit: Facilitating Human Potential Through Rhythm*. Tempe: White Cliffs Media, 1998.
Husserl, Edmund. *On the Phenomenology of the Consciousness of Internal Time*. Translated by James B. Brough. Dordrecht: Kluwer Academic Publishers, 1991.
Hyde, Lewis. *The Gift: Imagination and the Erotic Life of Property*. New York: Vintage Books, 1983.
Hyde, Lewis. *Trickster Makes This World: Mischief, Myth, and Art*. New York: Farrar, Straus and Giroux, 1998.
Iacoboni, Marco. *Mirroring People: The New Science of How We Connect with Others*. New York: Farrar, Straus and Giroux, 2008.
Ingold, Tim. *The Perception of the Environment: Essays on Livelihood, Dwelling and Skill*. London: Routledge, 2000.
Jaques, Susan. "Dance as Embodied Social Action." In *Postmodern Culture and Dance*, 123–144. Pittsburgh: University of Pittsburgh Press, 1991.
Johnstone, Keith. *Impro: Improvisation and the Theatre*. London: Faber & Faber, 1979.
Jung, Carl G. *The Archetypes and the Collective Unconscious*. Translated by R.F.C. Hull. 2nd ed. Princeton: Princeton University Press, 1998.

Katz, Richard. *Boiling Energy: Community Healing Among the Kalahari Kung*. Cambridge: Harvard University Press, 1982.
Kauffman, Stuart A. *The Origins of Order: Self-Organization and Selection in Evolution*. Oxford: Oxford University Press, 1993.
Kawagley, Angayuqaq O. *A Yupiaq Worldview: A Pathway to Ecology and Spirit*. Prospect Heights: Waveland Press, 1995.
Kaye, Nick. *Site-Specific Art: Performance, Place and Documentation*. London: Routledge, 2000.
Keali'inohomoku, Joann. "An Anthropologist Looks at Ballet as a Form of Ethnic Dance." In *What Is Dance?*, 533–549. Oxford: Oxford University Press, 1980.
Kimmerer, Robin Wall. *Braiding Sweetgrass: Indigenous Wisdom, Scientific Knowledge and the Teachings of Plants*. Minneapolis: Milkweed Editions, 2013.
King, Thomas. *The Truth About Stories: A Native Narrative*. Toronto: House of Anansi Press, 2003.
Kleinman, Arthur. *The Illness Narratives: Suffering, Healing, and the Human Condition*. New York: Basic Books, 1988.
Kohn, Eduardo. *How Forests Think: Toward an Anthropology Beyond the Human*. Berkeley: University of California Press, 2013.
Ladefoged, Peter. *Vowels and Consonants: An Introduction to the Sounds of Languages*. Oxford: Blackwell, 2001.
Lakoff, George, and Mark Johnson. *The Embodied Mind: The Bodily Basis of Meaning, Imagination, and Reason*. New York: Basic Books, 1999.
Leder, Drew. *The Absent Body*. Chicago: University of Chicago Press, 1990.
Lee, Meewon. "Kamunguk: The Mask-Dance Theatre of Korea." PhD diss., University of Pittsburgh, 1984.
Lefebvre, Henri. *Rhythmanalysis: Space, Time, and Everyday Life*. New York: Continuum, 2004.
Le Guin, Ursula K. "We Live in Capitalism. Its Power Seems Inescapable. So Did the Divine Right of Kings." Speech at National Book Awards, 2014.
Lehmann, Hans-Thies. *Postdramatic Theatre*. Translated by Karen Jürs-Munby. London: Routledge, 2006.
Linklater, Kristin. *Freeing the Natural Voice*. New York: Drama Book Publishers, 1976.
Lomax, Alan. *Folk Song Style and Culture*. New Brunswick: Transaction Books, 1978.
Lovelock, James. *Gaia: A New Look at Life on Earth*. Oxford: Oxford University Press, 1979.
Lowenstein, Tom. *Ancient Land: Sacred Whale: The Inuit Hunt and Its Rituals*. New York: North Point Press, 1993.
Mâche, François-Bernard. *Un demi-siècle de musique... et toujours contemporaine*. Paris: L'Harmattan, 2000.
Madson, Patricia R. *Improv Wisdom: Don't Prepare, Just Show Up*. New York: Bell Tower, 2005.
Malinowski, Bronislaw. *Argonauts of the Western Pacific*. London: Routledge, 1922.
Malkki, H. Liisa. *Purity and Exile: Violence, Memory, and National Cosmology Among Hutu Refugees in Tanzania*. Chicago: University of Chicago Press, 1995.
Mark, Margaret, and Carol S. Pearson. *The Hero and the Outlaw: Building Extraordinary Brands Through the Power of Archetypes*. New York: McGraw-Hill, 2001.
Massey, Doreen. "A Global Sense of Place." *Marxism Today* 38 (1991): 24–29.
Mauss, Marcel. *The Gift: Forms and Functions of Exchange in Archaic Societies*. Translated by Ian Cunnison. New York: W.W. Norton & Company, 1967.
McLean, Mervyn. *Maori Music*. Auckland: Auckland University Press, 1996.

Merchant, Carolyn. *Reinventing Eden: The Fate of Nature in Western Culture*. New York: Routledge, 2003.
Merleau-Ponty, Maurice. *Phenomenology of Perception*. Translated by Colin Smith. London: Routledge, 2012. Originally published 1945.
Merzenich, Michael. *Soft-Wired: How the New Science of Brain Plasticity Can Change Your Life*. San Francisco: Parnassus Publishing, 2013.
Miller, Richard. *The Structure of Singing: System and Art in Vocal Technique*. New York: Schirmer Books, 1986.
Munn, Nancy D. "The Cultural Anthropology of Time: A Critical Essay." *Annual Review of Anthropology* 21 (1992): 93–123.
Myers, Thomas. *Anatomy Trains: Myofascial Meridians for Manual and Movement Therapists*. Edinburgh: Churchill Livingstone, 2001.
Napoleon, Harold, Eric Christopher Madsen, and Alaska Native Knowledge Network. *Yuuyaraq: The Way of the Human Being: With Commentary*. Fairbanks: Alaska Native Knowledge Network, University of Alaska, Fairbanks, 1996.
Nelson, Edward W. *The Eskimo About Bering Strait*. Bureau of American Ethnology Annual Report 18. Washington: Smithsonian Institution, 1899.
Nelson, Richard K. *Hunters of the Northern Ice*. Chicago: University of Chicago Press, 1969.
Nelson, Richard K. *Personal Conversation*. 25 October Fairbanks: Alaska, 1990.
Oliveros, Pauline. *Deep Listening: A Composer's Sound Practice*. Lincoln: iUniverse, 2005.
Ong, Walter J. *Orality and Literacy: The Technologizing of the Word*. London: Methuen, 1982.
Osiński, Zbigniew. *Grotowski and His Laboratory*. Translated by Lillian Vallee and Robert Finlay. New York: PAJ Publications, 1986.
Pallasmaa, Juhani. *The Eyes of the Skin: Architecture and the Senses*. Chichester: John Wiley & Sons, 2005.
Palmer, Parker J. *The Courage to Teach: Exploring the Inner Landscape of a Teacher's Life*. San Francisco: Jossey-Bass, 1998.
Panksepp, Jaak. *Affective Neuroscience: The Foundations of Human and Animal Emotions*. Oxford: Oxford University Press, 1998.
Peek, Philip M., ed. *African Divination Systems: Ways of Knowing*. Bloomington: Indiana University Press, 1991.
Plsek, Paul E., and Trisha Greenhalgh. "Complexity Science: The Challenge of Complexity in Health Care." *BMJ* 323, no. 7313 (2001): 625–628.
Poincaré, Henri. "Sur le problème des trois corps et les équations de la dynamique." *Acta Mathematica* 13 (1890): 1–270.
Porges, Stephen W. *The Polyvagal Theory: Neurophysiological Foundations of Emotions, Attachment, Communication, and Self-Regulation*. New York: Norton, 2011.
Portelli, Alessandro. *The Death of Luigi Trastulli and Other Stories: Form and Meaning in Oral History*. Albany: SUNY Press, 1991.
Postman, Neil. *Technopoly: The Surrender of Culture to Technology*. New York: Knopf, 1992.
Putnam, Robert D. *Bowling Alone: The Collapse and Revival of American Community*. New York: Simon & Schuster, 2000.
Riccio, Thomas. "A Message from Eagle Mother: The Messenger's Feast of the Inupiat Eskimo." *TDR: The Drama Review* 37, no. 1 (1993): 115–146. https://doi.org/10.2307/1146274
Riccio, Thomas. "Unpublished Fieldwork Notes and Interviews With! Xuu and Khwe Bushmen Healers." Kalahari Desert (1994).

Riccio, Thomas. "N!ngongaio: People Come Out of Here Making a New Story with the! Xuu and Khwe Bushmen." *TheatreForum* 10 (1997): 45.
Riccio, Thomas. "Dead White Zombies, Dallas, TX." *TheatreForum* 8 (2014) 66–80.
Riccio, Thomas. "Holy Bone: Coming into Knowing." Unpublished manuscript, 2017.
Riccio, Thomas. "Huan Nuoyuan: Exorcism and Transformation in Miao Ritual Drama." *TDR : Drama Review* 63, no. 2 (2019): 78–101. https://doi.org/10.1162/dram_a_00836
Riccio, Thomas. Wiesna Mond-Kozlowska "Rhythm Reality." In *Rhythm Steps of Africa*, 125–142. Krakow: Jagiellonian University Press, 2010.
Riccio, Thomas. *Sophia Robot, Post-Human Being*. London: Routledge, 2024.
Riccio, Thomas. "Shadows in the Sun: Context, Process, and Performance in Ethiopia." *New Theatre Quarterly* 28, no. 3 (2012): 272–295. https://doi.org/10.1017/S0266464X12000450
Rosaldo, Renato. *Culture and Truth: The Remaking of Social Analysis*. Boston: Beacon Press, 1989.
Rosaldo, Renato. "Cultural Citizenship and Educational Democracy." *Cultural Anthropology* 9, no. 3 (1994): 402–411.
Rose, Deborah B. *Nourishing Terrains: Australian Aboriginal Views of Landscape and Wilderness*. Canberra: Australian Heritage Commission, 1996.
Schechner, Richard. *Between Theater and Anthropology*. Philadelphia: University of Pennsylvania Press, 1985.
Schechner, Richard. *Performance Studies: An Introduction*. London: Routledge, 2002.
Schipper, Kristofer. *The Taoist Body*. Translated by Karen C. Duval. Berkeley: University of California Press, 1993.
Schmidtbleicher, Dietmar. "Training for Power Events." In *Strength and Power in Sport*, edited by Paavo V. Komi, 381–395. Oxford: Blackwell Scientific Publications, 1992.
Schön, Donald A. *The Reflective Practitioner: How Professionals Think in Action*. New York: Basic Books, 1983.
Schutz, Alfred. *The Phenomenology of the Social World*. Evanston: Northwestern University Press, 1967.
Scott, James C. *Seeing Like a State: How Certain Schemes to Improve the Human Condition Have Failed*. New Haven: Yale University Press, 1998.
Senge, Peter M. *The Fifth Discipline: The Art and Practice of the Learning Organization*. New York: Doubleday, 1990.
Shepard, Paul. *Coming Home to the Pleistocene*. Washington, DC: Island Press, 1998.
Shi, Jinhe. Personal communication. Translated by Peng Jingquan. 6 January 2001, Ban Li Village, Xiangxi Tujia-Miao Autonomous Prefecture, China.
Shusterman, Richard. *Body Consciousness: A Philosophy of Mindfulness and Somaesthetics*. Cambridge: Cambridge University Press, 2008.
Simpson, Leanne B. *As We Have Always Done: Indigenous Freedom Through Radical Resistance*. Minneapolis: University of Minnesota Press, 2017.
Smart, Jennifer. "The Futility of Existence." Texas Arts+Culture, 26 November 2014.
Smart, Lauren. "Dead White Zombies' Immersive Theater is Reborn with Karaoke Motel." *Dallas Observer*, 25 November 2014.
Smith, Linda T. *Decolonizing Methodologies: Research and Indigenous Peoples*. London: Zed Books, 1999.
Sobel, David. *Place-Based Education: Connecting Classrooms & Communities*. Great Barrington: Orion Society, 2004.

Somé, Malidoma P. *Ritual: Power, Healing and Community*. New York: Penguin, 1993.
Somé, Malidoma P. *Of Water and the Spirit: Ritual, Magic, and Initiation in the Life of an African Shaman*. New York: Tarcher/Putnam, 1994.
Spencer, Robert F. *The North Alaskan Eskimo: A Study in Ecology and Society*. Bureau of American Ethnology Bulletin 171. Washington: Smithsonian Institution, 1959.
Spivak, Gayatri C. "Can the Subaltern Speak?" In *Marxism and the Interpretation of Culture*, edited by Cary Nelson and Lawrence Grossberg, 271–313. Urbana: University of Illinois Press, 1988.
Spolin, Viola. *Improvisation for the Theater*. Evanston: Northwestern University Press, 1963.
Spradley, James P. *Participant Observation*. New York: Holt, Rinehart and Winston, 1980.
Stacey, Ralph D. *Complex Responsive Processes in Organizations: Learning and Knowledge Creation*. London: Routledge, 2001.
Stanislavski, Constantin. *An Actor Prepares*. Translated by Elizabeth Reynolds Hapgood. New York: Theatre Arts Books, 1936.
Steyerl, Hito. *Duty Free Art: Art in the Age of Planetary Civil War*. London: Verso, 2017.
Sundberg, Johan. *The Science of the Singing Voice*. DeKalb: Northern Illinois University Press, 1987.
Suzuki, Shunryu. *Zen Mind, Beginner's Mind*. New York: Weatherhill, 1970.
Tarnas, Richard. *The Passion of the Western Mind*. London: Pimlico, 1991.
Taussig, Michael. *Mimesis and Alterity: A Particular History of the Senses*. New York: Routledge, 1993.
Taylor, Diana. *The Archive and the Repertoire: Performing Cultural Memory in the Americas*. Durham: Duke University Press, 2003.
Truax, Barry. *Acoustic Communication*. 2nd ed. Westport: Ablex Publishing, 2001.
Tsing, Anna L. *The Mushroom at the End of the World: On the Possibility of Life in Capitalist Ruins*. Princeton: Princeton University Press, 2015.
Tsing, Anna L. *Arts of Living on a Damaged Planet: Ghosts and Monsters of the Anthropocene*. Minneapolis: University of Minnesota Press, 2017.
Tuan, Yi-Fu. *Topophilia: A Study of Environmental Perception, Attitudes, and Values*. Englewood Cliffs: Prentice-Hall, 1974.
Tuan, Yi-Fu. *Space and Place: The Perspective of Experience*. Minneapolis: University of Minnesota Press, 1977.
Tuck, Eve, and K. Wayne Yang. "Decolonization is Not a Metaphor." *Decolonization* 1, no. 1 (2012): 1–40.
Tulving, Endel. "Episodic Memory: From Mind to Brain." *Annual Review of Psychology* 53, no. 1 (2002): 1–25.
Turner, Edith. *Among the Healers: Stories of Spiritual and Ritual Healing Around the World*. Westport: Praeger, 2006.
Turner, Victor. *The Ritual Process: Structure and Anti-Structure*. Chicago: Aldine de Gruyter, 1969.
Turner, Victor. *Dramas, Fields, and Metaphors: Symbolic Action in Human Society*. Ithaca: Cornell University Press, 1974.
Urry, John. *Mobilities*. Cambridge: Polity Press, 2007.
Valentin. Interview with the author. 28 July 2018, São Paulo, Brazil.
Varela, Francisco J., Evan Thompson, and Eleanor Rosch. *The Embodied Mind: Cognitive Science and Human Experience*. Cambridge: MIT Press, 1991.

Venuti, Lawrence. *The Translator's Invisibility: A History of Translation*. London: Routledge, 1995.
Vygotsky, Lev S. *Mind in Society: The Development of Higher Psychological Processes*. Cambridge: Harvard University Press, 1978.
Watts, Duncan J, and Steven H. Strogatz. "Collective Dynamics of 'Small-World' Networks." *Nature* 393, no. 6684 (1998): 440–442.
Wheatley, Margaret J. *Leadership and the New Science: Learning About Organization from an Orderly Universe*. San Francisco: Berrett-Koehler Publishers, 1992.
White, Gareth. "On Immersive Theatre." *Theatre Research International* 37, no. 3 (2012): 221–235.
Whitehouse, Harvey. *The Ritual Animal: Imitation and Cohesion in the Evolution of Social Complexity*. Oxford: Oxford University Press, 2021.
Willerslev, Rane. *Soul Hunters: Hunting, Animism, and Personhood among the Siberian Yukaghirs*. Berkeley: University of California Press, 2007.
Wilson, Lindsey. "The Rebirth of Dead White Zombies." *Texas Arts+Culture*, 1 November 2014.
Wilson, Shawn. *Research Is Ceremony: Indigenous Research Methods*. Halifax: Fernwood Publishing, 2008.
Wilson, Sule G. *The Drummer's Path: Moving the Spirit with Ritual and Traditional Drumming*. Rochester: Destiny Books, 1992.
Yunkaporta, Tyson. *Sand Talk: How Indigenous Thinking Can Save the World*. New York: HarperOne, 2020.

INDEX

Note: *Italicized* pages refer to figures.

Aboriginal Australian songlines 43–44
Abram, D. 5
Abramović, M. 226
acoustic ecology 338
Addis Ababa University 112
agayuliyararput 78, 301
aggression 138–143; circle pulls 141; finger follow 139; life boat 141–142; master-slave 143; pig in the middle 140; pull apart 140–141; resistance 139–140; Zambia journal 142
Alaska 1, 2, 9, 19, 68, 69, 71, 79, 87, 184, 213, 236, 298, 303, 306, 316, 319, 324, 325, 339–341, 344, 345, 349–350, 353, 374, 388, 391, 397
Alexander, E. 213
Alexander, J. 128
alternate breathing 55
alternate nostril 54
alternative moves 96–97
amoeba tag 126
Andegna ("the first") 375
animal character 405
animism/animist worldview 78
anthropocene 2, 88, 410
anthropocentric 3–4, 6, 9, 20–21, 79, 154, 198, 418; egotism 43; performance 9; systems 80
antigravity responses 335
Apple 157
Arat Kilo Theatre 384
archetypal: ensembles 289–294, *294*; spaces 277–278
archetype(s) 275–289; average guy/gal 284–285; the caregiver 287; the creator 287–288; elemental 278–289; exploration 65, 66; explorer 280–281; as foundation of cultural identity 276; hero 282–283; innocent 279–280; the jester 286; the lover 285–286; magician 283–284; outlaw 283; the ruler 288–289; sage 281–282
arctic hysteria 354
Aristotle 87, 366
arrogance space 154, 155, 198
artificial intelligence (AI) 2, 5–6, 157
Attali, J. 324, 373–374
Aufhebung 370
Augé, M. 154; *see also* non-place
aura space 273–274, *274*
awakenings 392–394, *394*

babble 84–85
balance 86–98; alternative moves 96–97; body and 88; body's inner landscape 203–204; and centered

awareness 91–92; cycles and 201–202; floating *95*, *96*; move and center 97–98, *98*; one leg 94; as practice 206–207; process 204–211; rhythm and 202–203; and roll around *90*, 90–91; sound 200–204
balanced space 208
balancing with others 207
bar mitzvahs 43
Bartra, R. 15
Beck, R. 5, 354
"Behavioral Sink" theory 246
being watched 144–145, *145*
Bell, M. 438
Benson, H. 331
Berhanu, L. 265–266
Berlin, N. 301
Beuys, J. 328–329
Bey, H. 162
big-foot 116–117
Black Lives Matter movement 321
blind movement 239
BLOB 106–107
Boal, A. 6, 133, 244, 259, 392
body 13; presence (performance); somatic practice; balance and 88; natural forces and 44; shapes 105–106; *see also* embodiment
body as a performance instrument 8, 23, 152, 340
body as script 301–313; Elemental Forms 303–305, *305*, 307–310, *310*; hieroglyphics 310–311; personal vocabulary 311–313
body-environment relationship 20–21
body parts: breathing 57–58; drama 215, *216*; exercise alternative 47; exploring 46–47; isolating 46–47; scenario 47–49
body part scenario 47–49; dance 50–51; drama 51; separating sound 50; sounding 47–49
Bogart, A. 117, 246
breath 204
breathing dramaturgy 372
Breuer, L. 366, 367, 369
Bridle, J. 11, 203
bringing place to the body 339–346; the architecture 343; somatic archaeology 344–346
Brook, P. 167
Brown, S. 338
Brown, W. 346
Burkert, W. 6–7
Burkina Faso journal 38

Cage, J. 327
Calhoun, J. B. 246
California Institute of Integral Studies (CIIS) 32
Campbell, J. 366, 367
capitalism 317, 326
capturing space 130
cardinal vowel production 338
case studies: Dead White Zombies (DWZ) 420–441; (W)hole 420–425, *422*, *425*; Holy Bone 431–441, *440*, *441*; kaRaoKe MoTeL 425–431, *426*
centered awareness: and balance 91–92
ceremony 347, 366, 378–379
chaos play 364
character 252–255; animal 405; gesture exploration 252–253, 406–407; from life 254–255; living the 408–409; multispecies 403–404; personality and 401; place 400–409; planetary crises and 403–404; postdramatic 402–403; and situation 258–259; as technology 401–402; towards Cosmocentric performance 404–405; warm-up 407–408
checking in space 198–200
China (Guizhou Province) 2, 190, 306, 325, 347, 384, 391, 431
choreographic rhythms 61
circadian rhythm 53
circle 21–25, 86; center of the 209; definitions 110, 159–160; engage in any activity and 161–162; functions 23; fundamental significance 21; the gauntlet 170–172; inherent fragility 23; liminal space and 24; in mathematics 22; modern civilization and 21–22; nice/nasty 169–170; represents 22; as a sacred space 24; sound 194; sound game 164; statements 165; symbolizes 21; talking 164; as the term 21; victim 177
circle-cycle principles 23, 24
circle definitions 205

circle pulls 141, 177
Clark, A. 332
climate crisis 157
climate destabilization 79
color spaces 271–272; group 273
column of light 209–210
Commedia Dell'arte warm-up 333
communities of place 4–5
Community Health Awareness Puppets (CHAPS) 267–269
community of place 78–81
competitive plasticity 335
compositions 146–147
consciousness technology 401, 418, 425
consciousness touch 229–230, *231*
consumerism 157
Cosmocentric 3–4, 11–12, 21, 78–81, 111, 199, 202, 314, 317, 318, 345–346, 354, 355, 401, 434
Cosmocentric performance 4, 7–9, 370–371, 401
Cosmocentric practice 11–12
Cosmocentric way, the 7–9, 11, 12, 111, 157, 204, 320, 372
Cosmocentric worldview 21
cosmological practice: narrative structure 372; performance as 372–373; rehearsal process 372; spatial dynamics 372; temporal experience: 372–373
COVID-19 pandemic 157
creating performance 358–365; chaos play 364; creating a moment 361–362; directing 358–359; ideas 360–361; necessary dialogue 365; onwards 359; storytelling 362
Creative Circle 25, 163, *163*
Creative Groups 382–383
cross-cultural performance work 102
crossing the floor 113; Siberian journal 114–116
cultural displacement 88
cultural narratives 182
cultural rhythm 185–186
cultural Zombification 3
cultures: cultural rhythm 185–186; physical language 188; rhythm chant 184–185; rhythm expression 183, *184*; rhythmic movement 186–187; rhythm memory 189–190; shaman rhythm 190–191, *191*; from Siberian journal 187; space 182–191
Culver, J. 433
cycles: balance and 201–202; space and 158–163; Stockholm journal 160–161

Dakis, J. 289
Damasio, A. 332
Dead White Zombies (DWZ) 3, 198–199, 246, 277, 289, 290, 299, 305, 325, 345, 351, 364, 382, 384, 386, 391, 393, 413, 416, 418; case studies 420–441; DP92 246, *246*, 364; (W)hole 420–425, *422*, *425*; Holy Bone 404, 431–441, *440*, *441*; immersive work and 416; industrial space as performance venue 421, 432; kaRaoKe MoTeL 425–431, *426*; methodology 416; performances 325, 345; post-disciplinary performance collective 3; site-specific performance 198, *387*; T.N. B. 386, *387*; trance 393
death 400–401
The Death of Character: Perspectives on Theater After Modernism (Fuchs) 400
decentralizing 80
deep listening 10
deep structural patterns 335
diaphragm breathing 54–55
Dietrich, A. 390–391
Diné (Navajo) Kinaaldá ceremony 43
DirectorsLab Chicago journal 65, *66*
disorientation spins 130–131
Dissociative Identity Disorder (DID) 289–290
Drama Therapy 32
Dylan, B. 295
dynamic enactment of place 318

Eagle-Wolf Dance 301
Eco, U. 374–375
edification by puzzlement and cryptic potency 358
elemental archetypes 278–289
Elemental Forms 303–305, *305*; body as script 303–305, *305*; sample 307–310, *310*
element warm-up 336

Eliade, M. 60, 276, 372
Eliade,. M. 366
Emandulo, Kwasa Group 27, *29*
embodiment /embodied knowledge 4, 16–17
emergent consciousness: markers of 10–11
empty space 154–155
environmental rhythms, ritual preparation 353–354
environment theater 353–354
Estonian National Theatre School 33
Ethiopian journal: hallucination 375–376; narrative spaces 265–266; shadows and humans 232–235, *235*
exocerebrum 15
expansion: radiance 270–271; space 269–275
exploration: blind movement 239; creating spaces 242–243; defining space 246–247, *247*; gesture 252–253; monster within 256–257; movie posters 243–244; out of body 240–241; photos 244–245; space 238–246; spaces within 241–242
explorations 7, 35, 110–120; big-foot 116–117; body phase 124; climbing the mountain 114; crossing the floor 113; falling down 119–120; grids 117; log-rolling 114; poles 117–118
exploring 238–246
extra sight 358
eye-to-eye contact 218–219

Facebook 157
falling down 119–120
Feldenkrais, M. 335
Feld, S. 338
the feminine 371–372
Festival of Native Arts (Alaska) *178*
finger follow 139
fireman lift relay 127–128
first communions 43
FITMO festival (Ouagadougou) 38–39
floating 95, *96*
Flores, A. 438
the fly 108–109
flying 135–136, *136*
focus and mindfulness 120–125; the journey 122–123; listening to feelings 121; Siberian journal 123–124; walking meditation 121
Form Fatigue 3
Frazer, J. G. 221
Freud, S. 275
Fuchs, E. 400

Gaffney, M. 339
Gaia hypothesis 11
Galen, C. 431
Gardner, S. 437
Gaster, T. H. 366
the gauntlet 170–172; Makanda Mahlanu 172, *173*; Tuma Theatre journal 173–174
gentle movement patterns 335
gesture exploration 252–253, 406–407
gifting and exchange 235–238
Girard, R. 221
Girma, K. 375
giving and talking $ 130
globalization 10, 157, 276, 317, 320, 326
The Golden Bough (Frazer) 221
Goodman, F. 392
Google 157
go-tag 127
Got Ojas 438
Gray, P. 332
the great transition 10–11
Greenland Inuit at Tukak' Teatret 68
Greenspan, J. *see* Alexander, J.
Gresham-Lancaster, S. 423, 434
grids 117
Grotowski, J. 6, 131, 222, 302, 395, 410, 416; *see also* poor theater, via negativa
group color spaces 273
group interview 389
group lift 134, *135*
group situation 259–260
group story 267
guidance 220–221

Hadebe, E. 393
hallucination 373–385; Creative Groups 382–383; Ethiopia journal 375–376; facilitating 383–385; orality 376–377; qayaq script *377*; !Xuu and Khwe Bushmen 378–381, *381*
Hanson, D. 6
Hanson Robotics 6, 277

heartbeats *55*, *56*
Hegel, G. W. F. 370, 373
Heidegger, M. 402
Hennigan, B. 428
hieroglyphics 310–311, 413–414
HIV/AIDS prevention 236
(W)hole case study 420–425, *422*, *425*
Holsonback, H. 434
Holy Bone case study 431–441, *440*, *441*; Agent Zay: tests 436; Cadmus: remediation 436–437; cycles of life 434; John: photos 440; Kapsay: the map 433–434; Kei: stone 435; Manta: song 438; Que and Bung: confrontation 437; Rem: maps 437; Sa'bu: transference 439; Tan: anxiety 434–435; Urd: acceptance 439; Vig: listening 438–439; Zun: elixir 438
Huan Nuoyuan ritual 347, *348*
hugs 216–217, *217*
Huizinga, J. 125
Hull, A. 338
human-centered bias 154
human measurement of reality *53–54*
hunter-gatherers 323; Indigenous 19–20; reality 79, 80–81
Hyde, L. 236, 326, 328

Iacoboni, M. 333
imagination warm-up 336–337
Imipashi (the spirits) *175*, 176, 237
immersion as place-making strategy 415–416
immersive performance 416
Indigenous hunter-gatherers 19–20
indigenous knowledge systems 11, 80, 374
indigenous performance 355, 404
indiki 202
individual and group meditation *56–57*
initiation 319, 358
interstitials 129–131; capturing space 130; disorientation spins 130–131; giving and talking $ 130; "X" ercise 129–131
interview 211–212, *212*
interviewing: group 389; guidelines 388, *389*; peculiar things 390; performance-focused 385–390
Inupiaq peoples 7, 9, 12; *see also* Alaska
Inupiat dance 342
Inupiat Eskimo 24
Inupiat performance 302, 306

Jaques, S. 335
Jeremiah, D. 386
the journey 122–123
Jumbo, P. 354–355
Jung, C. G. 22, 275–276, 366

Kalahari 2, 10, 20, 24, 45, 316, 325, 330–331, 378, 391
Kalahari Desert 10, 330, 378
Kamyonguk mask exploration *297*
Karamu House Theatre (Cleveland) 113
kaRaoKe MoTeL case study 425–431, *426*
Kasra, M. 434
Katz, R. 391
Kenya journal, narrative spaces 267–269
Khan, A. 439
Khwe Bushmen 2, 20, 24, 44–46, 329, 378–381, 391; *see also* Kalahari
kinesthetic awareness 335
Kiviq 236
Korean journal 100–101; self-mask 296–297
Korean National University of Arts (KNUA) 100, 294, 298

Ladefoged, P. 338
Landau, T. 117
language: circle sound game 164; circle statements 165; counting to 20 167–168; endings and beginnings 165; never stop 166–167; small stories 166; space of 164–168; talking circle 164
Lao Tzu 118; on Ying Yang 94
leading walks 41–42
Lee, D. 299
Lehman, J. 440
Lehmann, H.-T. 382, 400, 402, 403, 409
Lesheim, L. 118
Lévi-Strauss, C. 370
life boat 141–142
life transitions 43
liminality/liminal 326; *see also* initiation; threshold

liminal space 24, 301–302, 391
liminal (transitional) stage 426
Linklater, K. 338
listening 30–31
Lomax, A. 61
The Lord of the Rings 276
Lovelock, J. 11
Lowenstein, T. 24
Lul Theatre 164

Mâche, F.-B. 339
Makanda Mahlanu 172, *173*
mantra 82–84
Mark, M. 278
Marvel 276
masks 294–301; creating mask character 300; *Kamyonguk* mask exploration *297*, 298–299; modern mask characters 299; self-mask 296–297
mass extinction 79
master-slave 143
Mauss, M. 78
McDonald, B. 437
Merleau-Ponty, M. 4
Merzenich, M. 335
Metamorphosis Theatre Ritual Preparation 62, 349
MeToo movement 321
Meyer, C. 378
Miao communities 2, 10, 12, 185, 190, 201, 237, 277, 306, 325, 347, 370–372, 377, 384, 391, 418, 431; *see also* China (Guizhou Province)
Miao ritual tradition 370–371
Microsoft 157
Miller, A. 226
mimesis 221–235; mirror sequence 222–225, *225*; Siberian journal 226–228; touch 228–229
mime warm-up 334
mirror neuron activation 333
mirror sequence 222–225, *225*
Mizner, K. 438
more-than-human relationships 20, 21, 204–205
more-than-human world 6, 11–12, 418
Mountain, P. 377
mountain, the 14–15
movement flow 104–105
movement to music 74–75
moving 99–110; the fly 108–109; Korean journal 100–101; make a move 99; making my move 102–104; movement flow 104–105; Siberian journal 109–110; stop-start 107; triangles 108
moving with other beings 76–77, *77*
moving with others 75–76
Mtonga, M. 355
multispecies character and the posthuman 403–404
"musilanguage" system 338
Musonda, L. 175
Myers, T. 335
myth, performance and 366

Namibia 378
Napolean, H. 78–79
narrative(s): place 322–339; as space *versus* narrative of place 323; trickster performing in-between 325–329
narrative-as-practice 324
narrative-as-representation 324
narrative space(s) 155–156, 264–265; Ethiopian workshop journal 265–266; group story 267; Kenya journal 267–269
Natal Performing Arts Council (South Africa) 172
National Theatre of Kenya 112
National Theatre of Sweden 160
National Theatre School in Krakow 112
National Theatre School journal, trickster performing in-between 328
Native American medicine wheel 43
Nelson, R. 19–20
neo-colonialism 326
neoteny 335
neural flexibility 332
neural networks 338
neurological roles of play 332–335
Noh tradition 306
non-dual awareness 78
non-human intelligences 20
non-place 154; *see also* Auge, Marc; loss of place; spaces without place
non-Western performance 418–419

object: imaginary 247; imagination 247–248; space 247–252
Objective Drama 302
observation: drawing faces 214–215; eye-to-eye contact 218–219; guidance 220–221; hugs 216–217, *217*; interview 211–212, *212*; partner 213; person opposite 219–220; space 211–221
one leg balance 94
ontological assurance 400
Oquilluk, W. 301
orality 376–377, *377*
oral tradition 315
organic reactions 131
Organic Theatre (Chicago) 2, 71, 339
out of body 240–241

Pallasmaa, J. 429
Panksepp, J. 332
participatory performance 341–343
patterns 146
Pearson, C. 278
performance: aggression and 138–143; as algorithm 294; as cosmological practice 372–373; as engagement with the living world 6–7; knowledge 44; and myth 366; narrative spaces 264–265; as a place and a process of evolution 44; place through 317–318; space 264–269; stand-off 93–94; stand-up 89
performance-focused interviewing 385–390
performer(s): body part scenario 47–49; flying 135–136, *136*; leading walks 41–42; name 67–68; smiling 42–43; what's going on? 132–133
performing 144–151; being watched 144–145, *145*; compositions 146–147; patterns 146; push, pull, spin 147–149; Tanzania journal 150–151; three movements 149–150
performing myth 365–369
personality: characters and 401
personal vocabulary 311–313
phenomenology 4
photos 244–245
physical expressions 43
physical language culture 188; from St. Petersburg journal 188–189
physiological coherence 331
picture thinking 370
pig in the middle 140
place 314–418: ancient features 418–419; as balanced system 205–206; becoming earthlings 320–321; bringing place to the body 339–346; characters 400–409; contemporary crisis 319–320; creating performance 358–365; hallucination 373–385; immersion, ritual, and the sacred 415–418; interviewing 385–390; loss of 316–317; making sounds 338–339; narratives 322–339; performance as transformation 321–322; phase 314–316; rehearsal 409–415; and space 313; spaces without 317; space *vs.* 154, 157; theoretical frames 365–373; through performance 317–318; trance 390–400; *see also* loss of place; place-based knowledge; space *vs.* place distinction
place as a container 313, 314
place as enacted /dynamic enactment of place 317–318
place-based beings 316, 320–321
place-based communities 316, 318
place-based cultures 5
place-based knowledge 316, 419
place-based performance 6, 7, 79, 80, 87
place-consciousness 339
place embodiment 155
planetary consciousness 10
planetary evolution 10
play 125–129; alternative tags 126; chaos 364; fireman lift relay 127–128; neurological roles of 332–335; scorpion 128–129; tag 125–126
playful warm-up 334
poetics of space 318
political corruption 236
polyrhythmic place 355
Pondicherry University 215, *216*
poor theater 6
Porges, S. 332
postdramatic characters and Indigenous epistemology 402–403
postdramatic theater 400, 402, 403

post-liminal (integrative) stage 426
poverty of resolution 370–371
practical engagement 6
pre-liminal (preparatory) stage 426
presence (performance) 323
pride and prejudice 168–177; the gauntlet 170–172; nice/nasty 169–170
process, space: balance as practice 206–207; balanced space 208; balancing with others 207; center of the circle 209; column of light 209–210; contemporary challenges 206; Ethiopian journal 210–211; place as balanced system 205–206; reciprocity and exchange 204–205
programmable techno body 403
progress report 414–415
psychic space 239–240
psychological healing 275
psychophysical 131–134; free contact 132; what's going on? 132–133; whole body communication 133–134
pull apart 140–141
push, pull, spin 147–149
Putin, V. 349

Quinceañeras 43

Radcliff, C. 437
radiance: aura space 273–274, *274*; color spaces 271–272; group color spaces 273; a tree 274–275
Radin, P. 327
Rayle, B. 438
reaching warm-up 335
reactive strength 335
reciprocity and exchange process 204–205
rehearsal: emotional journey 411–412, *412*; hieroglyphics 413–414; performance journey 415; place 409–415; progress report 414–415; stepping out 412–413
relational ontology 402
resistance 139–140
resource depletion 79
restorative rhythm 52
revolution 53
rhythm(s) 52–54, 86; awareness 52, 64; balance and 202–203; chant 184–185; choreographic 61; circadian 53; dialog 181–182; earth's 53; expression 179, 183, *184*; of light and dark 53; as living practice 355–356; memory 189–190; music physically 70–71; from natural world 178; number mix-up 71–72; polyphonic 52; reality 178; restorative 52; sequence 180–181; shifting 52; space and 177–182; steady 52–53; story dancing 73, *74*; technology for 178
rhythmic drumming 395
rhythmic movement 60–68; animal husbandry 61; breathing a lie 64; creation and meltdown 63; extractors 61; heartbeat movement 62; irrigation 61; meltdown 63; moving with other beings 76–77, *77*; moving with others 75–76; to music 74–75; plow agriculture 61; producers 61; sticks 72–73
ritual 206, 315
ritual performance 293, 341, 417
ritual preparation 341, 346–357; Alaska 349–350; applications 349–352, *352*; collaborative process 348–349; Dallas 351–352; environmental rhythms 353–354; Ethiopia 351; polyrhythmic place 355; purpose of 347; rhythm as living practice 355–356; Russia 351; Siberia 350; sticks 353; three traditional Zulu movements 357; Tundra's heartbeat 354–355; Zambia 350–351; Zulu journal 356
rotation 53

sacred in secular contexts 417–418
sacred space 24, 152, 184–185, 237
Sakha National Theatre (Siberia) 35, 392; walking attitudes exploration at *37*
Sakha people 2, 31, 32, 35, 125, 179, 182, 187, 190, 195, 203, 226–227, 325, 343, 344, 350, 391, 392, 395, 405, 422, 433, 435; *see also* Siberia; Yakutsk
Sakha Republic 35
Sakha shaman's drum 178, *179*
San peoples. *see* Khwe Bushmen; !Xuu Bushmen

Sardaana 36
Schechner, R. 5, 6, 401–402, 409
Schipper, K. 201, 370, 371
Schumann's Resonance 5, 354
scorpion 128–129
self-mask 295–296; Korean journal 296–297
shadows and humans 231–235; Ethiopian journal 232–235, *235*
shamanism/shamanic practice 188, 198, 420. *see also* indigenous performance; ritual; spirit medium
shaman rhythm 190–191, *191*
sharemony 25
sharing 255–264; monster within 256–257; the situation 256–257
Shi Jing He 10, 347
Siberia 2, 31, 35, 179, 182, 190, 325, 350, 391, 422, 433, 435; *see also* Sakha Republic; Yakutsk
Siberian journal 36–37, 83–84; balance and centered awareness 92–93; cultures from 187; focus and mindfulness 123–124; mimesis 226–228; moving 109–110; trust 137–138; vocal seasons 195
Simpson, L. B. 403
the situation 256–257; group 259–260; one word 261–263; questions and answers 260–261; three lines 263–264
skills exchange 238
smiling 42–43
social engagement system 332
social media 2, 157
social robotics 5–6
Socrates 87
somatic flexibility 332
somatic resonance 338–339
somatic systems 335–337
somatic wisdom and practice 318–319
Somé, M. P. 7, 79, 355
Sophia robot 6, 401, 402, 404, 405
sound(s) 192–198; animal 339; balance 200–204; basic vowel 338; body of 81–82; bursting with song 196; chest and torso percussion 338; circle 194; conducting 196–197; deep stomach 338; guidance 197–198; head voice and nasal resonance 338–339; mantra 82–84; of other beings 85–86; place making 338–339; technological space 192–193; touching 81; vocal seasons 193–194
soundscape composition 339
South Africa 2, 25, 68, 172, 202, 325, 392
space 151–155; age 156–158; archetypal 277–278; body as script 301–313; character 252–255; checking in 198–200; circle definitions 159–160; creating 242–243; cultures 182–191; cycles and 158–163; defining 246–247, *247*; empty and 155; expansion 269–275; exploration 238–246; within exploration 241–242; gifting and exchange 235–238; of language 164–168; mimesis 221–235; narrative 155–156; object 247–252; observation 211–221; performance 264–269; performer about 157; and place 313; place *vs.* 154, 157; pride and prejudice 168–177; process 204–211; psychic 239–240; sharing 255–264; sound 192–198
space (performance) 157–158
spaces without place 317
space *vs.* place distinction 313
spatial diversity 157
spatial vertigo 157, 314, 317
spell of the sensuous 5, 11
spirit medium 201, 384
spirit/spirits 347, 353, 370, 377, 385, 392–393, 403, 422, 424
standing relaxed 13–14
stand-off 93–94
stand-up performance 89
Star Wars 276, 367
Steyerl, H. 3
sticks 68–70, *70*; rhythmic movement 72–73; ritual preparation 353
stop-start 107
story dancing 73, *74*, 302, 341
storytelling 323–324, 362, 374–375; by !Xuu Bushmen 323
St. Petersburg journal, physical language culture from 188–189
Sundberg, J. 338

tag(s) 125–126; alternative 126; amoeba 126; go-tag 127
talking circle 164
Tanzania journal, performance workshop 150–151
Taoism 369–370
Taoist ritual tradition 370–371
Taylor, D. 376–377
Techdigenous 403, 404; *see also* consciousness technology; indigenous knowledge systems
technological displacement 79; space 192–193; spirit 6
Temporary Autonomous Zone 162
theoretical frames 365–373; the feminine 371–372; mythic self 368–369; performing myth 365–369; poverty of resolution 370–371; unifying opposites 369–370
therukoothu (South India) *178*, 417–418
three movements 149–150
threshold 391, 416, 418, 440
time and space 30
T.N. B. development by Dead White Zombies 386, *387*
Toolie, J. 1, 9, 12, 30, *178*
touch 228–229; consciousness 229–230, *231*
Traditional Korean Performance 294
training observational awareness 213
trance 390–400; awakenings 392–394, *394*; dance 397–399; Dead White Zombies 393; ethical considerations and safety 396–400; Ethiopia 393; movement into 397; postures 302; rhythmic drumming 395; sharing 399–400; Siberia 392; techniques for inducing 395–396; Zulu 392
triangles 108
trickster performing in-between 325–329; importance of adopting 326; National Theatre School journal 328; role of 327; somatic resonance 338–339; somatic systems 335–337; warming up, creating heat 329–332, *331*; !Xuu Bushmen healing process 329–332, *331*
Truax, B. 339
trust 134–138; disembodied head 137; flying 135–136, *136*; group lift 134, *135*; Siberian journal 137–138
Tuma Theatre (Alaska) 9, 298, 303, 339, 345–346; Elemental Forms and 306
Tuma Theatre journal, the gauntlet 173–174
Tundra's heartbeat 354–355
Turner, V. 24, 391, 426
Twelve Moons 294, 299, *300*

Umbanda practitioners 385
unifying opposites 369–370
University of Alaska Fairbanks 19, 262, 289, 339
University of Dar es Salaam 150
University of Ouagadougou 39
University of Texas 240

Valentin 385–386
Van Gennep, A. 24, 426
Van Itallie, J.-C. 362–363
via negativa 6, 410
victim circle 177
virtual worlds 2
visualization exploration *58–60*, *60*
vocal seasons 193–194; Siberian journal 195

walking 31–33, 39; attitudes 34–35; on Earth 33; exploration during a workshop at the DirectorsLab, Chicago *32*; light 38; like a dancer 32; like an animal 31; move sexy 33; in space 33; tense 40
walking attitudes 39; exploration *37*; exploration at Sakha National Theatre (Siberia) *37*
warming up, creating heat: Commedia Dell'arte warm-up 333; element warm-up 336; imagination warm-up 336–337; mime warm-up 334; playful warm-up 334; reaching warm-up 335; social activation 332–335; somatic resonance 338–339; somatic systems 335–337
Watts, A. 431
Ways of Being (Bridle) 203
Weir, H. 435
Werle, A. 439

white light 58–60, *60*
whole body communication 133–134
Willerslev, R. 401
Wilson, S. G. 23
winding path 7
world awareness 17–18
world beat 354
Wu Zheng Niam 347

"X" ercise 129–131
!Xuu Bushmen 2, 20, 24, 44–46, 78, 329, 378–381, *381*, 391; hallucination 378–381, *381*; healer 201; healing process 329–332, *331*; journal 44–46; storytelling by 323; trance and 391; *see also* Kalahari Desert; San peoples

Yakutsk 7, 115, 343
Yunkaporta, T. 7, 204–205, 324, 327
Yup'ik peoples 1, 5, 7, 9, 30, 33, 68, 73, 78, 185, 196, 202, 301–303, 305–307, 315, 316, 319, 340, 341–344, 391; dance 342–343; performance 302
Yuuyaraq 78

Zambia journal, aggression 142
Zarrilli, P. 131
Zhang Quifeng 384
zoomusicology 339
Zulu 2, 25–29, 33, 56, 84, 179, 202, 229, 242, 257, 303, 378, 391, 392; *see also* South Africa
Zulu journal, ritual preparation 356

For Product Safety Concerns and Information please contact our EU representative GPSR@taylorandfrancis.com
Taylor & Francis Verlag GmbH, Kaufingerstraße 24, 80331 München, Germany

www.ingramcontent.com/pod-product-compliance
Ingram Content Group UK Ltd.
Pitfield, Milton Keynes, MK11 3LW, UK
UKHW021306260926
475702UK00048B/700
9781041168010